Nanocellulose and Sustainability

Sustainability: Contributions through Science and Technology

Series Editor: Michael C. Cann, Ph.D.
Professor of Chemistry and Co-Director of Environmental Science
University of Scranton, Pennsylvania

Preface to the Series

Sustainability is rapidly moving from the wings to center stage. Overconsumption of non-renewable and renewable resources, as well as the concomitant production of waste has brought the world to a crossroads. Green chemistry, along with other green sciences technologies, must play a leading role in bringing about a sustainable society. The **Sustainability: Contributions through Science and Technology** series focuses on the role science can play in developing technologies that lessen our environmental impact. This highly interdisciplinary series discusses significant and timely topics ranging from energy research to the implementation of sustainable technologies. Our intention is for scientists from a variety of disciplines to provide contributions that recognize how the development of green technologies affects the triple bottom line (society, economic, and environment). The series will be of interest to academics, researchers, professionals, business leaders, policy makers, and students, as well as individuals who want to know the basics of the science and technology of sustainability.

Michael C. Cann

Published Titles

Green Chemistry for Environmental Sustainability
Edited by Sanjay Kumar Sharma, Ackmez Mudhoo, 2010

Microwave Heating as a Tool for Sustainable Chemistry
Edited by Nicholas E. Leadbeater, 2010

Green Organic Chemistry in Lecture and Laboratory
Edited by Andrew P. Dicks, 2011

A Novel Green Treatment for Textiles:
Plasma Treatment as a Sustainable Technology
C. W. Kan, 2014

Environmentally Friendly Syntheses Using Ionic Liquids
Edited by Jairton Dupont, Toshiyuki Itoh, Pedro Lozano, Sanjay V. Malhotra, 2015

Catalysis for Sustainability: Goals, Challenges, and Impacts
Edited by Thomas P. Umile, 2015

Nanocellulose and Sustainability:
Production, Properties, Applications, and Case Studies
Edited by Koon-Yang Lee, 2017

Nanocellulose and Sustainability

Production, Properties, Applications, and Case Studies

Edited by

Koon-Yang Lee

CRC Press is an imprint of the
Taylor & Francis Group, an **informa** business

CRC Press
Taylor & Francis Group
6000 Broken Sound Parkway NW, Suite 300
Boca Raton, FL 33487-2742

Printed on acid-free paper

International Standard Book Number-13: 978-1-4987-6103-1 (Hardback)

Library of Congress Cataloging-in-Publication Data

Names: Lee, Koon-Yang.
Title: Nanocellulose and sustainability : production, properties, applications, and case studies / [edited by] Koon-Yang Lee.
Description: Boca Raton : CRC Press, [2018] | Series: Sustainability contributions through science and technology | Includes bibliographical references and index.
Identifiers: LCCN 2017036755| ISBN 9781498761031 (hardback : alk. paper) | ISBN 9781351262927 (ebook)
Subjects: LCSH: Cellulose fibers. | Cellulose. | Nanostructured materials. | Green technology.
Classification: LCC TS1544.C4 N36 2018 | DDC 628--dc23
LC record available at https://lccn.loc.gov/2017036755

Visit the Taylor & Francis Web site at
http://www.taylorandfrancis.com

and the CRC Press Web site at
http://www.crcpress.com

Contents

Preface....................vii
About the Editor....................ix
Contributors....................xi

Chapter 1 Process Modelling and Techno-Economic Evaluation of an Industrial Airlift Bacterial Cellulose Fermentation Process....................1

Fernando Dourado, Ana Isabel Fontão, Marta Leal, Ana Cristina Rodrigues and Miguel Gama

Chapter 2 Production of Cellulose Nanofibres....................17

Antonio Norio Nakagaito, Kentaro Abe and Hitoshi Takagi

Chapter 3 Preparation of Cellulose Nanocrystals: Background, Conventions and New Developments....................27

Eero Kontturi

Chapter 4 Oxidative Chemistry in Preparation and Modification on Cellulose Nanoparticles....................45

Tuomas Hänninen and Akira Isogai

Chapter 5 Crystallinity and Thermal Stability of Nanocellulose....................67

Alba Santmartí and Koon-Yang Lee

Chapter 6 Crucial Interfacial Features of Nanocellulose Materials....................87

Andreas Mautner, Minna Hakalahti, Ville Rissanen and Tekla Tammelin

Chapter 7 Nanocellulose-Based Membranes for Water Purification: Fundamental Concepts and Scale-Up Potential....................129

Aji P. Mathew, Peng Liu, Zoheb Karim and Jessica Lai

Chapter 8 Applications of Nanocellulose as Optically Transparent Papers and Composites....................147

Franck Quero

Chapter 9 Application of Nanocellulose as Pickering Emulsifier..................... 175

Isabelle Capron

Chapter 10 Upgrading the Properties of Woven and Non-Woven (Ligno)Cellulosic Fibre Preforms with Nanocellulose..................... 197

Marta Fortea-Verdejo and Alexander Bismarck

Chapter 11 Cellulose-Based Aerogels... 217

Jian Yu and Jun Zhang

Chapter 12 Production of Cellulose Nanocrystals at InnoTech Alberta.............269

Tri-Dung Ngo, Christophe Danumah and Behzad Ahvazi

Index...289

Preface

Nanometre-scale cellulose fibres, or nanocellulose, are emerging materials for various advanced applications. Nanocellulose can be obtained through two approaches: bottom–up or top–down. In the bottom–up approach, cellulose is produced by the fermentation of low molecular weight sugars using bacteria from the *Acetobacter*, later renamed as *Komagataeibacter* species. These cellulose fibres, more commonly known as bacterial cellulose, are inherently nanofibrillar. With regards to nanocellulose produced using the top–down approach, (ligno)cellulosic biomass such as wood pulp are either treated with strong ultrasound, passed through high-pressure homogenisers or stone grinders to reduce the size of these fibres to the nanometre scale. These wood-derived nanocellulose is more commonly known as cellulose nanofibres, nanofibrillated cellulose or microfibrillated cellulose.

This book has been compiled in response to the growing interest of nanocellulose-based materials. It consists of 12 chapters dealing with the various aspects of production, properties and applications of nanocellulose. This book also includes case studies in the form of the prospect of commercial production of bacterial cellulose and large-scale production of cellulose nanocrystals. Each chapter can be regarded as a self-standing chapter.

Chapter 1, 'Process modelling and techno-economic evaluation of an industrial airlift bacterial cellulose fermentation process', covers the production of bacterial cellulose. A case study in the form of techno-economic analysis of the production of bacterial cellulose using an industrial airlift bioreactor is also discussed.

Chapter 2, 'Production of cellulose nanofibres', discusses the major production routes of wood-derived cellulose nanofibres. Both mainstream and novel and more recent cellulose nanofibre extraction processes are covered in this chapter.

Chapter 3, 'Preparation of cellulose nanocrystals: Background, conventions and new developments' covers the preparation and properties of nanosized cellulose rods derived from the acid hydrolysis of native cellulosic fibres.

Chapter 4, 'Oxidative chemistry in preparation and modification on cellulose nanoparticles', summarises various oxidative techniques to produce nanocellulose.

Chapter 5 'Crystallinity and thermal stability of nanocellulose', summarises the crystal structure of bacterial cellulose and cellulose nanofibres. The thermal stability of various types of nanocellulose reported in literature can also be found in this chapter.

Chapter 6, 'Crucial interfacial features of nanocellulose materials', discusses the various surface and interfacial properties of nanocellulose. The specific surface area and surface charges, surface reactivity, interaction of nanocellulose surface with various molecules are also discussed.

Chapter 7, 'Nanocellulose-based membranes for water purification: Fundamental concepts and scale-up potential', covers the concept of using nanocellulose as membranes. Case studies in the form of scaling-up the nanocellulose membrane production are also covered in this chapter.

Chapter 8, 'Applications of nanocellulose as optically transparent papers and composites', provides an overview of nanocellulose-based optically transparent materials.

Chapter 9, 'Application of nanocellulose as Pickering emulsifier', discusses the use of nanocellulose as stabilisers for emulsions. This chapter also covers the fundamentals of particle-stabilised emulsions, otherwise known as Pickering emulsions.

Chapter 10, 'Upgrading the properties of woven and non-woven (ligno)cellulosic fibre preforms with nanocellulose', covers the application of nanocellulose as a binder for various types of natural fibres to produce non-woven fibre preforms.

Chapter 11, 'Cellulose-based aerogels', discusses the properties and applications of nanocellulose-based aerogels. Various techniques to produce nanocellulose-based aerogels are also covered in this chapter.

Chapter 12, 'Production of cellulose nanocrystals at InnoTech Alberta', covers the pilot-scale production of cellulose nanocrystals in Canada. Various engineering challenges associated with the production scale-up of cellulose nanocrystals are discussed.

I would like to thank all contributors, who provided excellent chapters for this book. I am forever grateful for their commitment and contribution to this book. Without them, this book will never be realised. I would also extend my appreciation to the publishing staff at Taylor & Francis Group/CRC Press for their help. Special thanks go to Dr. Melanie Lee for creating the graphics for the front cover.

About the Editor

Dr. Koon-Yang Lee currently leads the Future Materials Group at the Department of Aeronautics, Imperial College London, United Kingdom. His research focuses on the manufacturing and development of novel polymeric materials with a focus on tailoring the interface between two (or more) phases to bridge the gap between chemistry, materials science and engineering. His area of expertise includes the design, manufacturing and characterisation of nanocellulosic structures and nanocellulose-reinforced polymers, nanocellulose-enhanced, natural fibre-reinforced hierarchical composites, foam-templated macroporous polymers and recycling/upcycling of waste materials.

Contributors

Kentaro Abe
Research Institute for Sustainable Humanosphere
Kyoto University
Kyoto, Japan

Behzad Ahvazi
Alberta Innovates-Technology Futures
Edmonton, Alberta, Canada

Alexander Bismarck
Polymer and Composite Engineering (PaCE) Group, Faculty of Chemistry, Institute for Materials Chemistry and Research
University of Vienna
Vienna, Austria

and

Polymer and Composite Engineering (PaCE) Group, Department of Chemical Engineering
Imperial College London
London, United Kingdom

Isabelle Capron
Institut Nationale de la Recherche Agronomique (INRA)
Nantes, France

Christophe Danumah
Alberta Innovates-Technology Futures
Edmonton, Alberta, Canada

Fernando Dourado
Centre of Biological Engineering
University of Minho
Braga, Portugal

Ana Isabel Fontão
Centre of Biological Engineering
University of Minho
Braga, Portugal

Miguel Gama
Centre of Biological Engineering
University of Minho
Braga, Portugal

Minna Hakalahti
High Performance Fibre Products
VTT Technical Research Centre of Finland Ltd
Espoo, Finland

Tuomas Hänninen
Department of Forest Products Technology
Aalto University
Espoo, Finland

Akira Isogai
Department of Biomaterials Science
The University of Tokyo
Tokyo, Japan

Zoheb Karim
MoRe Research AB
Örnsköldsvik, Sweden

and

Wallenberg Wood Science Centre
Stockholm, Sweden

Eero Kontturi
Department of Bioproducts and Biosystems, School of Chemical Engineering
Aalto University
Espoo, Finland

Jessica Lai
Department of Chemical Engineering
Imperial College London
London, United Kingdom

Marta Leal
Centre of Biological Engineering
University of Minho
Braga, Portugal

Koon-Yang Lee
The Composites Centre, Department of Aeronautics
Imperial College London
London, United Kingdom

Peng Liu
Division of Materials and Environmental Chemistry
Stockholm University
Stockholm, Sweden

Aji P. Mathew
Division of Materials and Environmental Chemistry
Stockholm University
Stockholm, Sweden

Andreas Mautner
Polymer and Composite Engineering (PaCE) Group, Institute for Materials Chemistry and Research
University of Vienna
Vienna, Austria

Antonio Norio Nakagaito
Graduate School of Technology, Industrial and Social Sciences
Tokushima University
Tokushima, Japan

Tri-Dung Ngo
Alberta Innovates-Technology Futures
Edmonton, Alberta, Canada

Franck Quero
Departamento de Ciencia de los Materiales, Facultad de Ciencias Físicas y Matemáticas
Universidad de Chile
Santiago, Chile

Ville Rissanen
High Performance Fibre Products
VTT Technical Research Centre of Finland Ltd
Espoo, Finland

Ana Cristina Rodrigues
Centre of Biological Engineering
University of Minho
Braga, Portugal

Alba Santmartí
The Composites Centre, Department of Aeronautics
Imperial College London
London, United Kingdom

Hitoshi Takagi
Research Institute for Sustainable Humanosphere
Kyoto University
Kyoto, Japan

Tekla Tammelin
High Performance Fibre Products
VTT Technical Research Centre of Finland Ltd
Espoo, Finland

Marta Fortea-Verdejo
Polymer and Composite Engineering (PaCE) Group, Faculty of Chemistry, Institute for Materials Chemistry and Research
University of Vienna
Vienna, Austria

Jian Yu
Chinese Academy of Sciences Key Laboratory of Engineering Plastics, Chinese Academy of Sciences Research/Education Center for Excellence in Molecular Sciences, Institute of Chemistry
Chinese Academy of Sciences
Beijing, China

Jun Zhang
Chinese Academy of Sciences Key Laboratory of Engineering Plastics, Chinese Academy of Sciences Research/Education Center for Excellence in Molecular Sciences, Institute of Chemistry
Chinese Academy of Sciences
and
University of Chinese Academy of Sciences
Beijing, China

1 Process Modelling and Techno-Economic Evaluation of an Industrial Airlift Bacterial Cellulose Fermentation Process

Fernando Dourado, Ana Isabel Fontão, Marta Leal, Ana Cristina Rodrigues and Miguel Gama

CONTENTS

1.1 Introduction 1
1.2 Process Simulation 3
1.3 Cost Estimations 3
1.4 Process Simulation of Bacterial Cellulose Production 4
1.5 Conclusion 13
References 13

1.1 INTRODUCTION

Bacterial cellulose (BC) is an exopolysaccharide produced by *Komagataeibacter xylinus* (formerly known as *Gluconacetobacter xylinus*), a gram negative and strictly aerobic bacterium that belongs to the acetic acid bacteria (AAB) family, Acetobacteraceae. BC has the same chemical composition as that of vegetable cellulose. However, it is deprived of lignin, pectin, hemicelluloses and other biogenic compounds [1,2]. Among the several genera of this family, the *Acetobacter* and *Komagataeibacter* (*Gluconacetobacter*) genus are the most notable cellulose producers [3–8]. BC is an outstanding biomaterial with unique properties, including high water holding capacity, high crystallinity, ultrafine fibre network, high tensile strength in the wet state and the possibility to be shaped into three-dimensional (3D) structures during synthesis. These properties have allowed to propose a wide range of applications in human and veterinary medicine, odontology, pharmaceutical industry, acoustic and filter membranes, biotechnological devices and in the food and paper industry [9–16].

Many developments have occurred in practical applications of BC in the biomedical field, with Bioprocess®, XCell®, Biofill®, Dermafill® and Gengiflex® as examples of patented products, with wide applications in surgery and dental implants (Gengiflex®) and especially for wound dressings (Bioprocess®, XCell®, Biofill®, Dermafill®). Indeed, in clinical trials, BC has been shown to be a superior product, as compared to conventional wound dressings, in retaining exudate, reducing pain, accelerating reepithelialisation and healing times, reducing wound infection rates, facilitating wound inspection (due to the high transparency of the thin BC membranes), and in reducing scarring. Gengiflex® has been used to treat periodontal diseases, in dental implants, in guided bone regeneration alone or in association with osteointegrated implants, proving to be a good alternative for guided tissue regeneration [17–21]. Several other potential applications in the biomedical field have been reported [11,13,22–27].

Due to its unique properties and structure, BC has been shown to have important applications in a variety of food formulations where the low usage levels, the lack of flavour, the foam stabilisation ability and stability under broad pH, temperature, and freeze–thaw conditions are needed. BC could effectively be used as a low-calorie additive, thickener, stabiliser, texture modifier, pasty condiments and as a vegetarian foodstuff [28–33]. In Asian countries, BC is produced and commercialised under the name 'Nata de coco' and is obtained from the fermentation of coconut water by cellulose-producing AAB, using traditional fermentation methods. By far the most common uses of nata de coco include low-calorie sweetened desserts, fruit salads and high-fibre foods [34–36]. In the western world, Cetus Co. (Berkeley, California) and Weyerhaeuser Co. (Seattle, Washington) have used a patented, genetically modified *Acetobacter* strain for the fermentation of BC under agitated culture (by deep-tank fermentation technique). The obtained BC, Cellulon®, was aimed to be used as a food stabiliser and thickener [13,37]. By mid-1990s, Kelco, Inc. (Lexington, Kentucky) purchased the BC business from Weyerhaeuser and launched the product as PrimaCels, which was also aimed to be used in the food industry. Currently, with the exception of nata de coco, there are no commercial BC products for food applications.

A fermentative process affording high BC yields at low capital and operating costs will allow the release onto the market of a product with a range of potential applications that exceeds the biomedical niche market. The economic feasibility of BC production is directly dependent on its productivity (e.g. BC yield). Several fermentation technologies have been experimented using specific fermentation media, overproducing mutant strains, using agitated, airlift, membrane and horizontal bioreactors. Two fermentation methods have been traditionally explored to improve product yield: the static and the agitated culture. Under static culture conditions, BC production occurs at the air–water interface, where the assembly of reticulated crystalline ribbons results in a gel or pellicle. The growth of the pellicle occurs downward, until the entrapped cells become inactive and eventually die, due to oxygen and nutrient limitations. An alternative approach to BC production is submerged fermentation through aerated (airlift) or agitated cultivation (stirred tank). An advantage of the stirred tank reactor is its ability to prevent the heterogeneity of the culture broth by strong mechanical agitation, whereas its drawback is its high energy cost for generating the mechanical power. Contrarily, the energy cost of an airlift reactor is typically one-sixth of that of a stirred tank reactor. However, the agitation power of an airlift reactor is limited, resulting in low fluidity of the culture broth, especially at high cellulose concentrations. Both agitation and aeration systems result in cellulose-negative mutants (non-cellulose producers), highly branched, three-dimensional, reticulated BC structure (thus limiting BC applications). With membrane bioreactors, the major drawbacks include the high operating costs and the difficulty in collecting the cellulose from the reactors [15,16,38–40].

1.2 PROCESS SIMULATION

For a successful development of a plant design, plant location and layout, the availability of raw materials, construction materials, structural design, utilities, buildings, storage, material handling, safety, waste disposal, taxations, patents, transportation and markets needs to be considered. Process simulation software has been used since the early 1960s to first emulate industrial processes that operate under continuous transient behaviour. The evolution of process modelling technology strongly impacted on process engineering. Current software can be used to modulate an entire industrial process, execute material and energy balances, estimate equipment size and energy needs, calculate the demands for labour and utilities over time, performing cost analysis and assessing the environmental impact of a given industrial process. Early stage cost estimation plays a critical role in both rising and already established companies in assessing the need for financial investment in a given process or unit procedure. However, cost simulations at such early stage of development may yield an overestimated capital investment by up to 50% of the final investment, as most of the economic data available in the process simulation software are based on price indexes at a given year of data acquisition. To account for the time value of money, these indexes are adjusted through specific mathematical functions to provide updated price estimations of equipments, buildings, labour and so on. However, these indexes should be regarded only as general price estimates [41–43].

1.3 COST ESTIMATIONS

The net profitability, briefly described as the total income minus total expense, and the business sustainability are the key drivers of any plant design simulation and ultimately condition the decision of investing in the construction of a new processing plant, or expanding or modifying an existing one. The assessment, procurement and proper allocation of capital investment must thus be made in any project design. The total capital investment (TCI) required to set up a project includes two major components: fixed capital investment and working capital. The first component, which can be further divided into direct and indirect costs, briefly corresponds to the capital necessary to construct a ready-to-work processing plant; this involves buildings, equipment, machinery, land and associated construction expenses. The working capital corresponds to the additional necessary investment, above the fixed capital, to start and maintain the operation process, especially during the first months (or years) of plant activity. This capital allows to cover salaries, utility bills (such as electricity or gas, fuels), raw materials and other supplies. Due to its specific purpose (ensuring liquidity of the firm for a certain period), working capital is included as the second component of the TCI. Along with capital investment, an estimation of the operating costs (or production costs) must also be made. These costs, which are further divided into direct production costs, fixed charges and general expenses, represent the expenses incurred during plant operation and product selling and are usually expressed on an annual basis. These expenses include raw materials, labour, transportation and other miscellaneous operations, utilities, royalties, research and development, financing, advertising, administrative services and product disposal [43,44]. Further details on the cost structure will be provided below, along with the economic analysis of the BC production simulation.

1.4 PROCESS SIMULATION OF BACTERIAL CELLULOSE PRODUCTION

To explore the process and economics of a computer-simulated large-scale production of BC by agitated/aerated conditions, a survey of the literature was done to gather information on the type of strains, culture media (CM) and fermentation conditions. Table 1.1 summarises the collected data.

For the simulation of the BC fermentation conditions, the use of a commercial strain allows to obviating regulatory issues when considering the implementation of an industrial (commercial) bioprocess. Among the several existing possibilities (Table 1.1), the paper from Cheng et al. [54] was selected to retrieve the relevant data for the simulation. Briefly, as reported in that paper, *Acetobacter xylinum* BPR2001 strain was cultivated in a 20 L modified airlift reactor, using a modified Hestrin and Schramn (HS) culture medium. The modification of the fermenter consisted of using perforated pipes with holes of 1 mm in diameter as the sparger (to allow aeration and agitation), placed at the bottom of the draft tubes. From the fermentation experiments, the optimum BC yield was determined to be 7.72 g/L (dry mass) after 72 h fermentation. This must be recognised as an optimistic production yield, especially

TABLE 1.1
Summary of a Literature Survey on BC Production with Different Strains and Culture Conditions (Agitated and Aerated)

	Medium Composition			Culture Conditions				BC Yield	Productivity	
Strain	Carbon	Nitrogen	Additives	pH	T/°C	Time	Fermenter	g/L	$\times 10^{-2}$ g/L/h	References
BPR-2001 (ATCC700178)	5% w/v Fructose	2% v/v CSL	1.5% w/v CMC; complex medium	5.0	30	120 h	2 L PCS bioreactor	13.0	10.8	[45]
	4% w/v Fructose		Agar (0–1% w/v); complex medium			70 h	10 L jar fermenter	12.8	18.3	[46]
			–			3 days	1 L jar fermenter	7.7	10.7	[47]
	3.9% w/v Fructose		Different substrates; complex medium			3 days	Bioshaker BR-3000L	7.5	10.4	[48]
	4% w/v Fructose		Agar (0.1% w/w);			5 days	50 L airlift	8.7	7.3	[49]
			Xanthan (0.06% w/w); complex medium	4.5		67 h	50 L airlift	3.8	5.7	[50]
	7% w/v Fructose		Oxygen-enriched air; complex medium	5.5		48 h	50 L internal-loop airlift	10.4	21.7	[51]
	2% w/v Fructose	8% v/v CSL	3 g/L Polyacrylamide-*co*-acrylic acid; complex medium	5.0	28	7 days	Flask shaken 175 rpm	6.5	3.9	[52]
	5% w/v Fructose	2% v/v CSL	0%–1% w/v CMC; 0.2%–0.5% w/v MCC; 0.2%–0.5% w/v Agar; 0.2%–0.5% w/v Sodium alginate; complex medium		30	5 days	Flask	8.2	6.8	[53]
	2.0% w/v Glucose	0.6% v/v Yeast extract	Modified Hestrin and Schramn (HS) medium			72 h	20 L modified airlift	7.72	10.7	[54]

(*Continued*)

TABLE 1.1 (*Continued*)
Summary of a Literature Survey on BC Production with Different Strains and Culture Conditions (Agitated and Aerated)

	Medium Composition			Culture Conditions				BC Yield	Productivity	
Strain	Carbon	Nitrogen	Additives	pH	T/°C	Time	Fermenter	g/L	$\times 10^{-2}$ g/L/h	References
BRC5	3.5% w/v Fructose; 0.5% w/v Glucose	8% w/v CSL	5 g/L of carbon substrate supply	6.0	30	48 h	Fed-batch 5 L jar fermenter	6.8	14.2	[37]
	2% w/v Glucose		10% Oxygen enriched air; ethanol	5.5		50 h	5 L jar fermenter	15.3	30.6	[55]
Acetobacter sp. A9	2% w/v Glucose	0.5% w/v Yeast extract; 0.5% w/v polypeptone	0%–2% v/v Ethanol	6.5	30	8 days	200 rpm	15.2	7.9	[56]
	4% w/v Glucose	0.1% w/v Yeast extract; 0.7% w/v polypeptone	Complex medium	–		7 days	–	7.21	4.3	[57]
Ga. hansenii PJK (KCTC 10505BP)	1% w/v Glucose	1% w/v Peptone; 0.7% w/v yeast extract	0%–2% v/v Ethanol	5.0	30	5 days	200 rpm	2.31	1.9	[58]
			1% v/v Ethanol			48 h	Jar-fermenter 200 rpm	1.72	3.6	[59]
AJ 12368	5% w/v Sucrose	0.5% w/v Yeast extract	0.5% w/v Ammonium hydrogen sulphate; 0.3% w/v monopotassium phosphate; 0.0005% w/v magnesium sulphate heptahydrate	5.0	30	3 days + 3 days	Airlift + Static (petri dishes)	1.07	1.5	[60]

(*Continued*)

TABLE 1.1 (*Continued*)
Summary of a Literature Survey on BC Production with Different Strains and Culture Conditions (Agitated and Aerated)

	Medium Composition			Culture Conditions				BC Yield	Productivity	
Strain	Carbon	Nitrogen	Additives	pH	T/°C	Time	Fermenter	g/L	$\times 10^{-2}$ g/L/h	References
Acetobacter sp. V6	1% w/v Glucose	–	Ethanol, lactic acid, acetic acid, fumaric acid, pyruvic acid, succinic acid, inorganic salts	6.5	30	8 days	200 rpm	4.16	2.2	[61]
Ga. xylinus sp. St60-12	4% w/v Sucrose	4% w/v CSL	–	–	28	72 h	150 rpm	4.5	6.3	[62]
NUST4.1	2% w/v Glucose	2% w/v CSL	0%–0.1% w/v Sodium alginate	6.0	29	5 days	150 rpm	6.0	5.0	[63]
PJK (KCTC 10505BP)	1% w/v Glucose	Peptone (1% w/v); yeast extract (0.7% w/v)	1% v/v Ethanol	5.0	30	140 h	Jar-fermenter with spin filter	4.57	3.3	[64]
RKY5 (KCTC 10683BP)	2% w/v Glucose	0.5% w/v Peptone; 0.5% w/v yeast extract	–	–	30	96 h	Rotatory biofilm contactor	6.15	6.4	[65]

TABLE 1.2
Composition and Estimated Purchase Prices of the Main Raw Materials for BC Production

	Density (kg/L)	Culture Media (g/L)	Price (US$/ton)[a]
Glucose		20	640.00
Yeast extract		6	2,000.00
Lactic acid	1.2	9.6	1,000.00
Sodium dihydrogen phosphate ($12H_2O$)		2,7	200.00
Citric acid monohydrate		1.15	460.00
Water	1		2.50
Downstream washing			
NaOH		4	2,000.00

[a] An average price of the stock components was calculated based on values from different suppliers and literature.

for a large-scale airlift reactor. The composition of the CM, as extracted from the manuscript [54], is shown in Table 1.2.

The conceptual project and economic analysis of BC fermentation were done using Super-Pro Designer software, version 9.0, for Windows operating system. The plant was arbitrarily projected to produce 504 ton/year of BC (hydrated). Based on the yield of BC and by assuming that hydrated BC contains 99% water, the monthly CM requirements would be 54,404.15 L. Considering that each fermentation step lasts for 72 h, an airlift allowing a working volume of 5,440.41 L would be necessary. Figure 1.1 displays a flowsheet of the BC fermentation process, as retrieved from the software. Table 1.3 describes the main equipment (type and number of used equipment, their size and cost).

Inoculum propagation is usually achieved by successive propagations of biomass and CM at a ratio of 1:10 (biomass:CM); this procedure was also used in this work. For the sake of simplicity, propagations below 500 L were omitted in the design, as these can easily be done at laboratory scale. The inoculum from BPR2001 strain is to be transferred to the 737.85 L seed fermenter (SBR-101, Table 1.3, Figure 1.1, 'Inoculum Propagation' stage). In addition, a simplified version of the CM preparation and pasteurisation was chosen. For academic purposes, a single entry containing the mixture of the CM components (CM1) was fed to a storage tank (V-101) before pasteurisation (PZ-101). The pasteurised CM is then fed to the seed fermenter, which operates for 3 days at 30 °C for biomass growth. The bacteria and additional pasteurised CM (CM2) are then combined in the airlift fermenter (AFR-101, totalling a working volume of 5,440.41 L) for the fermentation, also operating at 30 °C for 3 days (Figure 1.1, 'Fermentation' stage).

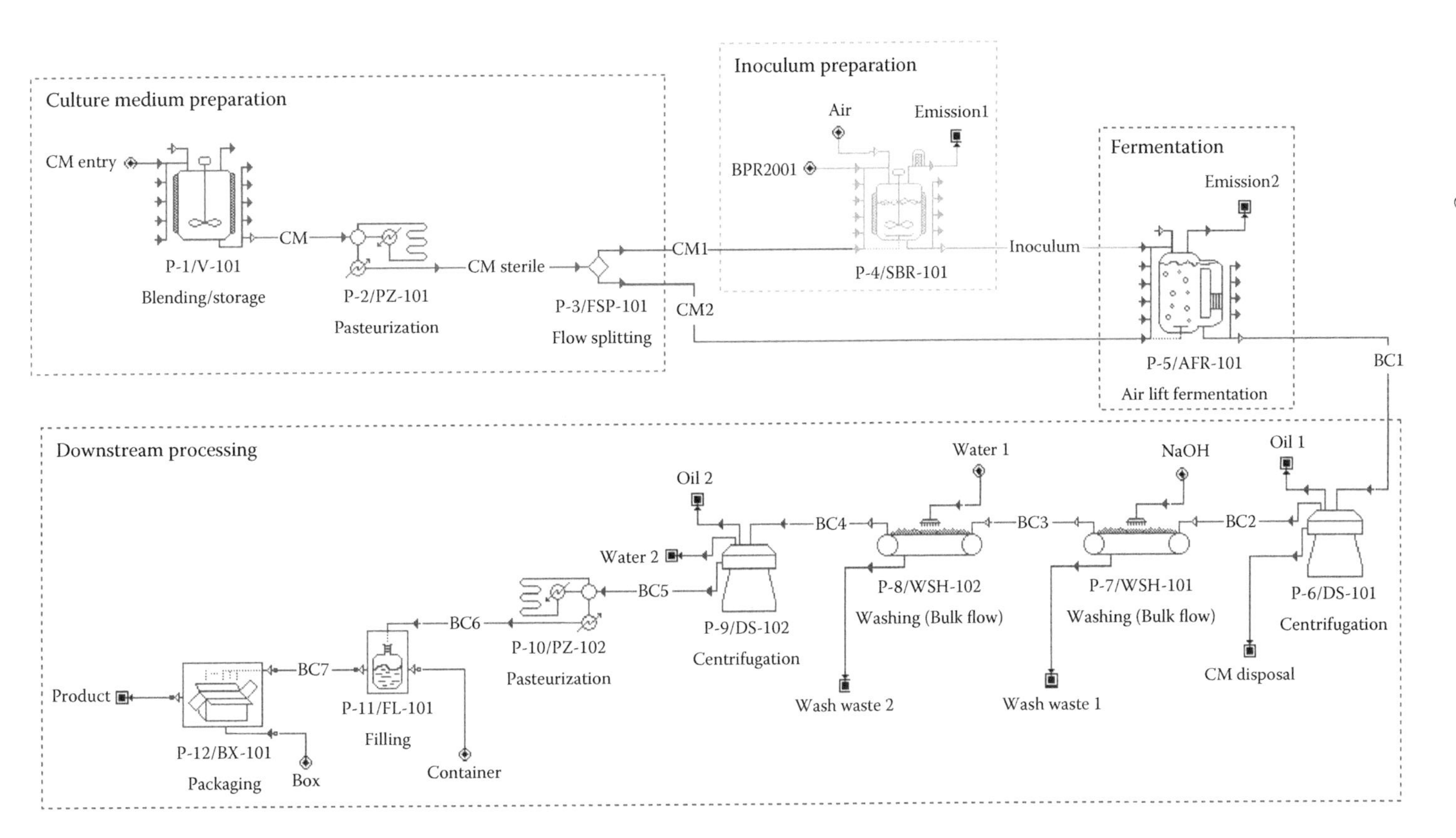

FIGURE 1.1 Super-Pro process flowsheet of the fermentative production of bacterial cellulose.

TABLE 1.3
Equipment Specifications and Costs (2016 prices)

Name	Type	Units	Size (Capacity)	Purchase Cost (US$/unit)
V-101	Blending tank	1	6,044.90 L	272,000.00
PZ-101	Pasteuriser	1	1,360.10 L/h	23,000.00
PZ-102	Pasteuriser	1	3,638.29 L/h	37,000.00
SBR-101	Seed bioreactor	1	737.85 L	679,000.00
AFR-101	Airlift fermenter	1	6,080.69 L	255,000.00
DS-101	Disc-stack centrifuge	1	21,890.48 L/h	1,941,000.00
DS-102	Disc-stack centrifuge	1	15,305.54 L/h	1,600,000.00
WSH-101, 102	Washer (Bulk Flow)	2	3,813.49 kg/h	20,000.00
FL-101	Filler	1	724.84 units/h	25,000.00[a]
BX-101	Packer	1		20,000.00[a]
Unlisted equipment				1,223,000.00
Total				6,115,000.00

[a] Estimate costs obtained from local companies. Remaining price estimates were provided by Super-Pro simulation software.

On fermentation ('Downstream Processing' stage), the resulting BC is collected, centrifuged (DS-101) and washed with NaOH and water (WSH-101, 102) to remove the bacteria and remnants of the culture medium. The purified BC is once more centrifuged (DS-102), pasteurised (PZ-102) and packed (in plastic bags and cardboard boxes; FL-101 and BX-101). Based on the inserted data (regarding unit operations, reaction kinetics, raw materials, desired BC production etc.), Super-Pro Designer can estimate the equipment size and cost (Tables 1.3 and 1.4) by using built-in cost correlations from data derived from a number of vendors and literature sources (data from some of the equipments were obtained from direct contact with several companies as these were not available in Super-Pro). The total cost of the equipment is US$6.12 million.

Table 1.4 summarises the resulting estimation of the capital investment costs, operating costs and profitability analysis of the BC fermentation process simulation. Some of the financial data were calculated based on information outlined by Peters and Timmerhaus [43]. The TCI for an industrial facility capable of producing 504 tons/year of BC was estimated to be near US$17.9 million, with direct costs (DCs) representing 70% of the TCI. DCs pertain to equipment and installation, piping, instrumentation, insulation, electrical facilities, building costs, yard improvements and auxiliary facilities and land. Indirect costs (30% of the TCI) consist of engineering and construction costs. Contingency charges (US$1.5 million) are extra costs added into a project budget to accommodate variations in the cost estimates.

TABLE 1.4
Cost Structure (in US$) and Profitability Analysis of the BC Fermentation Process

Estimation of Capital Investment Cost	
Direct costs (DCs)	**12,481,330.50**
Purchased equipment (PE)	6,115,000.00
Installation, including insulation and painting	1,212,634.38
Instrumentation and controls	1,010,528.65
Piping	808,422.92
Electrical	909,475.79
Buildings, process and auxiliary	1,010,528.65
Service facilities and yard improvements	404,211.46
Land	1,010,528.65
Indirect costs (ICs)	**3,072,007.10**
Engineering and supervision	848,844.07
Construction expense and contractor's fee	808,422.92
Contingency	1,414,740.11
Fixed capital investment (FCI)	**15,553,337.59**
Working capital (WC)	**2,333,000.64**
Total capital investment (TCI)	**17,886,338.23**
Estimation of the Annual Product Manufacturing Cost	
Manufacturing costs (MC)	**6,152,457.31**
Direct production cost (Variable costs)	**2,395,793.11**
Raw materials	34,892.95
Operating labour (OL)	806,000.00
Direct supervisory and clerical labour (DS & CL)	141,050.00
Utilities (Electricity, Steam, Chilled Water)	750,000.00
Maintenance and repairs (2%–10% FCI)	466,600.13[a]
Operating supplies (0.5%–1% FCI)	116,650.03[a]
Laboratory charges (10%–20% OL)	80,600.00[a]
Fixed charges	**2,908,474.13**
Depreciation (13% FCI machin. & equip. + 2%–3% FCI Buildings)	2,410,767.33[a]
Local taxes	388,833.44[a]
Insurances	108,873.36[a]
Plant overhead costs (50%–70% OL, DS & CL, and M & R)	**848,190.08**
General expenses (GE)	**1,831,816.91**
Administrative costs (2%–6% of TPC)	300,000.00[a]
Distribution and selling costs (2%–20% of TPC)	262,500.00[a]
Research and development costs (5% of TPC)	375,000.00[a]
Financing (interest) (0%–10% TCI)	894,316.91
Total product cost (*TPC*): MC + GE	**7,984,274.22**
Profitability[a] (504,000 kg/year, at a selling price of US$25/kg)	
Total income: Selling price × quantity of product	12,600,000.00
Gross income: Total income − Total product cost	4,615,725.78

(*Continued*)

TABLE 1.4 (*Continued*)
Cost Structure (in US$) and Profitability Analysis of the BC Fermentation Process

Estimation of the Annual Product Manufacturing Cost	
Taxes: 30%–40% Gross income	1,615,504.02
Net profit: Gross income − Taxes	3,000,221.75
Rate of return: Net profit*100/TCI	7%
Payout period: FCI/(Net profit + depreciation)	5 years

[a] Calculated as described in Peters and Timmerhaus [43].

This capital allows to compensate for unpredictable expenses, minor process changes, price fluctuations and estimating errors.

The annual manufacturing (product or production) costs (total product costs [TPC]), totalling almost US$8 million, include elements that contribute directly to the production costs (such as direct operating costs, fixed charges and plant overhead costs) and general expenses. Operating labour (OL; 34%) and utility costs (31%) are the most representative costs of the direct production costs (which total US$2.4 million). Fixed charges, totalling US$3 million, relate to the physical plant in itself, thus unaffected by the productivity levels. These include depreciation, local taxes, insurances and rent. Depreciation (70% of the fixed charges) is a time-dependent operating cost that represents a fixed capital loss mostly due to equipment and facilities wear out and obsolescence. Plant overhead costs (US$848,190.08) include charges for services that are not attributable to the cost of the product, such as medical service, safety and protection, storage facilities, plant superintendence, cafeteria, janitorial services, administrative and accounting services. The general expenses (US$1.3 million) cover the management costs to develop new processes (research and development). Administrative costs include salaries for administrators, accounting, legal support and computer support, as well as office supply and equipment, administrative buildings and so on. The sum of the manufacturing costs (MC; 77% of the TPC) and general expenses (GE; 23% of the TPC) makes up the total product manufacturing costs (TPC; almost US$8 million).

Based on the aforementioned capital expenses, profitability analysis was done considering a market price for BC at US$25/kg (corresponding to packed BC as the final selling product with 99% moisture). BC is thus an expensive product, as compared with other celluloses that are recently reaching the market: US$15–20/kg dry weight for hydrocolloidal microcrystalline cellulose, US$50–100/kg dry weight for nanocellulose. The resulting net profit amounts to US$3 million per year. The return on investment (of 7%) is the ratio of profit to investment and measures how effectively a company converts the invested capital into profit (i.e. it represents the return per dollar invested). Finally, the payback period, representing the length of time necessary to recover the capital investment, was calculated to be 5 years.

1.5 CONCLUSION

Ever since its discovery, BC has been gaining significant attention from academia and industry in various research fields, which suggests that new potential market segments (aside from the biomedical one) are indeed waiting for a 'new product' (i.e. new properties) to be widely available. However, biotechnological processes can only be realised on controlling several operating parameters (such as fermentation time, temperature, pH, substrate pre-treatment, inoculum–substrate ratio). Extensive efforts have been devoted to determine the scientific and technological factors that mediate BC production. As observed in this chapter through process simulation, biotechnological processes are highly capital-intensive. Coupled to the low BC yields and despite the use of low cost substrate in this simulation, the high capital investment and high operating costs associated present a strong economic constraint to the commercialisation of BC at a 'low' cost. Data gathered here showed that although it is possible to devise an economically feasible biotechnological process for BC production, the high selling costs would indeed restrain BC to high-value niche markets.

REFERENCES

1. Oliveira RL, Silva Barud H, Assunção RMN, Silva Meireles C, Carvalho GO, Filho GR et al. Synthesis and characterization of microcrystalline cellulose produced from bacterial cellulose. *Journal of Thermal Analysis and Calorimetry*. 2011;106(3):703–709.
2. Yamada Y, Yukphan P, Vu HTL, Muramatsu Y, Ochaikul D, Tanasupawat S et al. Description of Komagataeibacter gen. nov., with proposals of new combinations (Acetobacteraceae). *The Journal of General and Applied Microbiology*. 2012;58(5):397–404.
3. Yamada Y, Hoshino K-I, Ishikawa T. Taxonomic. Studies of acetic acid bacteria and allied organisms. Part XI. The phylogeny of acetic acid bacteria based on the partial sequences of 16S ribosomal RNA: The elevation of the subgenus gluconoacetobacter to the generic level. *Bioscience Biotechnology and Biochemistry*. 1997;61(8):1244–1251.
4. Yamada Y, Hosono R, Lisdyanti P, Widyastuti Y, Saono S, Uchimura T et al. Identification of acetic acid bacteria isolated from Indonesian sources, especially of isolates classified in the genus gluconobacter. *The Journal of General and Applied Microbiology*. 1999;45(1):23–28.
5. Matsushita K, Inque T, Theeragool G, Trcek J, Toyama H, Adachi O. Acetic acid production in acetic acid bacteria leading to their 'death' and survival. In: Yamada M (Ed.), *Survival and Death in Bacteria*. Research Signpost, Kerala/India; 2005, pp. 169–181.
6. Sievers M, Swings J. Family II. Acetobacteraceae. In: Staley JT, Boone DR, Brenner DJ, De Vos Editor P, Goodfellow M, Krieg NR et al. (Eds.), *Bergey's Manual® of Systematic Bacteriology* (2nd ed.), Vol. 2C. New York: Springer-Verlag New York, LLC; 2005, pp. 41–94.
7. Kersters K, Lisdiyanti P, Komagata K, Swings J. The family acetobacteraceae: the genera acetobacter, acidomonas, asaia, aluconacetobacter, aluconobacter, and kozakia. In: Dworkin M, Falkow S, Rosenberg E, Schleifer K-H, Stackebrandt E (Eds.), *Prokaryotes* (3 ed.), Vol. 5. New York: Springer; 2006, pp. 163–200.
8. Cleenwerck I, De Wachter M, Gonzalez A, De Vuyst L, De Vos P. Differentiation of species of the family Acetobacteraceae by AFLP DNA fingerprinting: Gluconacetobacter kombuchae is a later heterotypic synonym of *Gluconacetobacter hansenii*. *International Journal of Systematic and Evolutionary Microbiology*. 2009;59(7):1771–1786.

9. Watanabe K, Eto Y, Takano S, Nakamori S, Shibai H, Yamanaka S. A new bacterial cellulose substrate for mammalian-cell culture—A new bacterial cellulose substrate. *Cytotechnology*. 1993;13(2):107–114.
10. Iguchi M, Yamanaka S, Budhiono A. Bacterial cellulose—A masterpiece of nature's arts. *Journal of Materials Science*. 2000;35(2):261–270.
11. Klemm D, Schumann D, Udhardt U, Marsch S. Bacterial synthesized cellulose—Artificial blood vessels for microsurgery. *Progress in Polymer Science*. 2001;26(9):1561–1603.
12. Svensson A, Nicklasson E, Harrah T, Panilaitis B, Kaplan DL, Brittberg M et al. Bacterial cellulose as a potential scaffold for tissue engineering of cartilage. *Biomaterials*. 2005;26(4):419–431.
13. Czaja W, Krystynowicz A, Bielecki S, Brown JRR. Microbial cellulose—The natural power to heal wounds. *Biomaterials*. 2006;27(2):145–151.
14. Czaja WK, Young DJ, Kawecki M, Brown RM, Jr. The future prospects of microbial cellulose in biomedical applications. *Biomacromolecules*. 2007;8(1):1–12.
15. Andrade FK, Pertile RAN, Dourado F, Gama FM. Bacterial cellulose: Properties, production and applications. In: Lejeune A, Deprez T (Eds.), *Cellulose: Structure and Properties, Derivatives and Industrial Uses*. New York: Nova Science Publishers; 2010, pp. 427–458.
16. Keshk SMAS. Bacterial cellulose production and its industrial applications. *Journal of Bioprocessing & Biotechniques*. 2014;04(02). doi: 10.4172/2155-9821.1000150.
17. Novaes AB, Jr., Novaes AB, Grissi MFM, Soares UN, Gabarra F. Gengiflex, an Alkali-Cellulose membrane for GTR: Histologic observations. *Brazilian Dental Journal*. 1993;4(2):65–71.
18. Novaes AB, Jr., Novaes AB. Bone formation over a TiAl6V4 (IMZ) implant placed into an extraction socket in association with membrane therapy (Gengiflex). *Clinical Oral Implants Research*. 1993;4(2):106–110.
19. Novaes AB, Jr., Novaes AB. Immediate implants placed into infected sites: A clinical report. *International Journal of Oral and Maxillofacial Implants*. 1995;10(5):609–613.
20. Novaes AB, Jr., Novaes AB. Soft tissue management for primary closure in guided bone regeneration: Surgical technique and case report. *International Journal of Oral and Maxillofacial Implants*. 1997;12(1):84–87.
21. dos Anjos B, Novaes AB, Meffert R, Barboza EP. Clinical comparison of cellulose and expanded polytetrafluoroethylene membranes in the treatment of class II furcations in mandibular molars with 6-month re-entry. *Journal of Periodontology*. 1998;69(4):454–459.
22. Fontana JD, De Souza AM, Fontana CK, Torriani IL, Moreschi JC, Gallotti BJ et al. Acetobacter cellulose pellicle as a temporary skin substitute. *Applied Biochemistry and Biotechnology*. 1990;24(1):253–264.
23. Bäckdahl H, Helenius G, Bodin A, Nannmark U, Johansson BR, Risberg B et al. Mechanical properties of bacterial cellulose and interactions with smooth muscle cells. *Biomaterials*. 2006;27(9):2141–2149.
24. Czaja WK, Young DJ, Kawecki M, Brown JRM. The future prospects of microbial cellulose in biomedical applications. *Biomacromolecules*. 2007;8(1):1–12.
25. Fu L, Zhang J, Yang G. Present status and applications of bacterial cellulose-based materials for skin tissue repair. *Carbohydrate Polymers*. 2013;92(2):1432–1442.
26. Laçin NT. Development of biodegradable antibacterial cellulose based hydrogel membranes for wound healing. *International Journal of Biological Macromolecules*. 2014;67:22–27.
27. Hong F, Wei B, Chen L. Preliminary study on biosynthesis of bacterial nanocellulose tubes in a novel double-silicone-tube bioreactor for potential vascular prosthesis. *BioMedical Research International*. 2015;2015:1–9.

28. Lapuz MM, Gallardo EG, Palo MA. The nata organism—Cultural requirements, characteristics and identity. *The Philippine Journal of Science*. 1967;96(2):91–109.
29. Okiyama A, Motoki M, Yamanaka S. Bacterial cellulose II. Processing of the gelatinous cellulose for food materials. *Food Hydrocolloids*. 1992;6(5):479–487.
30. Okiyama A, Motoki M, Yamanaka S. Bacterial cellulose IV. Application to processed foods. *Food Hydrocolloids*. 1993;6(6):503–511.
31. Lin KW, Lin HY. Quality characteristics of Chinese-style meatball containing bacterial cellulose (nata). *Journal of Food Science*. 2004;69(3):S107–S111.
32. Sheu F, Wang CL, Shyu YT. Fermentation of monascus purpureus on bacterial cellulose-nata and the color stability of Monascus-nata complex. *Journal of Food Science*. 2000;65(2):342–345.
33. Ochaikul D, Chotirittikrai K, Chantra J, Wutigornsombatkul S. Studies on fermentation of monascus purpureus TISTR 3090 with bacterial cellulose from *Acetobacter xylinum* TISTR 967. *KMITL Science and Technology Journal*. 2006;6(1):13–17.
34. Lapuz MM, Gallardo EG, Palo MA. The nata organism—Cultural requirements, characteristics, and identify. *The Philippine Journal of Science*. 1967;96(2):91–109.
35. Affairs OoI, Council NR. *Applications of Biotechnology in Traditional Fermented Foods: Report of an Ad Hoc Panel of the Board on Science and Technology for International Development*. Washington, DC: The National Academies Press; 1992.
36. Seumahu CA, Suwanto A, Hadisusanto D, Suhartono MT. The dynamics of bacterial communities during traditional nata de coco fermentation. *Microbiology Indonesia*. 2007;1(2):65–68.
37. Vandamme E, Baets S, Vanbaelen A, Joris K, Wulf P. Improved production of bacterial cellulose and its application potential. *Polymer Degradation and Stability*. 1998;59:93–99.
38. Chawla PR, Bajaj IB, Survase SA, Singhal RS. Microbial cellulose: Fermentative production and applications. *Food Technology and Biotechnology*. 2009;47(2):107–124.
39. Shah N, Ul-Islam M, Khattak WA, Park JK. Overview of bacterial cellulose composites: A multipurpose advanced material. *Carbohydrate Polymers*. 2013;98(2):1585–1598.
40. Lee KY, Buldum G, Mantalaris A, Bismarck A. More than meets the eye in bacterial cellulose: Biosynthesis, bioprocessing, and applications in advanced fiber composites. *Macromol Bioscience*. 2014;14(1):10–32.
41. Rouf SA, Douglas PL, Moo-Young M, Scharer JM. Computer simulation for large scale bioprocess design. *Biochemical Engineering Journal*. 2001;8(3):229–234.
42. Dimian AC, Bildea CS, Kiss AA. Process design project. In: Gani R (Ed.), *Integrated Design and Simulation of Chemical Processes*, Vol. 13. Amsterdam, The Netherlands: Elsevier Science; 2014, pp. 557–570.
43. Peters MS, Timmerhaus KD. Cost estimation. In: Peters MS, Timmerhaus KD (Eds.), *Plant Design and Economics for Chemical Engineers* (4th ed.). Singapore: McGraw-Hill; 1991. pp. 150–215.
44. Economic evaluation of projects. In: Dimian AC (Ed.), *Integrated Design and simulation of Chemical Processes* (1st ed.). Amsterdan, The Netherlands: Elsevier; 2003, pp. 571–604. http://www.sciencedirect.com/science/article/pii/S1570794603800391?via%3Dihub.
45. Cheng KC, Catchmark JM, Demirci A. Effects of CMC addition on bacterial cellulose production in a biofilm reactor and its paper sheets analysis. *Biomacromolecules*. 2011;12(3):730–736.
46. Bae S, Sugano Y, Shoda M. Improvement of bacterial cellulose production by addition of agar in a jar fermentor. *Journal of Bioscience and Bioengineering*. 2004;97(1):33–38.
47. Toyosaki H, Naritomi T, Seto A, Matsuoka M, Tsuchida T, Yoshinaga F. Screening of bacterial cellulose-producing acetobacter strains suitable for agitated culture. *Bioscience, Biotechnology, and Biochemistry*. 1995;59(8):1498–1502.
48. Bae SO, Shoda M. Production of bacterial cellulose by *Acetobacter xylinum* BPR2001 using molasses medium in a jar fermentor. *Applied Microbiology and Biotechnology*. 2005;67(1):45–51.

49. Chao Y, Mitarai M, Sugano Y, Shoda M. Effect of addition of water-soluble polysaccharides on bacterial cellulose production in a 50-L airlift reactor. *Biotechnology Progress*. 2001;17(4):781–785.
50. Chao Y, Ishida T, Sugano Y, Shoda M. Bacterial cellulose production by *Acetobacter xylinum* in a 50-L internal-loop airlift reactor. *Biotechnology and Bioengineering*. 2000;68(3):345–352.
51. Chao Y, Sugano Y, Shoda M. Bacterial cellulose production under oxygen-enriched air at different fructose concentrations in a 50-liter, internal-loop airlift reactor. *Applied Microbiology and Biotechnology*. 2001;55(6):673–679.
52. Joseph G, Rowe GE, Margaritis A, Wan W. Effects of polyacrylamide-co-acrylic acid on cellulose production by *Acetobacter xylinum*. *Journal of Chemical Technology and Biotechnology*. 2003;78(9):964–970.
53. Cheng K-C, Catchmark JM, Demirci A. Effect of different additives on bacterial cellulose production by *Acetobacter xylinum* and analysis of material property. *Cellulose*. 2009;16(6):1033–1045.
54. Cheng HP, Wang PM, Chen JW, Wu WT. Cultivation of *Acetobacter xylinum* for bacterial cellulose production in a modified airlift reactor. *Biotechnology and Applied Biochemistry*. 2002;35(Pt 2):125–132.
55. Yang Y, Park S, Hwang J, Pyun Y, Kim Y. Cellulose production by *Acetobacter xylinum* BRC5 under agitated condition. *Journal of Fermentation and Bioengineering*. 1998;85(3):312–317.
56. Son HJ, Heo MS, Kim YG, Lee SJ. Optimization of fermentation conditions for the production of bacterial cellulose by a newly isolated *Acetobacter* sp. A9 in shaking cultures. *Biotechnology and Applied Biochemistry*. 2001;33(Pt 1):1–5.
57. Heo M-S, Son H-J. Development of an optimized, simple chemically defined medium for bacterial cellulose production by *Acetobacter* sp. A9 in shaking cultures. *Biotechnology and Applied Biochemistry*. 2002;36:41–45.
58. Park JK, Jung JY, Park YH. Cellulose production by *Gluconacetobacter hansenii* in a medium containing ethanol. *Biotechnology Letters*. 2003;25(24):2055–2059.
59. Jung JY, Park JK, Chang HN. Bacterial cellulose production by *Gluconacetobacter hansenii* in an agitated culture without living non-cellulose producing cells. *Enzyme and Microbial Technology*. 2005;37(3):347–354.
60. Okiyama A, Shirae H, Kano H, Yamanaka S. Bacterial cellulose I. Two-stage fermentation process for cellulose production by *Acetobacter aceti*. *Food Hydrocolloids*. 1992;6(5):471–477.
61. Son H, Kim H, Kim K, Kim H, Kim Y, Lee S. Increased production of bacterial cellulose by *Acetobacter* sp. V6 in synthetic media under shaking culture conditions. *Bioresource Technology*. 2003;86(3):215–219.
62. Seto A, Saito Y, Matsushige M, Kobayashi H, Sasaki Y, Tonouchi N et al. Effective cellulose production by a coculture of *Gluconacetobacter xylinus* and *Lactobacillus mali*. *Applied Microbiology and Biotechnology*. 2006;73(4):915–921.
63. Zhou LL, Sun DP, Hu LY, Li YW, Yang JZ. Effect of addition of sodium alginate on bacterial cellulose production by *Acetobacter xylinum*. *Journal of Industrial Microbiology and Biotechnology*. 2007;34(7):483–489.
64. Jung JY, Khan T, Park JK, Chang HN. Production of bacterial cellulose by Gluconacetobacter hansenii using a novel bioreactor equipped with a spin filter. *Korean Journal of Chemical Engineering*. 2007;24(2):265–271.
65. Kim Y-J, Kim J-N, Wee Y-J, Park D-H, Ryu H-W. Bacterial cellulose production by *gluconacetobacter* sp. RKY5 in a rotary biofilm contactor. In: Mielenz J, Klasson KT, Adney W, McMillan J (Eds.), *Applied Biochemistry and Biotecnology*. New York: Humana Press; 2007, pp. 529–537.

2 Production of Cellulose Nanofibres

Antonio Norio Nakagaito, Kentaro Abe and Hitoshi Takagi

CONTENTS

2.1 Introduction 17
2.2 The Mainstream Cellulose Nanofibre Extraction Processes 17
2.3 Recent Developments in Cellulose Nanofibre Extraction 21
2.4 Conclusion 23
References 24

2.1 INTRODUCTION

Cellulose is the most abundant polysaccharide on earth and is synthesised by higher plants from carbon dioxide and water in a process driven by sunlight energy. It is produced annually in massive amounts and is naturally decomposed in a perfectly sustainable carbon neutral process. Cellulose forms the framework of the cell wall of plants and is always present in the form of a semi-crystalline structure of a few nanometres in diameter known as cellulose nanofibres. The crystalline portion possesses a Young's modulus of 138 GPa [1], and the tensile strength along the nanofibre length is estimated to be in the range of 1.6–3 GPa [2]. The molecular structure of cellulose consists of long chains of glucose rings connected without folding, very much resembling the way in which benzene rings are joined to form aramid molecules. Even the mechanical properties and density of cellulose nanofibres are also similar to those that of aramid fibres [3]. The only caveat is that cellulose nanofibres are always part of a complex biocomposite in the plant cell wall and have to be properly extracted and individualised in order to be exploited in intended applications.

2.2 THE MAINSTREAM CELLULOSE NANOFIBRE EXTRACTION PROCESSES

The biggest challenge to obtain cellulose nanofibres is the extraction from plant fibres. The nanofibres are originally the reinforcing phase of a complex biocomposite, and therefore their extraction must be carried out with the least possible damage by achieving nanoscale diameters but maintaining the long axial lengths to attain high aspect ratio. This chapter is by no means intended to be a comprehensive review of the subject but just a guide about the basic principles involved in cellulose nanofibrillation.

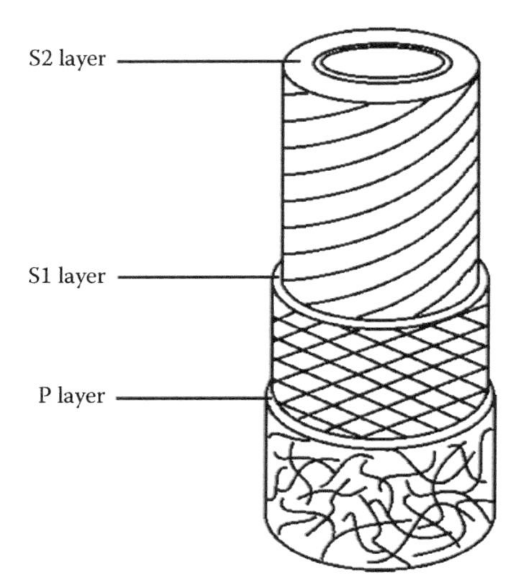

FIGURE 2.1 Layered structure of a wood pulp fibre cell wall.

Plant fibres are one of the different types of plant cells, and being so they possess cell walls in the form of hollow tubes made up of tinier fibrils embedded in matrix substances such as lignin, hemicelluloses and pectin. These reinforcing fibrils, also known as microfibrils, from now on will be referred to as cellulose nanofibres (CNFs). The structure of these cell walls is multilayered; for instance, in wood (Figure 2.1) the thickest layer (S2) has the CNFs aligned forming a helix, which is responsible for the longitudinal tensile strength of the fibre. There are two thinner outer layers (P and S1) with CNFs oriented rather randomly, wrapping around the spirally wound nanofibres. Therefore, the extraction of CNFs should start by removing the outer layers encasing the more oriented S2 layer. Among the various extraction processes proposed so far, most are mechanical and are summarised as the following methods.

In the beginning of the 1980s, a cellulose morphology called microfibrillated cellulose was developed by Turbak et al. [4]. It was initially intended to be used as a food additive and consisted of a new form of expanded high-volume cellulose with greatly expanded surface area, obtained by a homogenisation process. The fibrillation is carried out by a device called high-pressure homogeniser, which is in essence a spring-loaded valve (Figure 2.2). The wood pulp fibres are first treated by a disc refiner in order to peel off the outer P and S1 layers and to expose the more oriented helically wound nanofibres of the S2 layer. Next, a dilute aqueous suspension of refiner pre-treated fibres is injected under high pressure into the high-pressure homogeniser. The pressure opens the valve slightly; the fibres pass through a small slit being subjected to shear forces and acceleration and end up by colliding against the impact ring as shown in Figure 2.2. Once the fibres reach the outlet side of the device, they are subjected to a large pressure drop that expands and separates them into smaller fibril bundles. Because of the pressure drop, the valve shuts but soon opens again due to pressure build up in the inlet, repeating the cycle in a reciprocating motion. The process is iterated by passing the fibre suspension several times

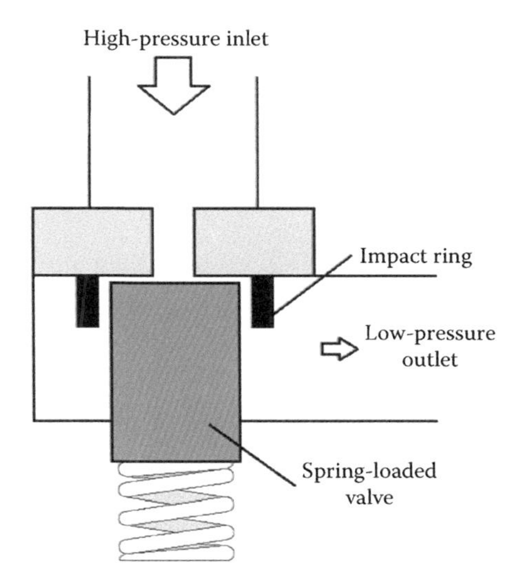

FIGURE 2.2 Schematic diagram of a high-pressure homogeniser.

through the high-pressure homogeniser. The fibrillation is accomplished by subjecting the fibres to shear and impact forces, accompanied by sudden pressure drop. This microfibrillated cellulose is partially nanofibrillated, with most nanofibre bundles having sub-micrometre diameters, and has been commercially available for a long time since its inception.

Much later in the late 1990s, another nanofibrillation process was proposed by Taniguchi and Okamura [5] using a grinder. A specialised grinder consisting of a stator and rotor grindstones (Figure 2.3) has the aperture adjusted to a point when the stones slightly touch each other's surface when dry. Then a dilute aqueous suspension of pulp fibres, usually below 1 wt.%, is poured into the grinder and as soon as the grindstones' surfaces get wet, the aperture is further closed by a few tenths of millimetre. Due to the hydrodynamic water pressure created by the spinning grindstone rotor, a tiny aperture is kept where the fibres slowly pass through and are subjected to shear and high pressure in the slit, suffering a sudden pressure drop when they leave the grindstone aperture. This pass is repeated several times as needed.

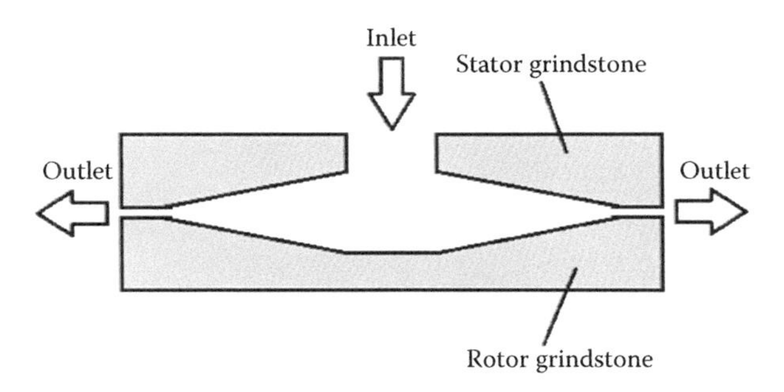

FIGURE 2.3 Schematic diagram of a grinder.

This process later received some important improvements that permitted the production of extremely fine and dimensionally uniform CNF. Abe et al. [6] proposed that fibres should be kept wet after pulping and before fibrillation by the grinder. If dried, CNFs would become bound to neighbouring nanofibres by hydrogen bonds that make posterior separation difficult without considerable damage. By the never-dried process, a single pass through the grinder produced nanofibres with diameters about 15 nm that are suitable for the production of optically transparent materials. Iwamoto et al. [7] showed that the severity of fibrillation by multiple passes through the grinder decreases the strength of the obtained nanofibres. Nanofibres produced by up to five passes through the grinder had finer morphology; however, the strength was significantly decreased due to both reduced crystallinity and degree of polymerisation of cellulose. They also clarified the beneficial role of hemicelluloses on nanofibrillation [8]. When pulp fibres are dried, the presence of hemicelluloses impedes the formation of irreversible hydrogen bonds between the CNFs. And by rewetting, hemicelluloses are plasticised by absorbing water, facilitating posterior nanofibrillation. The important role of hemicelluloses on nanofibrillation was later validated by Chaker et al. [9] using grass, sunflower stem and hardwood fibres as starting material.

Another major cellulose nanofibrillation process was described by Zimmermann et al. [10] who used a microfluidizer, a type of cell disruptor. As in other methods, a dilute aqueous suspension of fibres pre-treated by a high-performance disperser is introduced under high pressure into the microfluidiser. There are basically two designs of the device as seen in Figure 2.4. The Y-type separates the suspension into two streams that have to pass through microchannels where they are accelerated before being reunited by collision in an interaction chamber. The Z-type brings the highly pressurised suspension through a single microchannel where the flow is accelerated before colliding against a wall on the outlet. Similar to other processes,

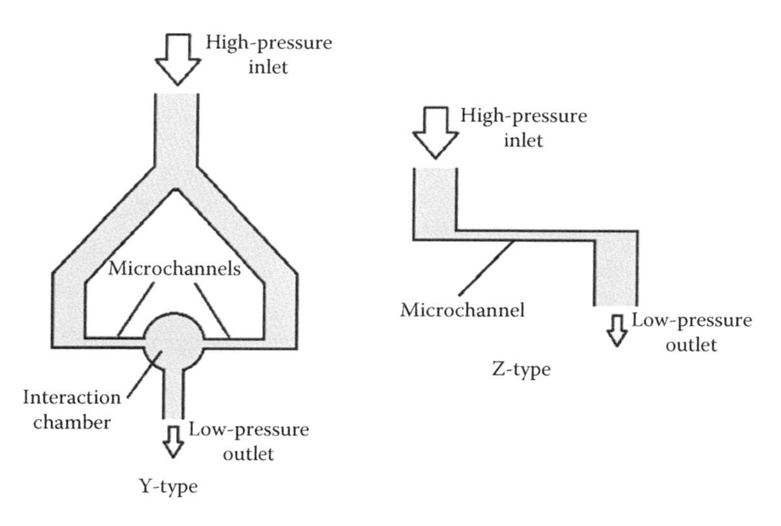

FIGURE 2.4 Schematic diagram of a microfluidiser.

the fibres are subjected to shear while passing through the microchannels, impact forces, and to pressure drop due to the pressure gap between inlet and outlet. This process too delivers high-quality CNFs. Improvements were implemented to this method such as the enzymatic pre-treatment of fibres described by Paakko et al. [11]. A mild enzymatic hydrolysis is intended to selectively dissolve the amorphous portions of cellulose before mechanical treatment to facilitate nanofibrillation.

In essence, all of the mainstream nanofibrillation processes are based on the application of shear and impact forces or pressure drop to the cellulosic fibres. Another common characteristic is the use of water as dispersion medium, as water swells the cellulosic fibres by its molecules bonding to the hydroxyl groups on the surface of CNFs, preventing the formation of interfibrillar hydrogen bonds. As the fibre suspensions are highly diluted, usually below 1 wt.% concentration, the amount of fibres processed is relatively small making the energy consumption high during nanofibrillation, whereas the nanofibre output yield is low, resulting in extremely costly CNFs. Other processes were later presented, but still as lab-scale operations. In cryocrushing [12], liquid nitrogen is used to freeze the water contained in the interstices of fibres that are disintegrated by a mortar and pestle. Another method uses ultrasonic cavitation in fibre aqueous suspensions to nanofibrillate [13]. The shockwaves produced by the collapse of microbubbles erode the surface of fibres, splitting them in the axial direction. Another process is the counter collision in water developed by Kondo [14], whose fibrillation principle derived many other variations of the method.

2.3 RECENT DEVELOPMENTS IN CELLULOSE NANOFIBRE EXTRACTION

Many other processes of CNF extraction were later developed but mostly by altering the pre-treatment of pulp fibres prior to mechanical fibrillation. Clever chemical approaches such as mild enzymatic hydrolysis, carboxymethylation, acetylation, oxidation, and other position-selective modification of the surface of CNFs have been applied extensively in order to facilitate posterior fibrillation, whether to reduce energy consumption or to improve the degree of nanofibrillation [15]. Among these chemical processes, the most successful process has been the one proposed by Saito et al. [16]. The method consists of oxidising the surface of the CNFs by a 2,2,6,6-tetramethylpiperidine-1-oxyl (TEMPO) radical-catalyzed process. Part of the hydroxyl groups on the surface of CNFs are converted into negatively charged sodium carboxylate groups loosening their mutual adhesion so that even the mild agitation by a blender is enough to obtain individualised CNFs of 3–5 nm in diameter.

Even though new nanofibrillation processes or variations of existing ones have been developed, still the production yields are low and the energy required is too high to be economically viable for large-scale production. Hence, as each extraction method delivers varying degrees of nanofibrillation, this should be determined by the end application. A partially nanofibrillated cellulose morphology with a broad distribution of diameters would be suitable as mechanical reinforcement in composites or as additives to strengthen papers, whereas a completely nanofibrillated morphology with diameters of few nanometres would be necessary to fabricate optically transparent composites. Cellulose morphologies should be able to satisfy the different application demands in terms of performance but also from the aspect of cost as well.

As most of these nanofibre extraction processes rely on highly specialised and costly devices, while the production yield is low and the energy consumption is high, a search for alternative more affordable processes is being continually pursued. One of these attempts was reported by Uetani and Yano [17], who used a kitchen blender to fibrillate never-dried wood pulp fibres. The process relies on the impact of fibres against the blender blades as a means to rupture the fibre cell walls to accomplish fibrillation. This idea is not completely new as an earlier patent by Galati [18] described the fibrillation of synthetic fibres by a kitchen blender. Nonetheless, the blending proved to be effective even in fibrillating more elaborate structures of plant fibres, showing promise as a lower cost nanofibre extraction at a laboratory scale. Uetani and Yano used a high-powered blender for food processing, which is capable of breaking down the parenchyma cells of vegetables and fruits to extract nutrients that eventually worked to break down the cell wall of fibres. The process is being improved by using agricultural crop residues instead of wood fibres [19] as the starting material and through modifications to the blender bottle to increase fibrillation efficiency. Preliminary results indicated that it is possible to obtain cellulose morphology similar to commercially available microfibrillated cellulose produced by the high-pressure homogeniser. Crop residues contain less lignin compared to wood, requiring fewer amounts of chemicals for pulping. In addition, another alternative cellulosic fibre source is cattle manure. The grass eaten by ruminants is mechanically reduced in size by mastication and later enzymatically hydrolyzed in the rumen in a way resembling the enzymatic pre-treatment described by Paako et al. [11]. The difference is that the pre-treatment is part of the nutritional process of cattle, and its by-product is a partially hydrolyzed cellulose pulp that requires less chemicals and less intensive mechanical nanofibrillation. By the proper choice of raw material and an affordable device, it would be possible to significantly decrease the overall cost of CNF extraction. Chaker et al. [9] demonstrated the possibility to extract high-quality CNF from sunflower stem fibres with a kitchen blender, especially if the hemicelluloses are left to facilitate fibrillation. Among agricultural waste products that contain cellulosic fibres, collenchyma and parenchyma cells are quite attractive for having simpler cell wall structures and hence are more suited to less intensive fibrillation treatments.

If a batch process using blenders limits its use to lab scale, ultrasonication [13] is an easily scalable process and can continuously treat fibre suspensions by the use of a flow cell. However, still a mechanical pre-treatment based on a blender to remove the outer P and S1 layers or even breakdown of the cell wall of fibres [20] would be necessary before the ultrasonication step. Another simple process is the one proposed by Abe [21], who was able to nanofibrillate once-dried pulp fibres using a beads mill by keeping the fibres in an alkali aqueous solution containing 8 wt.% NaOH during fibrillation. The concept is to swell the cellulose at the interfaces between nanofibres in order to break up the interfibrillar hydrogen bonds formed by drying and therefore to facilitate separation. The alkali concentration is carefully adjusted to avoid intra-crystalline swelling of cellulose that would result in crystalline structure conversion. Doing so, it is possible to produce fine 12–20 nm thick CNFs but keeping the native cellulose I crystal form.

Perhaps the fibrillation process closest to industrial scale is the one described by Suzuki et al. [22]. It consisted of using extrusion for both fibrillation and compounding. Never-dried pulp fibre pre-treated by a disc refiner was mixed with micrometre-sized polypropylene powder. The still wet mixture was then kneaded via a twin-screw extruder keeping the barrels cooled in order to attain fibrillation through shear forces. Afterwards, the fibrillated pulp and powdered polypropylene were melt-compounded with the excess moisture being evaporated through vents in the barrels. The composites obtained by injection moulding had the tensile modulus doubled and the strength increased by 50% when the fibre content was 50 wt.%. The cellulose morphology obtained by extrusion resembles the microfibrillated cellulose produced by a high-pressure homogeniser. The method has been optimised for nanofibrillation purposes as a possible mass production process of CNFs [23].

The newer extraction processes that have been reported lately are variations on existing methods. In a typical process, there are two main aspects that can be tweaked. The first one is the starting material where nanofibres are extracted from. There are three different kinds of plant cells available, namely sclerenchyma, collenchyma and parenchyma. The sclerenchyma cells comprise plant fibres in general, which are responsible for the mechanical support of the plant body. They are highly lignified and possess complex multilayered cell walls aiming mechanical stiffness. Collenchyma cells that consist of stalks are less lignified and are mostly produced by residues of cereal crops. Parenchyma cells are those intended for storage of nutrients and are made up of thin and flexible cell walls containing the least lignin among the different cell types. In that sense, instead of using wood pulp fibres as raw material, it would be advantageous to use collenchyma and parenchyma cells because of the lower cost as by-products, in addition to the less intensive fibrillation required [9]. Another important aspect is the pre-treatment prior to fibrillation. Depending on the raw material chosen, cost savings could be achieved by the use of lesser amounts of chemicals or less energy-intensive mechanical pre-treatments. If these two factors were properly settled, the final step of fibrillation could very much be facilitated, translating into lower energy consumption and related cost reduction.

2.4 CONCLUSION

The major processes of CNF extraction are fibrillation methods based on the application of shear or impact forces or sudden pressure drop to separate and individualise nanofibres. Although some methods produce extremely fine nanofibre morphologies, the high energy consumption, low yields and consequent overall high cost are the most serious impediments for mass market applications. A lower cost process more accessible to researchers worldwide could accelerate the development of new materials based on CNFs. Even though the nanofibrillation processes are limited and based on the same physical principles, optimisation is always confined to the choice of raw material and then to the improvements in pre-treatment and fibrillation methods. Unless a completely new fibrillation concept that is able to drastically reduce nanofibre production cost is conceived, morphologies with different degrees of nanofibrillation have to find the most suitable applications. A completely nanofibrillated morphology is not needed for most applications requiring mechanical reinforcement, whereas high-end materials with optical transparency depend directly on the nanosize uniformity of reinforcing elements. At this moment, the wider application of CNF is in need of a really low-cost extraction process.

REFERENCES

1. Nishino T, Takano K, Nakamae K. Elastic-modulus of the crystalline regions of cellulose polymorphs. *Journal of Polymer Science Part B: Polymer Physics.* 1995;33(11):1647–1651.
2. Saito T, Kuramae R, Wohlert J, Berglund LA, Isogai A. An ultrastrong nanofibrillar biomaterial: The strength of single cellulose nanofibrils revealed via sonication-induced fragmentation. *Biomacromolecules.* 2013;14(1):248–253.
3. Gordon JE. *The New Science of Strong Materials.* Princeton, NJ: Princeton University Press; 1976.
4. Turbak AF, Snyder FW, Sandberg KR. Microfibrillated cellulose, a new cellulose product: Properties, uses, and commercial potential. *Journal of Applied Polymer Science: Applied Polymer Symposium.* 1983;37:815–827.
5. Taniguchi T, Okamura K. New films produced from microfibrillated natural fibres. *Polymer International.* 1998;47(3):291–294.
6. Abe K, Iwamoto S, Yano H. Obtaining cellulose nanofibers with a uniform width of 15 nm from wood. *Biomacromolecules.* 2007;8(10):3276–3278.
7. Iwamoto S, Nakagaito AN, Yano H. Nano-fibrillation of pulp fibers for the processing of transparent nanocomposites. *Applied Physics A: Materials Science and Processing.* 2007;89(2):461–466.
8. Iwamoto S, Abe K, Yano H. The effect of hemicelluloses on wood pulp nanofibrillation and nanofiber network characteristics. *Biomacromolecules.* 2008;9(3):1022–1026.
9. Chaker A, Alila S, Mutjé P, Vilar MR, Boufi S. Key role of the hemicellulose content and the cell morphology on the nanofibrillation effectiveness of cellulose pulps. *Cellulose.* 2013;20(6):2863–2875.
10. Zimmermann T, Pöhler E, Geiger T. Cellulose fibrils for polymer reinforcement. *Advanced Engineering Materials.* 2004;6(9):754–761.
11. Pääkko M, Ankerfors M, Kosonen H, Nykänen A, Ahola S, Österberg M et al. Enzymatic hydrolysis combined with mechanical shearing and high-pressure homogenization for nanoscale cellulose fibrils and strong gels. *Biomacromolecules.* 2007;8(6):1934–1941.
12. Chakraborty A, Sain M, Kortschot M. Cellulose microfibrils: A novel method of preparation using high shear refining and cryocrushing. *Holzforschung.* 2005;59(1):102–107.
13. Zhao HP, Feng XQ, Gao HJ. Ultrasonic technique for extracting nanofibers from nature materials. *Applied Physics Letters.* 2007;90(7):073112.
14. Kondo T. New aspects of cellulose nanofibers. *Mokuzai Gakkaishi.* 2008;54(3):107–115.
15. Isogai A. Wood nanocelluloses: Fundamentals and applications as new bio-based nanomaterials. *Journal of Wood Science.* 2013;59(6):449–459.
16. Saito T, Nishiyama Y, Putaux JL, Vignon M, Isogai A. Homogeneous suspensions of individualized microfibrils from TEMPO-catalyzed oxidation of native cellulose. *Biomacromolecules.* 2006;7(6):1687–1691.
17. Uetani K, Yano H. Nanofibrillation of wood pulp using a high-speed blender. *Biomacromolecules.* 2011;12(2):348–353.
18. Galati CC, (Motion Control Industries, Inc.). Method of fibrillating fibers. In: *Patent US*, (Ed.), Charlottesville, VA: Motion Control Industries; 1989.
19. Nakagaito AN, Nakano K, Takagi H, Pandey JK. Extraction of cellulose nanofibers from grass by a domestic blender. *7th International Workshop on Green Composites*, Hamamatsu, Japan; 2012, pp. 6–8.

20. Chen WS, Abe K, Uetani K, Yu HP, Liu YX, Yano H. Individual cotton cellulose nanofibers: pretreatment and fibrillation technique. *Cellulose.* 2014;21(3):1517–1528.
21. Abe K. Nanofibrillation of dried pulp in NaOH solutions using bead milling. *Cellulose.* 2016;23(2):1257–1261.
22. Suzuki K, Okumura H, Kitagawa K, Sato S, Nakagaito AN, Yano H. Development of continuous process enabling nanofibrillation of pulp and melt compounding. *Cellulose.* 2013;20(1):201–210.
23. Ho TTT, Abe K, Zimmermann T, Yano H. Nanofibrillation of pulp fibers by twin-screw extrusion. *Cellulose.* 2015;22(1):421–433.

3 Preparation of Cellulose Nanocrystals

Background, Conventions and New Developments

Eero Kontturi

CONTENTS

3.1 Introduction 27
3.2 Background on Cellulose Nanocrystal Preparation 29
3.3 Current Paradigm: Sulphuric Acid Hydrolysis 30
  3.3.1 Effect of Reaction Conditions 31
  3.3.2 Effect of Source Material 34
3.4 Preparation of Cellulose Nanocrystals with Other Mineral Acids 35
3.5 Enzymatic Hydrolysis for Cellulose Nanocrystal Preparation 36
3.6 Oxidation Routes to Cellulose Nanocrystals 36
3.7 Esterification Routes to Cellulose Nanocrystals 38
3.8 Note on the Yield of Cellulose Nanocrystals 39
3.9 Conclusion 40
References 40

3.1 INTRODUCTION

To put it concisely, cellulose nanocrystals (CNCs) are rigid, nanosized rods derived from native cellulosic fibres (Figure 3.1). They are commonly deemed as single crystals of cellulose, but precise evidence as to whether they consist of completely crystalline cellulose is missing. CNCs were first reported in 1949 [1] and the first microscopy images appeared in 1951 [2] in papers by Rånby when transmission electron microscopy had sufficiently matured and broken out of its esoteric status, but presumably CNCs have already been in existence at least to some extent ever since acid hydrolysis has been applied on cellulose. It is the works of Rånby, however, which are generally regarded as the original works in the canon of CNCs. Thereafter, scattered literature entries appeared throughout 1950s to 1990s, but the advent of modern CNC research began in 1992 when Derek Gray's group in Montreal, guided with the intuition of Jean-Francoise Revol, reported a discovery that CNCs spontaneously formed chiral nematic liquid crystal phases in aqueous suspensions after exceeding a certain critical concentration [3]. The research involving CNCs later

FIGURE 3.1 **(See colour insert.)** 5 × 5 μm^2 atomic force microscopy image of cellulose nanocrystals prepared from Whatman No. 1 filter paper (cotton linter source).

gained significant momentum towards the latter half of the first decade of 2000s when nanosized cellulose started to attract the widespread attention that it still enjoys today. Here, however, it is impossible to pinpoint an explicit trigger that would be similar to the seminal papers by Rånby or Revol et al. [1–3]. Nevertheless, it is fair to say that the nanotechnology boom that started in 1990s caught up with the trend on renewable materials and green chemistry in around 2005–2007 to give rise to the prolific literature on nanosized cellulosics, including CNCs. The research on CNCs focuses largely on what can be categorised as materials science and/or technology. The scope is vast, ranging from basic composite materials to sophisticated responsive devices and optical as well as electronic applications. Proportionally speaking, the efforts that make use of the liquid crystal properties of CNCs have waned, but they still form an important part of the literature, particularly within the high end of the applications.

Isolation of CNCs from the fibre matrix can be considered a significant bottleneck in their utilisation. The current paradigm with sulphuric acid relies essentially on the method introduced by Rånby and developed further by Mukherjee and Woods nearly 70 years ago [4], with refinements and fine-tunings made by the Gray group in the 1990s [3,5]. It is an expensive method, consuming a lot of water whereby the possibility to completely recycle the acid is prevented. By and large, it is reasonable to state that the challenges posed by the sulphuric acid method have strongly impeded the industrial development of CNCs. Although the use of alternative mineral acids was already advocated by Rånby [1,2], the concept of preparing CNCs with totally different approaches has received substantial attention only within the past 5 years. The purpose of this chapter is to give an up-to-date review of this development with a bias on the fundamental structure of native cellulose that allows the preparation of CNCs. Multiple reviews on CNCs or nanocellulose in general exist and many of them give accounts on the basic preparation techniques [6–12]. None of them, however, deal with the issue from the fundamental perspective and they all give increasingly technical descriptions on the preparation techniques. Moreover, many references describe new preparation procedures, particularly with novel source materials, but often they lack important characterisation data. Yields or proper size distribution histograms for CNC dimensions, for example, are rarely reported. This review aims at addressing all these issues in a comprehensive manner. However, the list of references is not exhaustive and the aim is rather to provide a broad overview of what the current state of CNC preparation is and what are the factors that influence it from a fundamental point of view.

3.2 BACKGROUND ON CELLULOSE NANOCRYSTAL PREPARATION

Isolation from CNC is enabled by the morphological integrity of cellulose that exists in nature as microfibrils, that is, ordered agglomerates of some tens of cellulose chains that form the main mechanical scaffold of, for example, plant cells [13–16]. In the simplified presentation, the disordered or 'amorphous' segments of the microfibril are selectively degraded leaving the crystallites, that is, cellulose nanocrystals, intact (Figure 3.2a). On biosynthesis, microfibrils are directly formed because cellulose crystallises simultaneously with its molecular synthesis. The common perception is that the microfibrils are semi-crystalline, implying that highly ordered crystalline regions are frequently interrupted by disordered regions [17,18]. The dislocations are often referred to as 'amorphous regions' in the literature, but most of the fundamental research community working with cellulose agree nowadays that the term 'amorphous' is erroneous or at least misleading. A specific study on the issue points out that in at least ramie fibres, the length of the disordered regions spans only four to five anhydroglucose units or 1.5% of the total mass of the cellulose microfibrils [17], which is way too short a length for

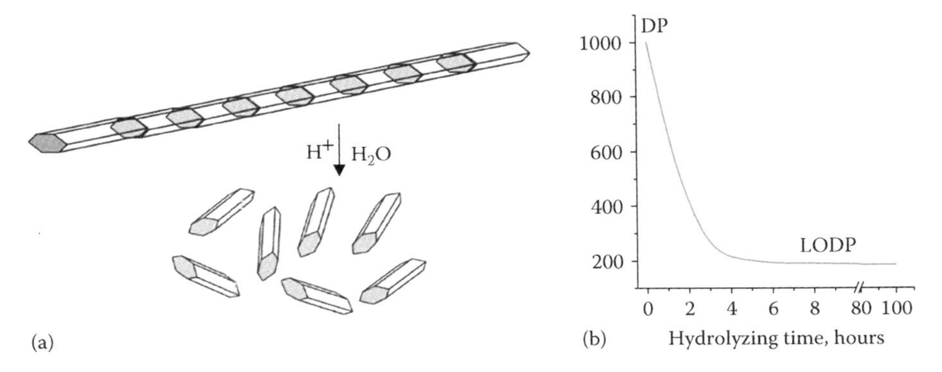

FIGURE 3.2 Principle behind preparing CNCs from native cellulose, (a) disordered segments in cellulose microfibrils are selectively degraded by acid hydrolysis, leaving crystallites intact, leading to isolation of CNCs and (b) during acid hydrolysis, the degradation of the disordered segments results in a rapid decrease in DP that subsequently reaches a nearly constant value (LODP) when only the recalcitrant crystallites are left.

bulky amorphous regions that are depicted in many schematic representations of cellulose microfibrils. Nor are the disordered regions similar to the genuine amorphous regions detected in several synthetic semi-crystalline polymers. In fact, the disordered segments in cellulose microfibrils can be viewed more like defects in the crystallites rather than representatives of amorphous polymers. For a more profound treatise on longitudinal disorder within cellulose microfibrils, the interested reader is referred to my recent review on the topic [19].

Whatever the actual physical mode of disordered cellulose may be, it is an undisputed fact that acid is able to selectively degrade the disordered segments in a native cellulose microfibril, leaving the crystallites intact. This selectivity is manifested by a phenomenon called levelling-off degree of polymerisation (LODP). Chronologically, when hydrolysing native cellulose with a strong acid, first the degree of polymerisation (DP) rapidly decreases before the degradation halts at a nearly constant value, the LODP (Figure 3.2b) [20,21]. This, the conventional wisdom holds, is the consequence of the longitudinal disorder in cellulose microfibrils: the disordered segments in microfibrils are selectively hydrolysed, whereas the extremely recalcitrant cellulose crystal is barely affected, thus leading to the LODP after the removal of disordered cellulose. The very same idea is behind the preparation of CNCs from native cellulose: disordered segments are hydrolysed with strong aqueous acid, leaving behind rigid crystalline rods, the CNCs (Figure 3.2a).

3.3 CURRENT PARADIGM: SULPHURIC ACID HYDROLYSIS

It is fair to say that probably >98% of the literature that deals with CNCs utilises sulphuric acid hydrolysis for their preparation. The reason for the use of sulphuric acid is simple: not only does it efficiently degrade cellulose but it also introduces sulphate half-esters on the surface of CNCs that are being isolated [6,7]. These sulphates bear a monovalent charge, resulting in colloidal stability for the CNC dispersions in water due to electrostatic repulsion.

Although sulphuric acid was used already in the 1950s for CNC preparation [2,4], the conditions for modern usage were consolidated in the 1990s by the Gray group [3,5]. Because of the extraordinary recalcitrance of cellulose, the sulphuric acid concentrations are generally very high around 63–66 wt.%, whereas the temperature is kept at moderately elevated levels between 40°C and 65°C. These conditions are also optimal for introducing a sufficient charge by sulphate esterification on the CNC surface. After the actual hydrolysis, the reaction is usually quenched in 10-fold volume of water, and the reaction mixture is purified first by repeated centrifugation and subsequently by dialysis. In the final step, the CNCs are properly dispersed by ultrasonic treatment and finally filtered in order to remove the non-nanosized fraction. Earlier, it was customary to perform an ion-exchange step with mixed bed resin, but this is usually omitted in the modern protocols. Recently, added purification steps with, for example, ethanol extraction have been added to the CNC preparation procedure in case high purity and reproducibility are in demand. In my group, for example, we first perform an ion exchange so that the sulphate groups bear a

Na^+ counterion, followed by freeze-drying, and subsequent ethanol extraction for 48 h, as described by Labet and Thielemans [22]. The extracted, dry CNC powder is then preserved in a vacuum desiccator and an appropriate amount is dispersed in water whenever needed. The counterion exchange is performed to facilitate the re-dispersion that is hindered with protonated CNCs [23]. We have noticed that purification is absolutely mandatory to obtain full reproducibility when surface sensitive methods have been used.

It is interesting to note how close the acid concentration is to the concentration of 72 wt.% used in analytical procedures that degrade even crystalline cellulose into monosaccharides, that is, glucose [24]. Altogether, the reaction window that leads to well isolated, properly sulphated CNCs is fairly narrow and indeed, relatively slight increases in the acid concentration would violate the integrity of the CNC. In other words, small changes in the harsh reaction conditions on CNC preparation affect their yield and the dimensions [5]. Likewise, the centrifugation step influences both the CNC yield [25] and the dimensions, and controlled centrifugation can be used to tune their dimensions [26]. A brief account on yield is presented at the end of this chapter. Comprehensive dimensional analysis, however, exists only for the CNCs that are prepared with sulphuric acid and will be referenced in Sections 3.3.1 and 3.3.2 that deal with the impact of reaction conditions and source materials.

3.3.1 Effect of Reaction Conditions

In the first studies on CNCs, Rånby and co-workers subjected wood sulphite pulp to relatively mild H_2SO_4 concentrations of 2.5 M (ca. 22 wt.%) at boiling temperatures [2]. Mukherjee and Wood noticed very soon afterwards that higher concentrations at milder temperatures were more beneficial for the efficient isolation of CNCs [4]. They adjusted the H_2SO_4 concentration to around 950 g/dm^3 (ca. 66 wt.%) and temperatures to around 20°C to 40°C with reaction times depending on the source material, laying the groundwork for modern CNC preparation protocols.

As said earlier, the group of Prof. Gray at McGill University in Canada consolidated the laboratory procedures for CNC preparation during the 1990s. In their seminal paper on chiral nematic liquid crystal structures, Revol et al. [3] introduced a method that made use of 65 wt.% sulphuric acid at 70°C for 30 min for bleached hardwood kraft pulp. A few years later, Whatman No. 1 filter paper from cotton linter source became the preferred substrate for CNC preparation, presumably because of its availability. The reaction conditions were placed under scrutiny for the first time since 1950s in a 1998 paper by Dong et al. [5], where the decaying length of CNCs as a function of extended reaction time was demonstrated. The length of CNCs appeared to reach a plateau mean value at ca. 180 nm after 120 min of hydrolysis. Proper size distribution histograms based on transmission electron microscopy (TEM) were also presented for the first time. The amount of sulphate groups was expectedly found to increase as a result of prolonged hydrolysis times. Because the sulphuric acid hydrolysis in

TABLE 3.1
CNC Lengths as Reported by Various Literature Sources

Cellulose Source and Reference	Hydrolysis Conditions	Mean Length of CNCs (nm)
Bleached sulphite pulp [2]	2.5 N H_2SO_4, 100°C, 6 h	46
Bleached sulphite pulp [28]	64 wt.% H_2SO_4, 45°C, 25 min	140–150
Bleached kraft pulp [29]	64 wt.% H_2SO_4, 45°C, 25 min	50–65
Bleached kraft pulp [30]	64 wt.% H_2SO_4, 70°C, 10 min	180
Avicell (wood based) [27]	65 wt.% H_2SO_4, 72°C, 30 min	118
Cotton linters [27]	65 wt.% H_2SO_4, 45°C, 30 min	163
Cotton linters [27]	65 wt.% H_2SO_4, 72°C, 30 min	128
Cotton linters [27]	64 wt.% H_2SO_4, 45°C, 45 min	226
Cotton linters [27]	64 wt.% H_2SO_4, 45°C, 240 min	177
Cotton linters [31]	64 wt.% H_2SO_4, 45°C, 45 min	103

CNC preparation is often aborted after 30–60 min, one can assume that the LODP value obtained by hydrolysis is incomplete in many modern accounts where CNCs have been prepared. With this in mind, it is plausible that most CNCs are not single crystals. Furthermore, different lengths have been reported for CNCs that are prepared under exactly the same or very similar reaction conditions. Table 3.1 highlights this contradiction.

A major effort to elucidate the impact of reaction conditions on CNC dimensions was published in an effort of researchers from centre de recherches sur les macromolécules végétales (CERMAV) (Grenoble, France) in 2008 [27]. The survey covered several source materials, but the most extensive investigation was performed on cotton linters (filter paper), showing the clear-cut influence of reaction temperature in 65 wt.% H_2SO_4 for 30 min reaction time on both the width and the length of the CNCs (Figure 3.3). Particularly, the variations in width attended to a thitherto unaddressed issue: most of the earlier accounts claim that the CNC diameters are monodisperse from the same source material, whereas the length is polydisperse. However, it is clear from many published micrographs and the dimensional analysis that in most cases the CNCs are present in lateral aggregates of two to four individual crystallites, and this fact was specifically studied in the cited work by Elazzouzi-Hafraoui et al. [27] (Figure 3.4).

The effect of reaction conditions on the sulphate group concentration has been studied relatively little. Nevertheless, it is clear that by tuning the reaction conditions from milder to harsher, the amount of sulphate groups can be increased, respectively [5]. A recent study by Beck et al. [32] pointed out that the sulphate group content did not vary more than 17% within 45°C to 65°C temperature range during hydrolysis. The reaction temperature did, however, affect the viscosity of the CNC suspension. Interestingly, the differences in viscosity were later ascribed to the presence of water-insoluble oligosaccharides that precipitated on the CNC surface at moderate (45°C) temperatures after quenching the hydrolysis with water, whereas at 65°C the oligosaccharides hydrolysed faster and they remained soluble [33].

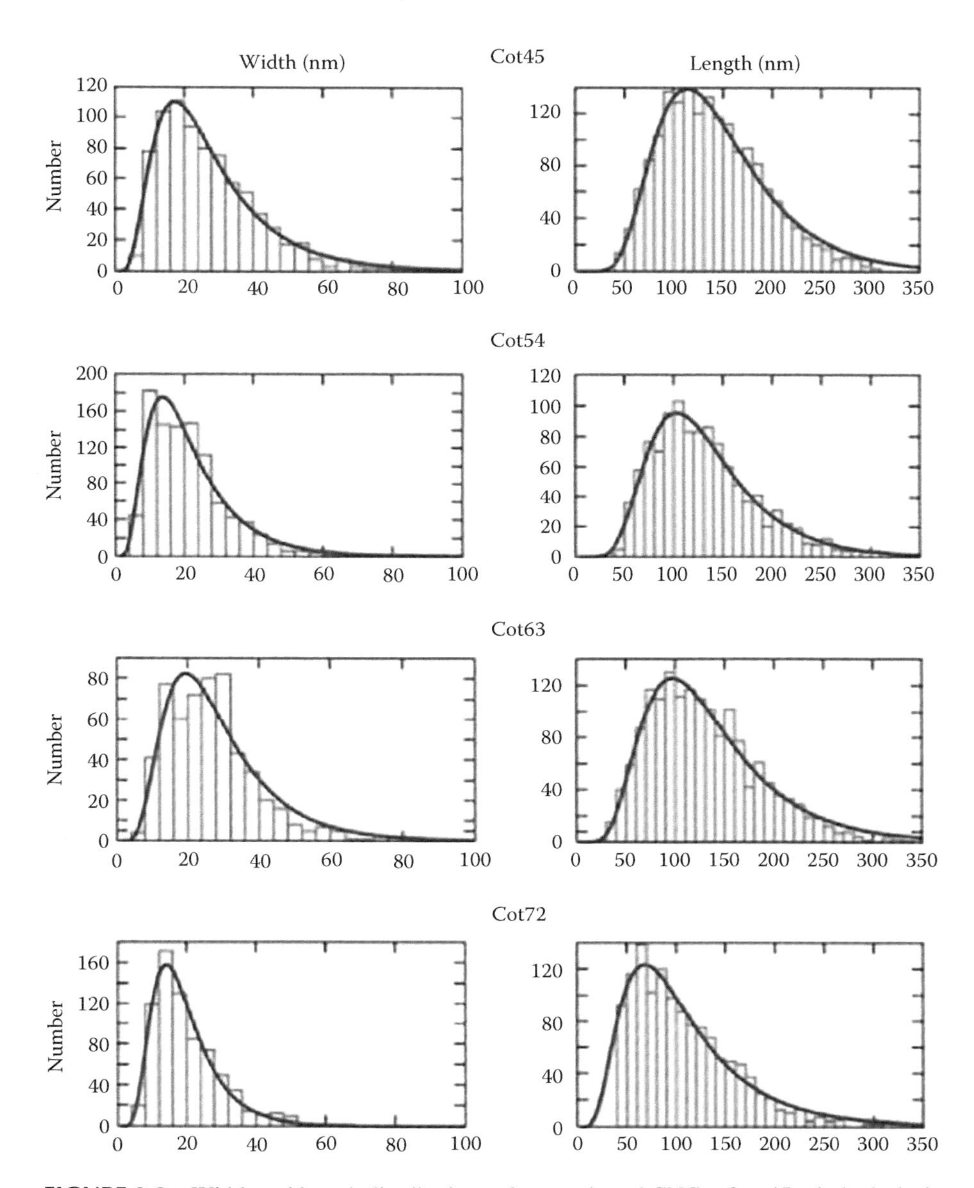

FIGURE 3.3 Width and length distributions of cotton-based CNCs after 45 min hydrolysis in 65 wt.% aqueous sulphuric acid at different temperature: 45°C, 54°C, 63°C and 72°C. (Reprinted with permission from Elazzouzi-Hafraoui, S. et al., *Biomacromolecules*, 9, 57–65, 2008. Copyright 2008 American Chemical Society.)

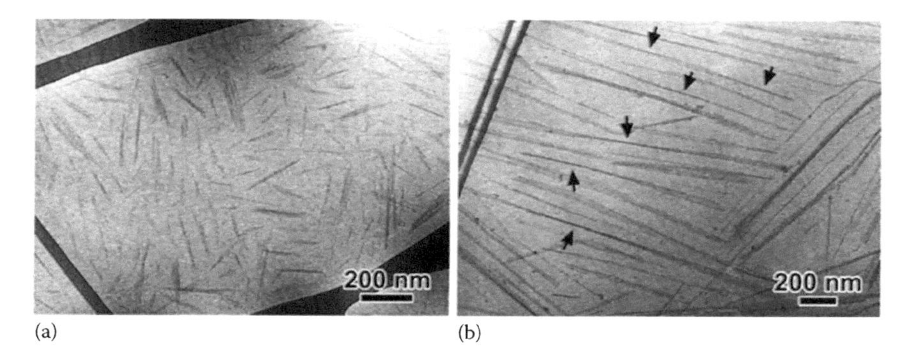

FIGURE 3.4 Cryoscopic transmission electron micrographs of (a) cotton-based CNCs and (b) tunicate-based CNCs, clearly exposing the lateral aggregates of several crystallites in dispersion. The arrows point to increased diffraction contrast in the areas where CNCs appear edge on. (Reprinted with permission from Elazzouzi-Hafraoui, S. et al., *Biomacromolecules*, 9, 57–65, 2008. Copyright 2008 American Chemical Society.)

TABLE 3.2
The Approximate Effect of Source Material on CNC Dimensions

Cellulose Source and Reference	Width (nm)	Length (nm)
Bleached wood pulp [27,29]	3–5	50–150
Cotton linters (Whatman No. 1 filter paper) [5,31]	6–8	50–300
Ramie [37]	6–8	100–300
Tunicate [27]	10–20	500–3,000
Bacterial cellulose [34–36]	10–15	500–1,500

3.3.2 Effect of Source Material

Source material is seen as the most decisive factor when selecting the desired dimensions for CNCs. For example, tunicate cellulose yields CNCs with lengths spanning a few micrometres, whereas wood-based CNCs have conspicuously short lengths with an average well under 100 nm (see the distinction in Figure 3.4) [7,8,27]. Table 3.2 demonstrates the well-known effect of source materials on the width and length of the CNC. Generally, the width is thought to be determined by the microfibril width in the source, whereas the length is connected to its LODP value. One should, however, be aware of the reservations raised in the previous passage that the CNC is usually a lateral aggregate of several crystallites and that the CNC preparation might be interrupted before it reaches the LODP value, rendering the length higher than that would be apparent from the LODP. Furthermore, bacterial cellulose is a special case where widely varying CNC dimensions have been reported [34–46]. This is probably due to the structure of the bacterial cellulose microfibril that is formed by elementary fibrils laterally aggregating into large flat ribbons of ca. 70–140 nm width and ca. 7 nm height. Moreover, the contrast between the disordered and the ordered (liquid crystal) phases after separation is particularly stark with CNCs from bacterial cellulose [34].

With wood-based starting materials (chemical pulps), the tree species has hardly any effect on the CNC dimensions, as attested by a study between pulps based on hardwood and softwood [28]. However, the accessibility plays a role, but counter-intuitively less accessible dried pulp hydrolyses faster to shorter CNCs than never-dried pulp [29]. This was hypothetically ascribed to tensions occurring in microfibrils on drying, resulting in more reactive disordered segments in microfibrils. These are among few examples of systematic studies on the source materials of CNCs. A well-cited publication on producing CNCs from microcrystalline cellulose with 63.5 wt.% sulphuric acid exists [38]. Because the LODP value has been reached already in the preparation of microcrystalline cellulose, the authors are actually presenting a way of extracting CNCs via sulphate esterification. The same method was applied by Araki et al. [39] on CNCs originally isolated via HCl hydrolysis.

All in all, the utilisation of different botanical sources represents one of the biggest trends in the present literature on CNCs. The methods to isolate CNCs from the more traditional sources described in Table 3.2 are already established, but numerous recent entries report sulphuric acid-based production of CNCs from straw holocellulose [40], sunflower stalks [41], oil palm trunk [42], conifer branches, needles and bark [43,44], grain straws [45], groundnut shells [46], soy hulls [47], *Miscanthus Giganteus* grass [48], *Posidonia oceanica* sea grass [49], tomato peels [50], potato peels [51], rice straw [52], grape skins [53], mengkuang leaves [54] and asparagus [55], to name a few. On the other hand, regenerated or mercerised cellulose in the form of cellulose II polymorph has been utilised as a source material for CNCs relatively little. Sulphuric acid hydrolysis of mercerised cotton fibres resulted in shorter CNCs compared with those produced from the corresponding native cotton fibres [56]. The crystallite width, however, did not change. Occasionally, partial formation of cellulose II under very specific conditions in aqueous sulphuric acid has been reported when preparing CNCs [57–58].

3.4 PREPARATION OF CELLULOSE NANOCRYSTALS WITH OTHER MINERAL ACIDS

Other than sulphuric acid, hydrolysis with other strong mineral acids has received sparing attention. HCl has been the most popular mineral acid [30,39], but its use has never gained widespread popularity because of the lack of charge on the resulting CNCs. Although sulphuric acid has the ability to introduce charged sulphate esters on the CNC surface, HCl hydrolysis results in virtually uncharged CNCs, which flocculate in water (and most other solvents) almost immediately [30]. Occasionally, HCl hydrolysis has been used as a pre-treatment for isolating CNCs via TEMPO-mediated oxidation or poly(ethylene glycol) grafting [59].

In addition to the conventional liquid/solid hydrolysis conditions, hydrolysis with gaseous HCl has recently been reported [60]. This method utilises adsorption of gaseous HCl on fibres that are always covered by a thin water layer under ambient conditions. The small amount of water is able to dissociate the acid, leading to very large local acid concentrations with the commencing hydrolysis quickly reaching the LODP. The resulting CNCs, dispersed in formic acid medium after extensive sonication, suffer from the same drawbacks with minimal charge as those prepared in liquid, but the CNC yields have been very high (>97%). In addition, cellulose crystallises during the degradation, slightly adding to the CNC yield. Moreover, the laborious purification steps with centrifugation and dialysis can be omitted when using acid vapour. The gas is also easier to recycle than a liquid solution of acid.

Hydrolysis with phosphoric acid, on the other hand, yields charged groups on the CNC surface via phosphate esterification. The phosphate groups are thermally far more stable than the autoxidative sulphate groups, which have prompted some authors to strongly encourage the route via phosphoric acid [61]. However, it has not been able to challenge the hegemony of sulphuric acid in CNC manufacturing.

Solid acid catalysts are attractive because they can be easily isolated from the reaction mixture and indeed some accounts on their usage for CNC preparation have been published. Possibly, the most successful attempt involved the use of phosphotungstic acid ($H_3PW_{12}O_{40}$) for the hydrolysis of bleached hardwood pulp [62]. The method appears effective although the CNC dimensions were somewhat larger than those prepared with the conventional sulphuric acid method. Moreover, the reaction times were fairly long (>15 h).

3.5 ENZYMATIC HYDROLYSIS FOR CELLULOSE NANOCRYSTAL PREPARATION

Besides acids, a host of other substances can be used to catalyse the degradation of cellulose. Cellulose-degrading enzymes, that is, cellulases, are a popular choice nowadays because of their prominent status in biofuel research, but they have not gained much recognition in the literature concerning CNCs. In principle, one of the cellulase types (trivial name endoglucanase) is selective towards disordered cellulose as opposed to the crystalline segments. With this in mind, Filson et al. [63] utilised monocomponent endoglucanase hydrolysis of recycled wood pulp (lignin content 1%) for isolating CNCs with yields running up to 40%. In another account, Satyamurthy et al. [64] took advantage of direct fungal hydrolysis with several generated cellulase species and managed to extract nanocrystals from cotton-based microcrystalline cellulose. Unfortunately, these studies lacked comprehensive dimensional analysis of the generated CNCs, but they nevertheless established a proof of concept that cellulose-degrading enzymes are able to isolate CNCs of some sort. More recently, CNCs from tunicate cellulose were isolated by a commercial multicomponent cellulase mixture, but the resulting crystals appeared more like fibrils in the subsequent atomic force microscopy images [65]. To summarise, it appears more difficult to undertake proper hydrolysis that exclusively leads to rigid CNC crystallites with enzymes than with acid.

3.6 OXIDATION ROUTES TO CELLULOSE NANOCRYSTALS

A popular choice for cellulose oxidation within nanocellulose preparation is to apply the TEMPO catalyst that stands for (2,2,6,6-tetramethylpiperidine-1-yl)oxyl radical. It is frequently used to catalyse the oxidation of OH groups in cellulose and in other hydroxyl-containing molecules and materials. TEMPO exclusively catalyses the oxidation of primary alcohols first into aldehydes and subsequently into carboxylic acid, usually with a hypochlorite oxidant (Figure 3.5a) [66,67]. As the extent

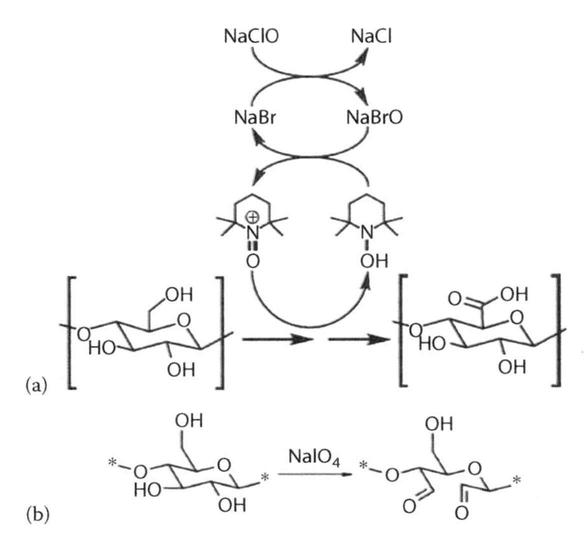

FIGURE 3.5 Reaction schemes for (a) TEMPO-mediated oxidation of cellulose under alkaline conditions with hypochlorite as the oxidising agent and (b) periodate oxidation of cellulose.

of TEMPO-mediated oxidation is low enough not to break the cellulose crystal itself but high enough for introducing significant charge on the crystal surface, it has extensively been utilised for isolation of cellulose nanofibres from plant-based fibres. The electrostatic repulsion caused by the carboxylic groups on the microfibril surfaces renders the required mechanical energy for nanofibre isolation very small indeed with acquired yields well over 90% [67].

With certain precautions, TEMPO-mediated oxidation appears as an attractive choice for CNC preparation as well, particularly when using microcrystalline cellulose as a starting material. Microcrystalline cellulose has already been hydrolysed to its LODP value, and one can expect that charges induced by TEMPO oxidation would enable the separation of CNCs from a microcrystalline matrix just like they enable the separation of nanofibres from a fibre matrix. However, a trial on TEMPO-mediated oxidation of microcrystalline cellulose resulted in a meagre 4% CNC yield [68]. In addition to CNCs, highly swollen microcrystals with elevated charge densities were produced. A more detailed study resulted in an optimisation of ca. 20% CNC yield from TEMPO-oxidised microcrystalline cellulose although the charge densities pointed towards complete oxidation of the crystallite surfaces [25]. The surprisingly low yield was ascribed to the inefficiency of centrifugation and ultrasonic treatment. A different take on oxidative CNC preparation was introduced by Li et al. [69], who first used formic acid for hydrolysis, followed by TEMPO oxidation for charge introduction. These efforts to exploit TEMPO-mediated oxidation for CNC preparation should not be confused with TEMPO-mediated oxidation of already prepared CNCs [59,70].

Periodate oxidation is another oxidation method that has been surveyed in connection with CNC preparation. It is selective towards the secondary alcohols in the anhydroglucose monomer of cellulose, thus oxidising the vicinal C2 and C3 hydroxyl groups to 2,3-dialdehyde units while simultaneously cleaving the C2–C3 bond (Figure 3.5b). Unlike TEMPO oxidation, periodate oxidation has the capability to break the cellulose crystal, and therefore it must be carefully implemented in an incomplete fashion when aiming at CNCs. Yang et al. [71] used periodate followed by chlorite oxidation that oxidised the aldehydes to carboxylic acid groups, enabling electrostatic stabilisation of the resulting CNCs. A follow-up work refined the method to partial periodate oxidation followed by a hot-water treatment and a fractionation accompanied with co-solvent addition, resulting in sterically stabilised CNCs [72]. The resulting CNCs had similar dimensions to those prepared by conventional sulphuric acid hydrolysis. In another study, partial periodate oxidation followed by reductive amination with butylamine isomers on the reactive aldehyde groups was utilised to gain hydrophobic CNCs [73].

A totally different route to CNCs via oxidation was introduced by Leung et al. [74] who utilised ammonium persulphate for a range of different cellulose substrates. The resulting CNCs possessed distinctly uniform diameters close to those of elementary microfibrils. The length of the CNCs was also surprisingly monodisperse in comparison with acid-hydrolysed CNCs. Furthermore, the yields were better than normally reported, running up to >80% with the conventional Whatman No. 1 filter paper substrate. Even the microcrystalline cellulose substrate gave a CNC yield of 65% with ammonium persulphate oxidation, which is at odds with the previously mentioned limitations with centrifugation and sonication observed with TEMPO oxidation. Overall, it appears that persulphate oxidation is one of the most promising methods that deserve to be pursued further in CNC production.

3.7 ESTERIFICATION ROUTES TO CELLULOSE NANOCRYSTALS

Esterification as a promising route to CNC production first received attention in 2009 in a study by Braun and Dorgan who used acid-catalysed Fischer esterification of cellulose hydroxyls by acetic acid and butyric acid while simultaneously degrading the disordered portions of cellulose by HCl [75]. The subsequent CNCs, prepared from cotton linters, were fairly similar in dimensions to those prepared by sulphuric acid hydrolysis. In addition, it was possible to disperse the CNCs into aprotic solvents such as toluene and ethyl acetate because of the acetylation of the surface hydroxyls. Acetylation was also used in an investigation with the cellulose solvent ionic liquid 1-ethyl-3-methylimidazolium acetate ([EMIM][OAc]) that was found both to degrade and to acetylate swollen (non-dissolved) cellulose, leading to CNCs directly from a wood substrate [76].

Dicarboxylic acids have also received fair attention within the literature on CNCs. With esterification of a single carboxyl group with a cellulose hydroxyl group, the second carboxyl group is able to remain unesterified and impart charge on the crystal, thus facilitating isolation of CNCs and yielding stable dispersions of those CNCs in water. In principle, complication in such procedure by cross-linking with esterification of the second carboxyl group is feasible but it has not been reported. Generally, the esterification by dicarboxylic acids for CNC production requires an additional accompaniment: mechanochemical energy input [77], swelling in deep eutectic solvents [78] or high concentrations (up to 80 wt.%) coupled with relatively high temperatures (80°C–120°C) [79]. Easy recovery of the solid-state acids from the reaction mixture is an added benefit of using the dicarboxylic acids.

3.8 NOTE ON THE YIELD OF CELLULOSE NANOCRYSTALS

Investigating the CNC yield can be considered a somewhat unexplored territory. Truly systematic studies on the effect of conditions on yield in the conventional sulphuric acid procedure have emerged only during the present decade, and they have a direct link to the commercialisation of CNCs [57,80–82]. It is evident and wholly expected that acid concentration and temperature profoundly affect the CNC yield and their dimensions [80]. Kinetic studies exposed that the standard 64 wt.% sulphuric acid concentration is not appropriate if yield is to be optimised. In fact, slightly milder concentrations between 58 and 62 wt.% resulted in superior yields (Figure 3.6) [81,82]. Of the more esoteric setups, incorporating the sonication treatment in situ with the hydrolysis affects the yield positively [83]. With other than sulphuric acid-based methods, yields are occasionally reported, but systematic approach is largely absent from such studies.

A weakness in most entries on CNC yield is the lack of consideration of steps such as centrifugation, sonication and filtration. Our recent investigation addresses these factors [25], but systematic quantification is yet to be done. The fact that the facile highest yields are achieved by methods with HCl vapour—something that omits the centrifugation steps completely—highlights the importance of these mechanical treatments on yield [60,84].

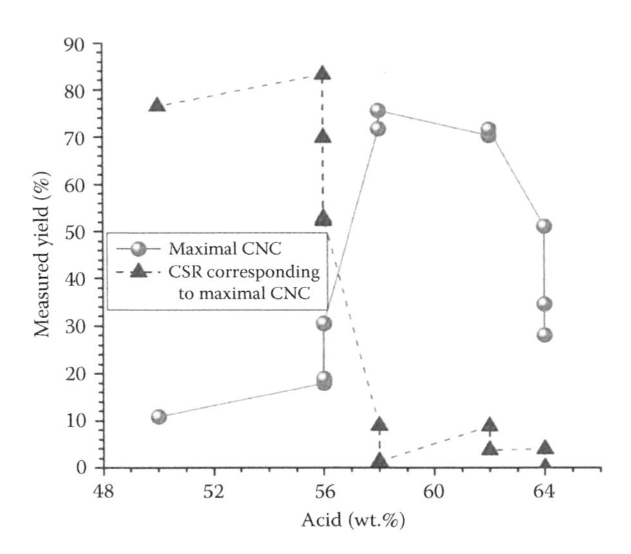

FIGURE 3.6 Influence of reaction conditions on CNC yield. CSR stands for cellulosic solid residue. (With kind permission from Springer Science+Business Media: *Cellulose*, 22, 2015, 1753–1762, Chen, L.I. et al.)

3.9 CONCLUSION

At present, method development for CNC preparation is at an important crossroads. The traditional route via sulphuric acid hydrolysis is being investigated in more and more detail and optimisations of yield, charge and specific crystallite size are currently under scrutiny. On the other hand, alternative routes to CNCs have emerged particularly within the past 5 years. Requirements for upscaling, purity, reagent recycling and water consumption will determine which method will ultimately gain the acceptance of academic and industrial communities. On another note, fundamental understanding of the cellulose microfibril, its exact size, shape and structure as well as its behaviour on treatments such as acid hydrolysis is necessary to access the full palette of preparing and utilising cellulosic nanomaterials.

REFERENCES

1. Rånby, B. G. Aqueous colloidal solutions of cellulose micelles. *Acta Chem. Scand.* 1949, 3, 649–650.
2. Rånby, B. The colloidal properties of cellulose micelles. *Discuss. Faraday Soc.* 1951, 11, 158–164.
3. Revol, J.-F., Bradford, H., Giasson, J., Marchessault, R. H., Gray, D. G. Helicoidal self-ordering of cellulose microfibrils in aqueous suspension. *Int. J. Biol. Macromol.* 1992, 14, 170–172.
4. Mukherjee, S. M., Woods, H. J. X-ray and electron microscope studies of the degradation of cellulose by sulphuric acid. *Biochim. Biophys. Acta* 1953, 10, 499–511.
5. Dong, X. M., Revol, J. F., Gray, D. G. Effect of microcrystalline cellulose preparation conditions on the formation of colloid crystals of cellulose. *Cellulose* 1998, 5, 19–32.
6. Fleming, K., Gray, D. G., Matthews, S. Cellulose crystallites. *Chem. Eur. J.* 2001, 7, 1831–1835.
7. Habibi, Y., Lucia, L. A., Rojas, O. Cellulose nanocrystals: Chemistry, self-assembly, and applications. *Chem. Rev.* 2010, 110, 3479–3500.
8. Eichhorn, S. J. Cellulose nanowhiskers: Promising materials for advanced applications. *Soft Matter* 2011, 7, 303–315.
9. Lin, N., Huang, J., Dufresne, A. Preparation, properties and applications of polysaccharide nanocrystals in advanced functional nanomaterials: A review. *Nanoscale* 2012, 4, 3274–3294.
10. Miao, C., Hamad, W. Y. Cellulose reinforced polymer composites and nanocomposites: A critical review. *Cellulose* 2013, 20, 2221–2262.
11. Brinchi, L., Cotana, F., Fortunati, E., Kenny, J. M. Production of nanocrystalline cellulose from lignocellulosic biomass: Technology and applications. *Carbohydr. Polym.* 2013, 94, 154–169.
12. Mariano, M., El Kissi, N., Dufresne, A. Cellulose nanocrystals and related nanocomposites: Review of some properties and challenges. *J. Polym. Sci., Part B: Polym. Phys.* 2014, 52, 791–806.
13. Nishiyama, Y. Structure and properties of the cellulose microfibril. *J. Wood. Sci.* 2009, 55, 241–249.
14. Fernandes, A. N., Thomas, L. H., Altaner, C. M., Callow, P., Forsyth, V. T., Apperley, D. C., Kennedy, C. J., Jarvis, M. C. Nanostructure of cellulose microfibrils in spruce wood. *PNAS* 2011, 108, E1195–E1203.

15. Oehme, D. P., Downton, M. T., Doblin, M. S., Wagner, J., Gidley, M. J., Bacic, A. Unique aspects of the structure and dynamics of elementary Iβ cellulose microfibrils revealed by computational simulations. *Plant Physiol.* 2015, 168, 3–17.
16. Reza, M., Bertinetto, C. G., Ruokolainen, J., Vuorinen, T. Cellulose elementary fibrils assemble into helical bundles in S1 layer of spruce tracheid wall. *Biomacromolecules* 2017, 18, 374–378.
17. Nishiyama, Y., Kim, U.-J., Kim, D.-Y., Katsumata, K. S., May, R. P., Langan, P. Periodic disorder along ramie microfibrils. *Biomacromolecules* 2003, 4, 1013–1017.
18. Müller, M., Czihak, C., Schober, H., Nishiyama, Y., Vogl, G. All disordered regions of native cellulose show common low-frequency dynamics. *Macromolecules* 2000, 33, 1834–1840.
19. Kontturi, E. Supramolecular aspects of native cellulose: Frigned-fibrillar model, levelling-off degree of polymerization, and production of cellulose nanocrystals. In: (Eds.) T. Rosenau and A. Potthast, *Advances in Cellulose Science and Technology: Chemistry, Analysis, and Applications*. Wiley-VCH, Weinheim, Germany: 2017.
20. Battista, O. A., Coppick, S., Howsmon, J. A., Morehead, F. F., Sisson, W. A. Level-off degree of polymerization. *Ind. Eng. Chem.* 1956, 48, 333–335.
21. Calvini, P., Gorassini, A., Merlani, A. L. On the kinetics of cellulose degradation: Looking beyond the pseudo zero order rate equation. *Cellulose* 2008, 15, 193–203.
22. Labet, M., Thielemans, W. Improving the reproducibility of chemical reactions on the surface of cellulose nanocrystals: ROP of ε-caprolactone as a case study. *Cellulose* 2011, 18, 607–617.
23. Beck, S., Bouchard, J., Berry, R. Dispersibility in water of dried nanocrystalline cellulose. *Biomacromolecules* 2012, 13, 1486–1494.
24. Sluiter A., Hames, B., Ruiz, R., Scarlata, C., Sluiter, J., Templeton, D., Crocker, D. Technical Report: NREL/TP-510-42618; National Renewable Energy Laboratory: Golden, CO, 2008.
25. Salminen, R., Reza, M., Pääkkönen, T., Peyre, J., Kontturi, E. TEMPO-mediated oxidation of microcrystalline cellulose: Limiting factors for cellulose nanocrystal yield. *Cellulose* 2017, 24(4), 1657–1667.
26. Bai, W., Holbery, J., Li, K. A technique for production of nanocrystalline cellulose with a narrow size distribution. *Cellulose* 2009, 16, 455–465.
27. Elazzouzi-Hafraoui, S., Nishiyama, Y., Putaux, J.-L., Heux, L., Dubreuil, F., Rochas, C. The shape and size distribution of crystalline nanoparticles prepared by acid hydrolysis of native cellulose. *Biomacromolecules* 2008, 9, 57–65.
28. Beck-Candanedo, S., Roman, M., Gray, D. G. Effect of reaction conditions on the properties and behavior of wood cellulose nanocrystal suspensions. *Biomacromolecules* 2005, 6, 1048–1054.
29. Kontturi, E., Vuorinen, T. Indirect evidence of supramolecular changes within cellulose microfibrils of chemical pulp fibers upon drying. *Cellulose* 2009, 16, 65–74.
30. Araki, J., Wada, M., Kuga, S., Okano, T. Flow properties of microcrystalline cellulose suspension prepared by acid treatment of native cellulose. *Colloid Surf. A* 1998, 142, 75–82.
31. Niinivaara, E., Faustini, M., Tammelin, T., Kontturi, E. Mimicking the humidity response of the plant cell wall by using two-dimensional systems: The critical role of amorphous and crystalline polysaccharides. *Langmuir* 2016, 32, 2032–2040.
32. Beck, S., Méthot, M., Bouchard, J. General procedure for determining cellulose nanocrystal sulfate half-ester content by conductometric titration. *Cellulose* 2015, 22, 101–116.
33. Bouchard, J., Méthot, M., Fraschini, C., Beck, S. Effect of oligosaccharide deposition on the surface of cellulose nanocrystals as a function of acid hydrolysis temperature. *Cellulose* 2016, 23, 3555–3567.

34. Hirai, A., Inui, O., Horii, F., Tsuji, M. Phase separation behavior in aqueous suspensions of bacterial cellulose nanocrystals prepared by sulfuric acid treatment. *Langmuir* 2009, 25, 497–502.
35. Vasconcelos, N. F., Feitosa, J. P. A., da Gama, F. M. P., Morais, J. P. S., Andrade, F. K., de Souza Filho, M. S. M., de Freitas Rosa, M. Bacterial cellulose nanocrystals produced under different hydrolysis conditions: Properties and morphological features. *Carbohydr. Polym.* 2017, 155, 425–431.
36. Sacui, I. A., Nieuwendaal, R. C., Burnett, D. J., Stranick, S. J., Jorfi, M., Weder, C., Foster, E. J., Olsson, R. T., Gilman, J. W. Comparison of the properties of cellulose nanocrystals and cellulose nanofibrils isolated from bacteria, tunicate, and wood processed using acid, enzymatic, mechanical, and oxidative methods. *ACS Appl. Mater. Interfaces* 2014, 6, 6127–6138.
37. Habibi, Y., Goffin, A.-L., Schiltz, N., Duquesne, E., Dubois, P., Dufresne, A. Bionanocomposites based on poly(ε-caprolactone)-grafted cellulose nanocrystals by ring-opening polymerization. *J. Mater. Chem.* 2008, 18, 5002–5010.
38. Bondeson, D., Mathew, A., Oksman, K. Optimization of the isolation of nanocrystals from microcrystalline cellulose by acid hydrolysis. *Cellulose* 2006, 13, 171–180.
39. Araki, J., Wada, M., Kuga, S., Okano, T. Influence of surface charge on viscosity behavior of cellulose microcrystal suspension. *J. Wood Sci.* 1999, 45, 258–261.
40. Jiang, F., Hsieh, Y.-L. Holocellulose nanocrystals: Amphiphilicity, oil/water emulsion, and self-assembly. *Biomacromolecules* 2015, 16, 1433–1441.
41. Fortunati, E., Luzi, F., Jiménez, A., Gopakumar, D. A., Puglia, D., Thomas, S., Kenny, J. M., Chiralt, A., Torre, L. Revalorization of sunflower stalks as novel sources of cellulose nanofibrils and nanocrystals and their effect on wheat gluten bionanocomposite properties. *Cabrohydr. Polym.* 2016, 149, 357–368.
42. Lamaming, J., Hashim, R., Sulaiman, O., Leh, C. P., Sugimoto, T., Nordin, N. A. Cellulose nanocrystals isolated from oil palm trunk. *Carbohydr. Polym.* 2015, 127, 202–208.
43. Moriana, R., Vilaplana, F., Ek, M. Cellulose nanocrystals from forest residues as reinforcing agents for composites: A study from macro- to nano-dimensions. *Carbohydr. Polym.* 2016, 139, 139–149.
44. Le Normand, M., Moriana, R., Ek, M. Isolation and characterization of cellulose nanocrystals from spruce bark in a biorefinery perspective. *Carbohydr. Polym.* 2014, 111, 979–987.
45. Oun, A. A., Rhim, J.-W. Isolation of cellulose nanocrystals from grain straws and their use for the preparation of carboxymethyl cellulose-based nanocomposite films. *Carbohydr. Polym.* 2016, 150, 187–200.
46. Bano, S., Negi, Y. S. Studies on cellulose nanocrystals isolated from groundnut shells. *Carbohydr. Polym.* 2017, 157, 1041–1049.
47. Flauzino Neto, W. P., Mariano, M., da Silva, I. S. V., Silvério, H. A., Putaux, J.-L., Otaguro, H., Pasquini, D., Dufresne, A. Mechanical properties of natural rubber nanocomposites reinforced with high aspect ratio cellulose nanocrystals isolated from soy hulls. *Carbohydr. Polym.* 2016, 153, 143–152.
48. Cudjoe, E., Hunsen, M., Xue, Z., Way, A. E., Barrios, E., Olson, R. A., Hore, M. J. A., Rowan, S. J. Miscanthus giganteus: A commercially viable sustainable source of cellulose nanocrystals. *Carbohydr. Polym.* 2017, 155, 230–241.
49. Bettiaeb, F., Khiari, R., Hassan, M. L., Belgacem, M. N., Bras, J., Dufresne, A., Mhenni, M. F. Preparation and characterization of new cellulose nanocrystals from marine biomass *Posedonia oceanica. Ind. Crops Prod.* 2015, 72, 175–182.
50. Jiang, F., Hsieh, Y.-L. Cellulose nanocrystal isolation from tomato peels and assembled nanofibers. *Carbohydr. Polym.* 2015, 122, 60–68.

51. Chen, D., Lawton, D., Thompson, M. R., Liu, Q. Biocomposites reinforced with cellulose derived from potato peel waste. *Carbohydr. Polym.* 2012, 90, 709–716.
52. Lu, P., Hsieh, Y.-L. Preparation and characterization of cellulose nanocrystals from rice straw. *Carbohydr. Polym.* 2012, 87, 564–573.
53. Lu, P., Hsieh, Y.-L. Cellulose isolation and core-shell nanostructures of cellulose nanocrystals from chardonnay grape skins. *Carbohydr. Polym.* 2012, 87, 2546–2553.
54. Sheltami, R. M., Abdullah, I., Ahmad, I., Dufresne, A., Kargarzadeh, H. Extraction of cellulose nanocrystals from mengkuang leaves (*Pandanus tectorus*). *Carbohydr. Polym.* 2012, 88, 772–779.
55. Wang, W., Du, G., Li, C., Zhang, H., Long, Y., Ni, Y. Preparation of cellulose nanocrystals from asparagus (*Asparagus officinalis* L.) and their applications to palm oil/water Pickering emulsion. *Carbohydr. Polym.* 2016, 151, 1–8.
56. Yue, Y., Zhou, C., French, A. D., Xia, G., Han, G., Wang, Q., Wu, Q. Comparative properties of cellulose nano-crystals from native and mercerized cotton fibers. *Cellulose* 2012, 19, 1173–1187.
57. Hu, T. Q., Hashaikeh, R., Berry, R. M. Isolation of a novel, crystalline cellulose material from the spent liquor of cellulose nanocrystals (CNCs). *Cellulose* 2014, 21, 3217–3229.
58. Sébe, G., Ham-Pichavant, F., Ibarboure, E., Koffi, A. L. C., Tingaut, P. Supramolecular structure characterization of cellulose II nanowhiskers produced by acid hydrolysis of cellulose I substrates. *Biomacromolecules* 2012, 13, 570–578.
59. Araki, J., Wada, M., Kuga, S. Steric stabilization of a cellulose microcrystal suspension by poly(ethylene glycol) grafting. *Langmuir* 2001, 17, 21–27.
60. Kontturi, E., Meriluoto, A., Penttilä, P. A., Baccile, N., Malho, J.-M., Potthast, A., Rosenau, T., Ruokolainen, J., Serimaa, R., Laine, J., Sixta, H. Degradation and crystallization of cellulose in hydrogen chloride vapor for high-yield isolation of cellulose nanocrystals. *Angew. Chem. Int. Ed.* 2016, 55, 14455–14458.
61. Espinosa, S. C., Kuhnt, T., Foster, E. J., Weder, C. Isolation of thermally stable cellulose nanocrystals by phosphoric acid hydrolysis. *Biomacromolecules* 2013, 14, 1223–1230.
62. Liu, Y., Wang, H., Yu, G., Yu, Q., Li, B., Mu, X. A novel approach for the preparation of nanocrystalline cellulose by using phosphotungstic acid. *Carbohydr. Polym.* 2014, 110, 415–422.
63. Filson, P. B., Dawson-Andoh, B. E., Schwegler-Berry, D. Enzymatic-mediated production of cellulose nanocrystals from recycled pulp. *Green Chem.* 2009, 11, 1808–1814.
64. Satyamurthy, P., Jain, P., Balasubramanya, R. H., Vigneshwaran, N. Preparation and characterization of cellulose nanowhiskers from cotton fibres by controlled microbial hydrolysis. *Carbohydr. Polym.* 2011, 83, 122–129.
65. Zhao, Y., Zhang, Y., Lindström, M. E., Li, J. Tunicate cellulose nanocrystals: Preparation, neat films and nanocomposite films with glucomannan. *Carbohydr. Polym.* 2015, 117, 286–296.
66. Saito, T., Kimura, S., Nishiyama, Y., Isogai, A. Cellulose nanofibers prepared by TEMPO-mediated oxidation of native cellulose. *Biomacromolecules* 2007, 8, 2485–2491.
67. Isogai, A., Saito, T., Fukuzumi, H. TEMPO-oxidized cellulose nanofibers. *Nanoscale* 2011, 3, 71–85.
68. Peyre, J., Pääkkönen, T., Reza, M., Kontturi, E. Simultaneous preparation of cellulose nanocrystals and micron-sized porous colloidal particles of cellulose by TEMPO-mediated oxidation. *Green Chem.* 2015, 17, 808–811.
69. Li, B., Xu, W., Kronlund, D., Määttänen, A., Liu, J., Smått, J.-H., Peltonen, J., Willför, S., Mu, X., Xu, C. Cellulose nanocrystals prepared via formic acid hydrolysis followed by TEMPO-mediated oxidation. *Carbohydr. Polym.* 2015, 133, 605–612.

70. Montanari, S., Roumani, M., Heux, L., Vignon, M. R. Topochemistry of carboxylated cellulose nanocrystals resulting from TEMPO-mediated oxidation. *Macromolecules* 2005, 38, 1665–1671.
71. Yang, H., Alam, N., van de Ven, T. G. M. Highly charged nanocrystalline cellulose and dicarboxylated cellulose from periodate and chlorite oxidized cellulose fibers. *Cellulose* 2013, 20, 1865–1875.
72. Yang, H., Chen, D., van de Ven, T. G. M. Preparation and characterization of sterically stabilized nanocrystalline cellulose obtained by periodate oxidation of cellulose fibers. *Cellulose* 2015, 22, 1743–1752.
73. Visanko, M., Liimatainen, H., Sirviö, J. A., Heiskanen, J. P., Niinimäki, J., Hormi, O. Amphiphilic cellulose nanocrystals from acid-free oxidative treatment: Physicochemical characteristics and use as oil-water stabilizer. *Biomacromolecules* 2014, 15, 2769–2775.
74. Leung, A. C. W., Hrapovic, S., Lam, E., Liu, Y., Male, K. B., Mahmoud, K. A., Luong, J. H. T. Characteristics and properties of carboxylated cellulose nanocrystals prepared from a novel one-step procedure. *Small* 2011, 7, 302–305.
75. Braun, B., Dorgan, J. R. Single-step method for the isolation and surface functionalization of cellulosic nanowhiskers. *Biomacromolecules* 2009, 10, 334–341.
76. Abushammala, H., Krossing, I., Laborie, M.-P. Ionic liquid-mediated technology to produce cellulose nanocrystals directly from wood. *Carbohydr. Polym.* 2015, 134, 609–616.
77. Tang, L., Huang, B., Yang, N., Li, T., Lu, Q., Lin, W., Chen, X. Organic solvent-free and efficient manufacture of functionalized cellulose nanocrystals via one-pot tandem reactions. *Green Chem.* 2013, 15, 2369–2373.
78. Sirviö, J. A., Visanko, M., Liimatainen, H. Acidic deep eutectic solvents as hydrolytic media for cellulose nanocrystal production. *Biomacromolecules* 2016, 17, 3025–3032.
79. Chen, L., Zhu, J. Y., Baez, C., Kitin, P., Elder, T. Highly thermal-stable and functional cellulose nanocrystals and nanofibrils produced using fully recyclable organic acids. *Green Chem.* 2016, 18, 3835–3843.
80. Hamad, W. Y., Hu, T. Q. Structure-process-yield interrelations in nanocrystalline cellulose extraction. *Can. J. Chem. Eng.* 2010, 88, 392–402.
81. Wang, Q., Zhao, X., Zhu, J. Y. Kinetics of strong acid hydrolysis of a bleached kraft pulp for producing cellulose nanocrystals (CNCs). *Ind. Eng. Chem. Res.* 2014, 53, 11007–11014.
82. Chen, L. I., Wang, Q., Hirth, K., Baez, C., Agarwal, U. P., Zhu, J. Y. Tailoring the yield and characteristics of wood cellulose nanocrystals (CNCs) using concentrated acid hydrolysis. *Cellulose* 2015, 22, 1753–1762.
83. Guo, J., Guo, X., Wang, S., Yin, Y. Effects of ultrasonic treatment during acid hydrolysis on the yield, particle size and structure of cellulose nanocrystals. *Carbohydr. Polym.* 2016, 135, 248–255.
84. Yu, H., Qin, Z., Liang, B., Liu, N., Zhou, Z., Chen, L. Facile extraction of thermally stable cellulose nanocrystals with a high yield of 93% through hydrochloric acid hydrolysis under hydrothermal conditions. *J. Mater. Chem. A* 2013, 1, 3938–3944.

4 Oxidative Chemistry in Preparation and Modification on Cellulose Nanoparticles

Tuomas Hänninen and Akira Isogai

CONTENTS

4.1 Introduction ... 45
4.2 Oxidation by Nitrogen Oxides ... 46
4.3 2,2,6,6-Tetramethylpiperidine-1-Oxyl-Mediated Oxidation ... 48
4.4 2-Azaadamantane *N*-Oxyl Oxidation ... 53
4.5 *N*-Hydroxypthalimide Oxidation ... 54
4.6 Periodate Oxidation ... 55
4.7 Industrial Possibilities of Oxidative Chemistry in Cellulose Nanoparticle Production ... 57
References ... 58

4.1 INTRODUCTION

The oxidation of cellulose has received a lot of attention during recent years as a tool for the preparation of nanocelluloses. Oxidation can be also used to modify cellulose for various applications ranging from therapeutics and pharmaceutics [1] to filtration [2–6]. In nature, polyglucuronic acids, which are oxidation products of cellulose, can rarely be found. For example, cellouronic acid can be obtained from the cell walls of certain bacteria or algae, and it is known to be excreted by the *Sinorhizobium meliloti* M5N1CS [7]. The oxidation of cellulose yields various kinds of products that can be used as a platform for more versatile chemicals [8], and its complete oxidation produces carbon dioxide and water [9]. However, we focus on more delicate oxidation methods in this chapter that can be used to prepare or modify cellulosic nanoscale particles. The oxidation of cellulose can be used to provide functional groups for chemical modification [10,11], to increase the water sorption of cellulosic materials [12], to protect the cellulose from acid hydrolysis [13], to increase fire retardancy [14–16] and to render cellulose more bioabsorbable [1]. Oxidation also significantly decreases the degradation rate of cellulose by cellulases and is dependent on the counterion of the carboxy group [17].

Oxidation of cellulose can proceed in two ways: by oxidation of hydroxy groups or by opening of the pyranose ring (e.g. periodate oxidation). During the oxidation, new functionalities, such as ketones, aldehydes and carboxy groups are introduced to the cellulose. Aldehyde groups are also present in native cellulose at the reducing end groups, and they can also be formed due to degradation of the cellulose polymer [18]. Oxidation methods can be divided into two groups: selective and non-selective methods. Although some non-selective oxidation methods have been successfully used in the preparation of nanoscale cellulosic particles, for example with ammonium persulphate [19], they commonly result in uncontrollable structures, severe degradation and eventually in the loss of fibrillar morphology. Selective methods are commonly more sensitive to the crystalline and fibrillary structure of cellulose and usually result in more defined chemical structures.

Oxidative treatments can be used to stabilise cellulosic particles electrostatically and sterically. Electrostatic stabilisation is based on interfibrillar repulsion that is caused by functional groups with like charge, which in the case of oxidised cellulose are carboxy groups; however, similar effects can also be achieved by cationisation. The increase in the amount of charged groups on the surface of the cellulose nanofibre (CNF) makes them easier to separate and provide the stability to the nanofibril suspension against aggregation in a similar manner to sulphate groups in cellulose nanocrystal (CNC) [20] or the carboxy groups of xylan in native CNF from hardwood pulp [21]. The content of charged groups, for example, in 2,2,6,6-tetramethylpiperidine-1-oxyl (TEMPO)-oxidised pulps is commonly and significantly higher compared to CNC (<0.4 mmol/g) [22] or CNF (0.04–0.07 mmol/g) [21,23]. High surface charge decreases the energy consumption that is required to mechanically disintegrate pulp fibres into nanosized particles. The mechanical preparation of microfibrillated cellulose (MFC) consumes about 700–1,400 MJ/kg, whereas TEMPO-oxidised cellulose nanofibrils (TOCNs) require only 7 MJ/kg [24]. The steric stabilisation is based on a layer of partially dissolved or highly swollen polymers on the surface of the nanoparticles. Such particles can be prepared, for example, by periodate oxidation [25]. Chemical changes in the structure of cellulose often render conventional analytical methods useless due to improved or hindered solubility or increased susceptibility to degradation. Chemical analyses of oxidised celluloses often require oxidative, reductive or other chemical modifications prior to analysis [26–28].

4.2 OXIDATION BY NITROGEN OXIDES

The preparation of oxidised cellulose using gaseous nitrogen dioxide was reported by Yackel and Kenyon in the 1940s [29]. Cellulose could be oxidised up to 15 wt.% of COOH content without affecting the original fibrous morphology or becoming fragile. Gaseous NO_2 oxidation has been shown to be highly selective to primary alcohols [30–34]. Oxidation with NO_2 can be performed in air or in non-polar solvents such as chloroform [33,35,36] or high-pressure CO_2 [37]. Although the method is often referred to as NO_2 oxidation, the actual oxidant is dinitrogen tetroxide. The equilibrium between NO_2 and N_2O_4 greatly favors the latter, and for example the equilibrium mixture at 21°C contains 0.08% NO_2 [34].

In gaseous or solution phase oxidation, NO_2 is absorbed chemically and physically on preferred sites, provided the moisture content of the cellulose is 5% to 10%. Primary alcohols of the cellulose are proposed to be oxidised via nitrite derivates to oximes or hydroxamic acid intermediates, which are hydrolyzed to carboxylic acid groups by concentrated acids present in the reaction mixtures. A proposed reaction scheme for the NO_2 oxidation of the primary alcohols of cellulose is as follows:

$$RCH_2OH + 2N_2O_4 \leftrightarrow RCOOH + 2HNO_3 + NO + \tfrac{1}{2}N_2$$

Processes used to prepare hemostats by circulating gaseous NO_2 across the fabrics were soon abandoned due to difficulties in handling the gas and the exothermic reaction, which led to the development of solvent-based oxidation systems.

Oxidation can be improved by dissolving cellulose in 85% phosphoric acid and by using sodium nitrite as an oxidant [33,38,39]. Despite the acid hydrolysis, high molar mass-oxidised products can be obtained by this method. During oxidation, the oxidation agent N_2O_3 that is generated forms foam within the viscous solution, which prevents the loss of the oxidising agents and enables high efficiency in the reaction [40]. The extent of $NaNO_2/H_3PO_4$ oxidation does not only depend on the reaction time but also on the degree of polymerization (DP) of the starting cellulose material. The carboxy group content has been reported to increase with the rising molecular weight of the starting material [38].

Cellulose oxidation can also be done in phosphoric acid by using sodium nitrate as a stoichiometric oxidant, and a catalytic amount of sodium nitrite is added to reduce the induction time [41]. Without the addition of sodium nitrite catalyst, the oxidation takes several hours to initiate. By using sodium nitrate as an oxidising agent, the formation of the toxic radical NO can be reduced threefold. The oxidation of cellulose also occurs with sodium nitrite in a mixture of nitric and phosphoric acids [42,43]. Varying the ratios of HNO_3 and H_3PO_4 affects the carboxy content and the yield only slightly. By replacing H_3PO_4 with H_2SO_4, the yield of the oxidation decreases significantly [43].

NO_2 oxidation can potentially be used for the preparation of cellulosic nanoparticles. Spectroscopic studies have shown that NO_2 oxidisation does not disrupt the crystalline structure of native or mercerised cellulose [36,44–46]. Acid hydrolysis in HNO_3 during NO_2 oxidation has been exploited to produce carboxylated submicron cellulose particles (Figure 4.1) from softwood sulphite pulp in a single-stage process [47]. Simultaneous hydrolysis and oxidation resulted in an oxidised pulp with up to 5.8 COOH wt.% that could be dispersed in submicron particles into water with a high-speed mixer equipped with a cutting blade. The fibrillation and gel formation properties of the oxidised pulp could be further improved by conversion of the carboxy groups into their sodium form. Oxidised pulps were reported to have increased crystallinity with only traces of nitrogen present in the pulp, possibly as nitrate esters. Yields from the process were comparable with the production of microcrystalline cellulose with mineral acids.

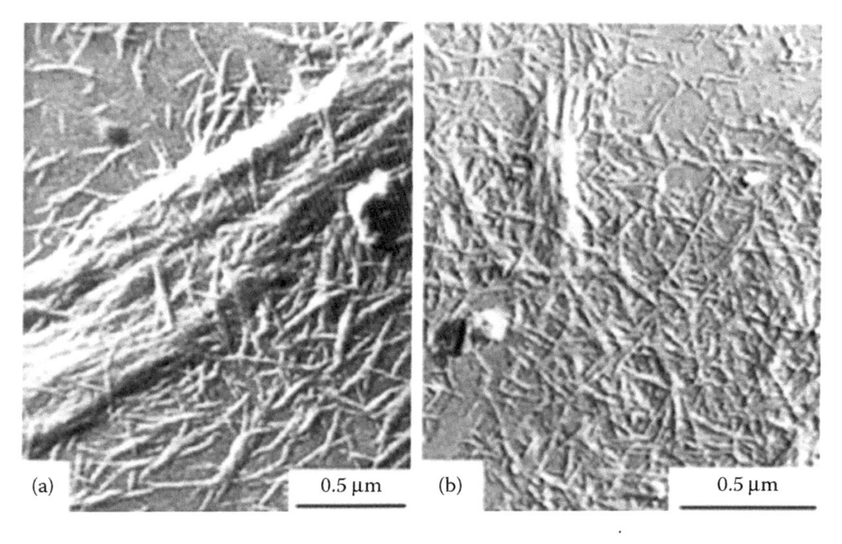

FIGURE 4.1 Electron micrographs of (a) HCMC and (b) NaCMC. (With kind permission from Springer Science+Business Media: Gert, E.V. et al., *Russ. J. Appl. Chem.*, 79, 2006, 1896–1901.)

4.3 2,2,6,6-TETRAMETHYLPIPERIDINE-1-OXYL-MEDIATED OXIDATION

TEMPO-mediated oxidation or in short, TEMPO oxidation, has recently received much attention industrially and scientifically as a pre-treatment for cellulosic materials in the production of CNF. TEMPO oxidation is a highly selective method to oxidise the primary alcohols of cellulose to aldehydes and subsequently to carboxy groups without disrupting fibrillar structure or crystallinity of native cellulose.

In an oxidation reaction, TEMPO acts as a mediator. The actual oxidant is the nitrosonium ion ($TEMPO^+$), which is reduced during oxidation to hydroxyamine (TEMPO-OH). $TEMPO^+$ can be used for oxidation in stoichiometric amounts; however, it can also be oxidised and regenerated by an oxidant in situ, which enables its use only in catalytic amounts. The regioselectivity is assumed to be caused by the four methyl groups in TEMPO, which sterically blocks access to the secondary alcohols [48].

Oxidation of alcohols by TEMPO oxidation was first presented by Semmelhack et al. [49]. Catalytic electro-oxidation using TEMPO was found to rapidly oxidise primary alcohols with high selectivity. Anelli et al. [50] reported on the highly selective oxidation of various primary alcohols using catalytic amounts of TEMPO with NaClO as the primary oxidant in mixture of water and dichloromethane. TEMPO oxidation was first applied to the oxidation of alcohols in carbohydrates by Davis and Flitsch [51]. Soon afterwards, the method was applied in several studies on various polysaccharides and natural fibres in an aqueous solvent system [48,51–58]. TEMPO-mediated oxidation of carbohydrates has been extensively covered in a review by Bragd et al. [59].

Although the TEMPO oxidation of natural fibres has been studied rather extensively, the aim of these studies had generally been to obtain water-soluble carbohydrates, and very little attention was paid to the water-insoluble fraction [54,55,60]. In the paper by Saito et al. [56], the water-insoluble fraction of TEMPO-oxidised cellulose (TOC) from various sources was reported to consist of well-defined CNFs with widths corresponding to individual nanofibrils and with lengths of several hundred nanometres. TOCNs were individualised with a mild mechanical treatment using only a Waring blender. After the publication [56], interest in the method increased rapidly.

The reaction scheme of the alkaline TEMPO/NaClO/NaBr system used by Saito et al. [56] is illustrated in Figure 4.2. In this method, catalytic amounts of TEMPO and NaBr are dissolved in suspension with pulp at pH 10, and the oxidation is initiated by the addition of the primary oxidant NaClO. The reaction can be easily controlled by monitoring the pH of the suspension. As the reaction proceeds and NaClO is consumed, the pH of the suspension increases continually and it has to be kept constant by adding NaOH. The pH becomes stable when all of the primary oxidant has been consumed or all of the accessible primary alcohols have been oxidised. The reaction is quenched by the addition of ethanol, which causes the rapid consumption of the remaining NaClO. The oxidised pulp maintains its fibre form, which makes washing and handling of the material easy before fibrillation. The degree of oxidation can be controlled by varying the amount of NaClO that is added [61].

Fibrils can be individualised from the TEMPO-oxidised pulp with a mild mechanical treatment. In dilute suspension, fibrils can be individualised from TEMPO-oxidised pulp with sufficiently high carboxylate content merely by using a magnetic stirrer [61]. Blenders or sonication can be used to obtain TOCN gels with higher concentration. Different kinds of fluidisers can be used for the preparation larger batches of TOCN. The width of the individualised TOCN is

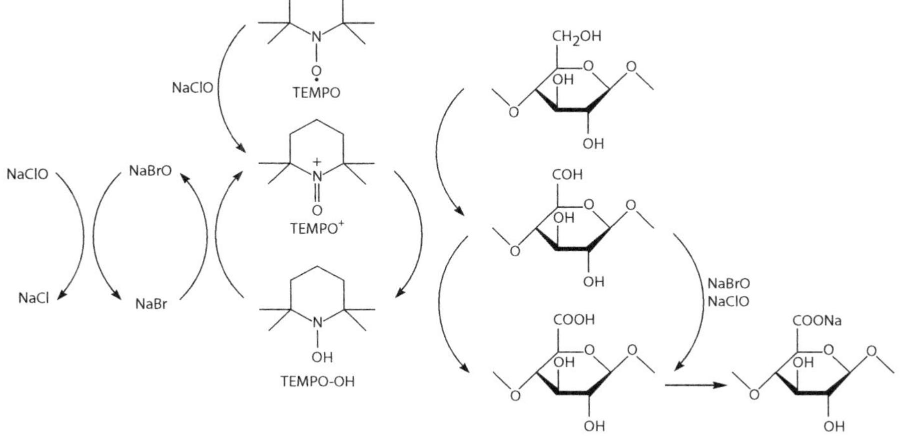

FIGURE 4.2 Scheme of alkaline TEMPO/NaClO/NaBr oxidation of cellulose.

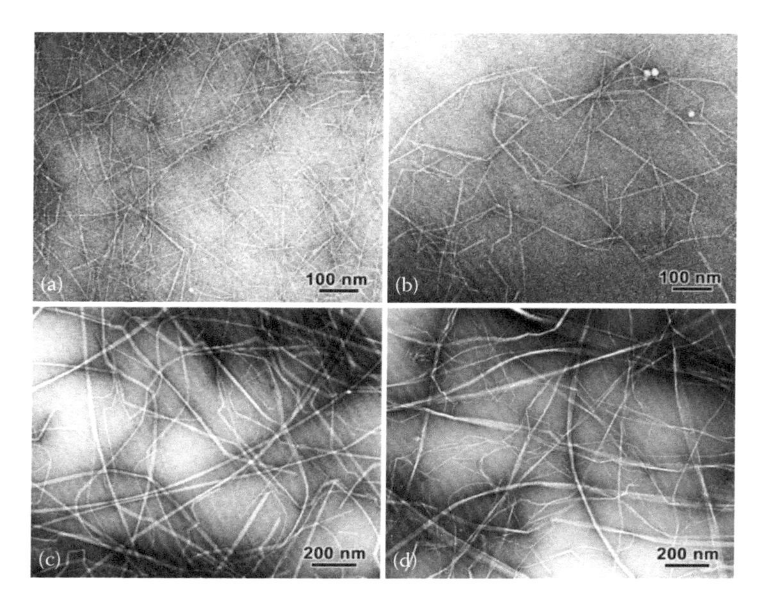

FIGURE 4.3 TEM images of uranyl acetate-stained TOCNs produced from (a) bleached sulphite wood pulp, (b) cotton, (c) tunicin and (d) bacterial cellulose. (Reprinted with permission from Saito, T. et al., *Biomacromolecules*, 7, 1687–1691, 2006. Copyright 2017 American Chemical Society.)

very consistent between different wood species. The width of the dry wood-based TOCN is about 2–2.7 nm [62–64] and in the water swollen state, it is 3–5 nm, depending on the analysis method [63,65]. The widths and lengths of the fibrils depend on the raw material, as seen from Figure 4.3. The length of the TOCNs prepared from softwood kraft pulp varies between 1,000 and 2,500 nm [66,67]. The mean strength of wood and tunicate nanofibrils has been determined to be 1.6–3 and 3–6 GPa by sonication-induced cavitation, respectively [68]. The elastic modulus of the tunicate TOCNs has been determined to be 145.2 ± 31.3 GPa [69].

The primary alcohols of the cellulose are efficiently oxidised to carboxylates via aldehyde intermediates. According to de Nooy et al. [48], the rate constant for the oxidation of hydroxy groups to aldehyde is smaller than the rate constant for the oxidation of aldehyde to carboxy group, which leads to the conclusion that in almost all cases, the oxidised samples contain more carboxy groups than aldehydes. Although the reaction kinetics of TEMPO oxidation is greatly affected by pH values, the ratio of the rate constants remains the same. Reaction rate has been observed to reach its maximum plateau above pH 10, and it rapidly decelerates as the pH decreases. Oxidised structures of cellulose, however, become severely prone to degradation at pH values over 10. Although cellouronic acids are only slightly degraded at pH below 10 [13], the aldehyde groups are highly prone to β-elimination in alkaline conditions [70]. Most severe degradation takes place during the first hour when oxidation is carried out under ambient conditions at pH 10 [71].

Oxidation rate can be increased with temperature; however, degradation of the cellulose becomes more prominent when the temperature is increased [48,72]. The degradation rate of the cellouronic acid has been reported to increase drastically at 60°C [13]. The oxidation rate can be enhanced sonocatalytically [73]; however, it also leads to degradation of cellulose, and in the worst cases, it leads to a significant decrease in the fibril length [28].

Cellulosic materials with native crystalline structures commonly maintain their fibre form after oxidation. The effects of the oxidation on fibre morphology depend on the extent of oxidation. As with mild conditions, the fibre length is affected only a little; samples oxidised to high carboxy contents disintegrate into fines very easily. This is partially not only due to the formation of carboxylate groups that enhance the fibrillation of the samples, but also due to the chemical degradation taking place [74]. TEMPO oxidation can also be performed on native CNF and amorphous cellulose films without disintegrating them, by controlling the pH or reaction time [75].

When oxidation is carried out on native celluloses that consist of the cellulose I crystalline structure, even with very harsh and extended oxidation times, very small water-soluble fractions are obtained [55,58,72]. TEMPO is unable to penetrate the crystalline regions of cellulose microfibrils, which leaves only the primary alcohols on the surface of the fibrils available for oxidation. The degree of oxidation correlates well with the calculated surface area of cellulose nanofibrils, which shows that TEMPO oxidation can be used to completely oxidise fibril surfaces [76]. TEMPO oxidation does not cause any changes in the crystalline structure of the cellulose allomorphs Iα and Iβ [77]. Cellulose Iα is less stable and therefore is considered to be more reactive than cellulose Iβ; however, in TEMPO oxidation there no differences were seen.

Water-soluble products can be obtained by oxidation of regenerated, mercerised, NH_3-treated or ball-milled samples [58,78]. Product of the oxidation in these cases is nearly homogeneous sodium (1-4)-β-D-anhydrogluronate, also known as cellouronic acid. Water-soluble co-polysaccharide with alternating anhydroglucose and glucuronic acid structures can be peeled from the surface of the native cellulose TOCs by using 20% NaOH solution [79].

Although TEMPO can also be activated by NaClO, the relatively low p*K*a value of HOCl formed limits the reaction rate of NaBr-free systems significantly. The p*K*a values of HOCl (7.5) and HOBr (8.7) define the applicable pH level during TEMPO activation [80]. At pH > p*K*a, the hypohalous acids increasingly exist as hypohalites that are inactive in the catalytic oxidation [80]. The conversion of TEMPO to $TEMPO^+$ is the slowest step in bromide-free oxidation of the alcohols by HOCl. The observations in study by Pääkkönen et al. [80] indicate that oxidation of TEMPO-OH to $TEMPO^+$ is much faster than its oxidative reaction with the primary alcohols of cellulose at mildly alkaline pH. At high pH, the reoxidation of TEMPO-OH becomes the rate-limiting reaction due to low concentration of HOCl.

Different kinds of nitroxyl radicals can be used to catalytically oxidise cellulose. Various TEMPO-analogous five-membered pyrrolidine and six-membered piperidine structures have been studied for the selective oxidation of cellulose [81–83]. Optimal oxidation pH varies for different TEMPO analogues [83]. 4-Acetamido-TEMPO

(4-AcNH-TEMPO) has been found to have higher oxidation efficiency in neutral and acidic conditions [81,82]; however, it is rather unstable and its degradation products have low or no oxidation tendency [84]. Optimal oxidation pH of 4-AcNH-TEMPO enables the efficient utilisation of TEMPO oxidation in neutral or acidic conditions. Performing the oxidation in neutral or acidic conditions decreases the depolymerization of the oxidation products by β-elimination, which takes place in alkaline conditions.

In the NaClO/$NaClO_2$/TEMPO system, the oxidation is done under neutral or weakly acidic conditions in order to prevent degradation. The reaction scheme, which is illustrated in Figure 4.4, differs from the one of alkaline oxidation. In neutral oxidation system, the $NaClO_2$ is the primary oxidant, and NaClO and TEMPO are added only in catalytic amounts. NaClO oxidises TEMPO to TEMPO$^-$, which is reduced to TEMPO-OH while rapidly oxidising primary hydroxy groups to aldehydes. Aldehydes are oxidised to carboxy groups by $NaClO_2$, forming NaClO that oxidises TEMPO-OH back to TEMPO$^-$. Fast oxidation of aldehydes by $NaClO_2$ decreases significantly by β-elimination and leads to an end product that does not contain any aldehyde groups [85,86].

High quantities of aldehydes remaining in the TOC are considered to be a disadvantage of the materials due to poor thermal stability, discolouration when heated and a tendency to form hemiacetal linkages between fibrils; however, they can be used as a template for selective chemical modification. The formation of hemiacetal linkages has been shown to be one of the main issues when redispersing dried TOC pulp. Oven drying has been shown to decrease

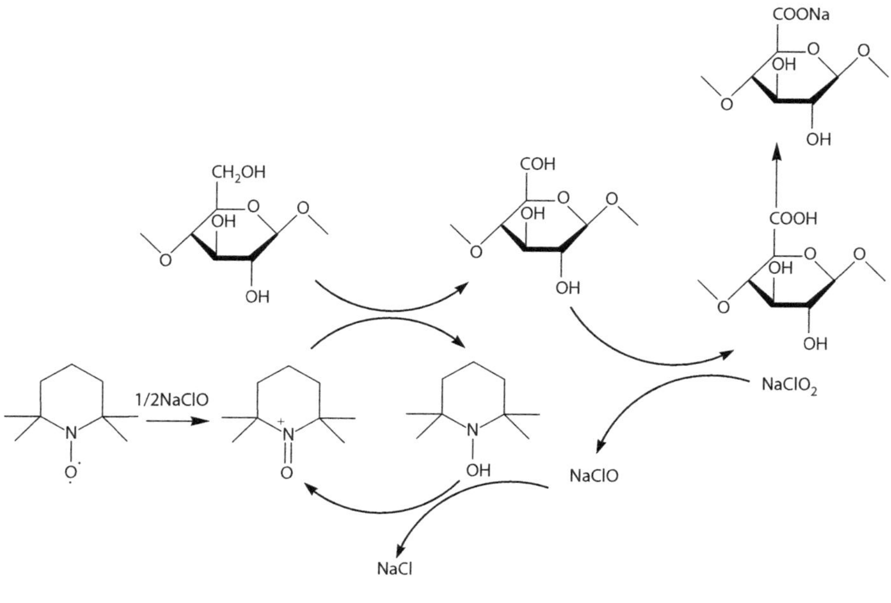

FIGURE 4.4 Reaction scheme of TEMPO/NaClO/$NaClO_2$ oxidation. (From Saito, T. et al., *J. Wood Sci.*, 56, 227–232, 2010.)

the nanofibrillation yield from 95% to 50%. However, after post oxidation with $NaClO_2$ the dispersibility of the pulp can increase significantly. When dispersion is done at elevated temperature (80°C), up to 70% yields can be achieved [87]. Hemicellulose content of the pulp, especially the xylan content, has been reported to affect negatively on the rate and the extent oxidation reaction and the dispersibility of TOCN [88,89].

A downside is that the low pH causes much slower reactivity of the system. At pH 6.8, it takes 3 days to achieve the nearly complete oxidation (~1.2 mmol/g) of softwood pulp with the TEMPO/NaClO/$NaClO_2$ system, and at pH 4.8 a carboxylate content of only about 0.8 mmol/g is reached in the same reaction time. By using 4-AcNH-TEMPO, the reaction rates at pH 4.8 can be significantly increased, and up to 1.4 mmol/g carboxylate content can be reached. The complete oxidation of bacterial cellulose has been reported to be achieved with reaction time of about 1 day at 65°C in neutral conditions [90]. TEMPO/NaClO/$NaClO_2$ oxidation can also be applied to regenerated celluloses in order to obtain cellouronic acid polymers with a high DP [82,91]. The water-insoluble fraction can be further utilised to produce cellulose II nanocrystals with high carboxylate content [92].

TEMPO oxidation can also be conducted by using a non-chemical method for activation of TEMPO. Activation of TEMPO without chlorous compounds provides an environmentally friendlier option; however, these methods often lack the efficiency of chemical activation. In electro-mediated TEMPO oxidation, TEMPO-OH compounds are repeatedly oxidised to $TEMPO^+$ under aqueous conditions by electrochemical energy. Carboxylate contents of up to 0.9 mmol/g can be achieved with electro-mediated TEMPO oxidation at neutral pH; however, the reaction time required to reach this level is 48 h, which makes the method very challenging at industrial scale [93,94]. Increasing the TEMPO concentration can be used for decreasing the reaction time; however, this increases the cost of the chemical [95].

A bio-mediated oxidation using enzymatic O_2/laccase/TEMPO oxidation system utilises laccase as a primary oxidant [96–100]. Laccase-based systems result in pulps with high amounts of aldehyde groups, and high carboxy content pulps can be achieved by redispersing the sample and repeating the oxidation [101]. The reaction rate has also been reported to increase with pulp consistency from 1% to 5% [102]. The reaction times, however, are very long for laccase-based method to be industrially viable.

4.4 2-AZAADAMANTANE *N*-OXYL OXIDATION

2-Azaadamantane *N*-oxyl (AZADO) or 1-methyl-AZADO oxidises efficiently not only C6-hydroxy groups but also C2- and C3-hydroxy groups to a certain extent with NaBr/NaClO system in water at pH 10. The AZADO oxidation is based on similar nitroxyl radical-mediated chemistry as TEMPO oxidation. The reaction scheme of AZADO oxidation is illustrated in Figure 4.5. AZADO-oxidised native cellulose samples contained a lower amount of ketone groups than TEMPO-oxidised samples (0–0.03 and 0.08 mmol/g, respectively); however, the ketone groups formed were

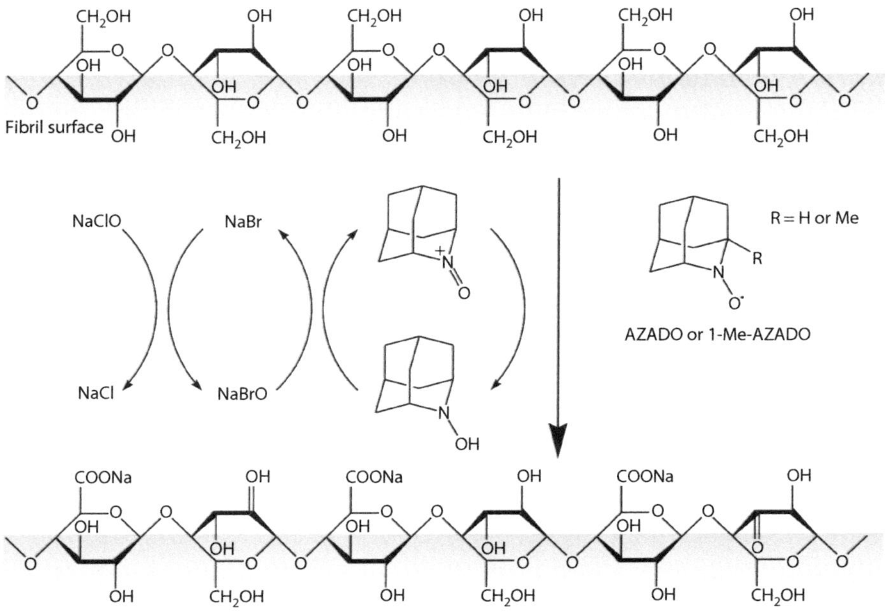

FIGURE 4.5 Reaction scheme of AZADO oxidation.

considered to be mostly at the C2 and C3 positions [103]. The oxidation time and the molar amounts of AZADO added can be significantly reduced using the AZADO/NaBr/NaClO system in comparison with the TEMPO system. AZADO or 1-methyl-AZADO contents in oxidation can be reduced to 1/32 and 1/16, respectively, than that of TEMPO to reach the same extent of oxidation in the case of TEMPO/NaClO/NaBr system. When AZADO oxidation is applied to regenerated cellulose, it results in 2,3,6-tricarboxylate cellulose [104].

4.5 *N*-HYDROXYPTHALIMIDE OXIDATION

Coseri et al. [105] have reported on another selective nitroxyl radical-mediated oxidation method for cellulose using *N*-hydroxypthalimide (NHPI) under aqueous conditions. The oxidation reaction proceeds via a phthalimide *N*-oxyl (PINO) radical intermediate. The reaction scheme of NHPI oxidation is illustrated in Figure 4.6. The method commonly involves the use of NaBr as co-catalyst and NaClO; however, NHPI has been shown to oxidise cellulose when NaBr has been replaced with $CuCl_2$ [106]. In addition, various other co-catalysts, for example, anthraquinone, O^2, $Pb(OAc)^4$, UV radiation, and their effects on oxidation rate have been studied [105,107–109]. In many of the studies, however, the degree of oxidation is about 0.3 mmol/g, which is below the level that enables individualisation of nanofibrils. In addition, severe degradation during oxidation has been reported [108]. It has been

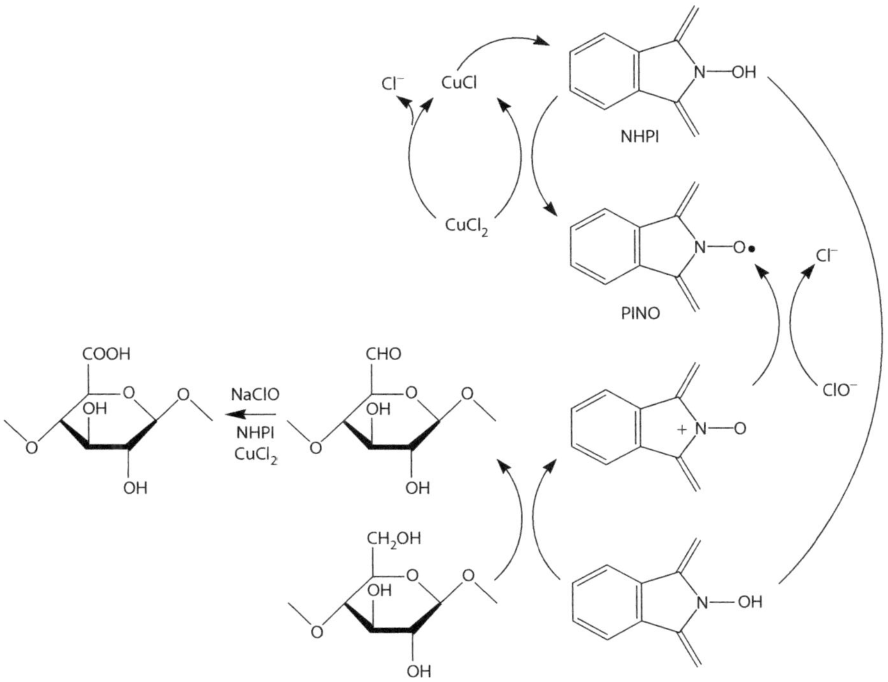

FIGURE 4.6 NHPI/co-catalyst/NaClO oxidation of cellulose in water at pH 10. (From Coseri, S. and Biliuta, G., *Carbohydr. Polym.*, 90, 1415–1419, 2012.)

proposed that where TEMPO is an efficient catalyst in a mild basic carbonate buffer, NHPI acts efficiently in a dilute acetic acid solution [110].

4.6 PERIODATE OXIDATION

Selectivity of periodate oxidation is based on cleavage of the glucopyranose ring at C2–C3 position resulting 2,3-dialdehyde cellulose (DAC), as illustrated in Figure 4.7. Similar structures have been achieved with lead(IV) tetraacetate oxidation; however, the method has received rather little attention due to its requirement for non-aqueous solvents [111]. Periodate oxidation has been used as a method to analyse the fine structure of polysaccharides, and later it has been used to produce metal absorbents with subsequent further oxidation to 2,3-dicarboxy groups by sodium chlorite [112–114]. Periodate oxidation of carbohydrates has been reviewed by Kristiansen et al. [115].

The rate and the extent of periodate oxidation can be easily controlled with concentration of periodate and reaction temperature [116,117]. Varma et al. [116] had proposed that in order to achieve greater extents of oxidation, it is preferable to utilise a higher concentration of periodate at 55°C. The oxidation is also shown to improve in the presence of inert metal salts, such as lithium, calcium, zinc and magnesium.

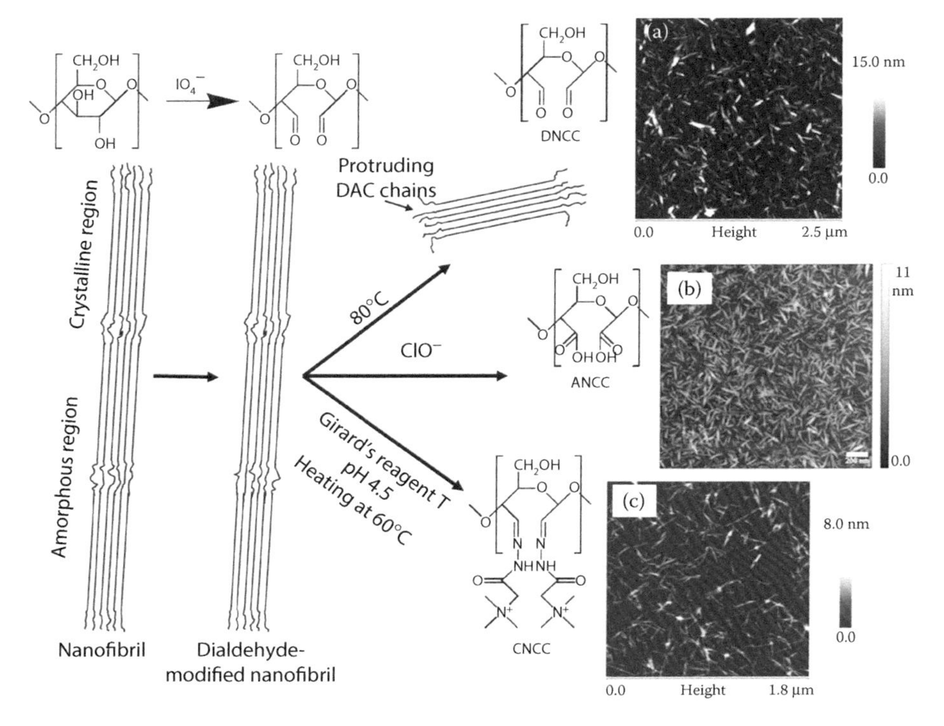

FIGURE 4.7 Reaction scheme for periodate oxidation with subsequent treatments to produce sterically (DCNC), anionically (ACNC) and cationically (CCNC) stabilised CNC. AFM height images from (a) DCNC, (b) ACNC and (c) CCNC samples are illustrated. (With kind permission from Springer Science+Business Media: Yang, H. et al., *Cellulose*, 20, 1865–1875, 2013; Yang, H. et al., *Cellulose*, 22, 1743–1752, 2015; Yang, H. and van de Ven, T.G.M., *Cellulose*, 23, 1791–1801, 2016.)

It has been speculated that the addition of inert salts causes a shift in Donnan equilibrium between pores and bulk, thus increasing the concentration of periodate in pores that leads to increased reactivity [10].

Unlike in the case of most of chemical reactions where the crystalline regions of the cellulose fibrils are mostly unmodified, periodate oxidation takes place in the crystalline regions in the early stage of oxidation [118]. A random mode of oxidation has been detected at low degrees of oxidation and as the degree is increased, the mode changes towards clustered oxidation [119]. Clustered oxidation has been analysed using X-ray diffraction and acid hydrolysis of oxidised samples, which resulted in the formation of short fragments of fibrils with the widths of the original material, indicating that oxidation had indeed proceeded through the crystalline regions [120]. An IR and wide-angle X-ray scattering (WAXS) studies by Calvini et al. [121] also support this mode of crystalline cellulose oxidation. Although the cellulose becomes more amorphous as the oxidation proceeds, the oxidation products differ from the actual amorphous cellulose [122].

At low degrees of oxidation, no degradation of the polymer takes place during the periodate oxidation. At high degrees, the cellulose degrades, becomes more compact and tends to form aggregates, which are likely to be associated by interchain hemiacetal bonds [118]. Subsequent reduction using $NaBH_4$ produced very flexible chains as hemiacetal cross links and ring structures were destroyed [123].

DAC is insoluble in water, but it can be dissolved in hot water. DAC suffers from minor degradation during hot water solubilisation, and storage in solution causes a slow decrease in molecular weight over time [124]. Water solubility and degradation of DAC can be used to advantage in the production of dialdehyde-modified nanocrystalline cellulose (DCNC). After the pulp is oxidised with $NaIO_4$ in ambient conditions in the presence of NaCl, the pulp suspension is heated to 80°C while constantly stirring for 6 h (Figure 4.7a). This initiates the degradation of the oxidised amorphous regions of the nanofibrils, leaving the crystalline regions mostly intact. Some of the cellulose chains from the amorphous regions remain partially at the ends of the crystals forming the so-called hairy nanocrystalloids. Oxidised 'hairs' remain in the dissolved state in water, providing steric stabilisation to DCNC. DCNC can be separated from DAC by adding a co-solvent, which is a poor solvent for DAC, such as propanol, with subsequent centrifugation. The dimensions and rod-like morphology of the DCNC are very similar to CNC that is prepared to conventional sulphuric acid hydrolysis. The steric stability of the DCNC is lost with time due to degradation of DAC [25]. The size and morphology on DCNCs can be controlled by the periodate reaction time. The average length of DCNC (26 h) is approximately 590 nm, whereas it decreases to 240 and 100 nm with 42 and 84 h reaction times. The widths of the DCNCs decreased in a similar manner from 8.7 to 5.6 nm [125].

DNCC can also be used to prepare electrostatically stabilised CNC by the introduction of anionically or cationically charged groups on the crystals. With a chlorite oxidation, the aldehyde groups can be selectively converted into carboxy groups, resulting in anionically charged CNC (ACNC) (Figure 4.7b). During chlorite oxidation, the DAC pulp is disintegrated into ACNC with only a very mild stirring. Dissolved dicarboxycellulose polymers can be removed from the suspension by ethanol-mediated coprecipitation, in a way similar to preparation of DCNC [126,127].

Cationic functionalisation is done by introducing the so-called Girard's reagent T, (2-hydrazinyl-2-oxoethyl)-trimethylazanium chloride, on the aldehyde groups through a Schiff base reaction [10] (Figure 4.7c) or by transforming ACNC through a bioconjugation reaction in which carboxy groups of ACNC are reacted with amine group with a suitable activator; however, the latter can be rather tedious way compared to the former method [10,126,128].

4.7 INDUSTRIAL POSSIBILITIES OF OXIDATIVE CHEMISTRY IN CELLULOSE NANOPARTICLE PRODUCTION

Utilisation of oxidative chemistries as a pre- or post-treatment in the preparation of cellulose nanoparticles of various grades broadens the field of applications on materials and can result in significant savings in energy consumed in production. Although NO_2 and periodate oxidations have been used earlier at industrial scale, currently TEMPO oxidation is the only method that has been applied industrially.

There are visible efforts by large companies such as Nippon Paper or UPM to utilise TEMPO in the industrial-scale production of nanocellulose, as witnessed by recent patenting activity in the area [129–134]. The feasibility of TEMPO oxidation in 30-ton annual production scale was started at Nippon Paper Groups, Iwakuni Mill in 2013. The first industrial scale TEMPO-oxidised pulp mill began its operation in 2017. The total investment in the facility was 1.6 billion Japanese Yen, and the production capacity of the mill was 500 tons per year [135]. Although TOCNs have been proven to have potential for various applications, currently its widespread use is limited. Nippon Paper has announced that its CNF will be used in the ink of ballpoint pens as a thickener [136]. Nippon Paper Industries has also produced TOCN sheet with a large amount of metal ions on the CNF surface that renders it antimicrobial and gives it deodorising effects [137]. Low-energy consumption and easy handling of TEMPO-oxidised pulp make the fibrillation of TOC possible even on-site for industrial applications.

REFERENCES

1. Bajerová, M., Krejčová, K., Rabišková, M., Gajdziok, J., Masteiková, R. (2009) Oxycellulose: Significant characteristics in relation to its pharmaceutical and medical applications. *Adv Polym Tech* 28:199–208.
2. Foglarova, M., Prokop, J., Milichovsky, M. (2009) Oxidized cellulose: An application in the form of sorption filter materials. *J App Polym Sci* 112:669–678.
3. Nemoto, J., Soyama, T., Saito, T., Isogai, A. (2012) Nanoporous networks prepared by simple air drying of aqueous TEMPO-oxidized cellulose nanofibril dispersions. *Biomacromolecules* 13:943–946.
4. Isobe, N., Chen, X., Kim, U.-J., Kimura, S., Wada, M., Saito, S., Isogai, A. (2013) TEMPO-oxidized cellulose hydrogel as a high-capacity and reusable heavy metal ion adsorbent. *J Hazard Mater* 260:195–201.
5. Fukuzumi, H., Fujisawa, S., Saito, T., Isogai, A. (2013) Transparent and high gas barrier films of cellulose nanofibers prepared by TEMPO-mediated oxidation. *Biomacromolecules* 14:1705–1709.
6. Hakalahti, M., Mautner, A., Johansson, L.-S., Hänninen, T., Setälä, H., Kontturi, E., Bismarck, A., Tammelin, T. (2016) Direct interfacial modification of nanocellulose films for thermoresponsive membrane templates. *ACS Appl Mater Inter* 8:2923–2927.
7. Elboutachfaiti, R., Delattre, C., Petit, E., Michaud, P. (2011) Polyuronic acids: Structures, functions and degrading enzymes. *Carbohydr Polym* 84:1–13.
8. Coseri, S. (2017) Cellulose: To depolymerize… or not to? *Biotechnol Adv* 35:251–266.
9. Birtwell, C., Ridge, B. P. (1928) The chemical analysis of cotton. The determination of cellulose by oxidation with chromic acid. *J Text Inst Trans* 19:T341–T348.
10. Yang, H., van de Ven, T. G. M. (2016) Preparation of hairy cationic nanocrystalline cellulose. *Cellulose* 23:1791–1801.
11. Lavoine, N., Bras, J., Saito, T., Isogai, A. (2017) Optimization of preparation of thermally stable cellulose nanofibrils via heat-induced conversion of ionic bonds to amide bonds. *J Polym Sci: Part A* 55:1750–1756.
12. Sjöstedt, A., Wohlert, J., Larsson, P. T., Wågberg, L. (2015) Structural changes during swelling of highly charged cellulose fibers. *Cellulose* 22:2943–2953.

13. Fujisawa, S., Isogai, T., Isogai, A. (2010) Temperature and pH stability of cellouronic acid. *Cellulose* 17:607–615.
14. Fukuzumi, H., Saito, T., Okita, Y., Isogai, A. (2010) Thermal stabilization of TEMPO-oxidized cellulose. *Polym Deg Stabil* 95:1502–1508.
15. Jiang, F., Han, S., Hsieh, Y.-L. (2013) Controlled defibrillation of rice straw cellulose and self-assembly of cellulose nanofibrils into highly crystalline fibrous materials. *RSC Advances* 3:12366.
16. Rohaizu, R., Wanrosli, W. D. (2017) Sono-assisted TEMPO oxidation of oil palm lignocellulosic biomass for isolation of nanocrystalline cellulose. *Ultrason Sonochem* 34:631–639.
17. Homma, I., Fukuzumi, H., Saito, T., Isogai, A. (2013) Effects of carboxyl-group counter-ions on biodegradation behaviors of TEMPO-oxidized cellulose fibers and nanofibril films. *Cellulose* 20:2505–2515.
18. Klemm, D., Philipp, B., Heinze, T., Heinze, U., Wagenknecht, W. (1998) Comprehensive cellulose chemistry, Vol. 1, *Fundamentals and Analytical Methods*, WILEY-VCH Verlag GmbH, Weinheim, Germany. http://onlinelibrary.wiley.com/book/10.1002/3527601929.
19. Leung, A. C. W., Hrapovic, S., Lam, E., Liu, Y., Male, K. B., Mahmoud, K. A., Loung, J. H. T. (2011) Characteristics and properties of carboxylated cellulose nanocrystals prepared from a novel one-step procedure. *Small* 7:302–305.
20. Habibi, Y., Lucia, L. A., Rojas, O. J. (2010) Cellulose nanocrystals: Chemistry, self-assembly, and applications. *Chem Rev* 11:3479–3500.
21. Tenhunen, T., Peresin, M. S., Penttilä P. A., Pere, J., Serimaa, R., Tammelin, T. (2014) Significance of xylan on the stability and water interactions of cellulosic nanofibrils. *React Funct Polym* 85:157–166.
22. Lin, N., Dufresne, A. (2014) Surface chemistry, morphological analysis and properties of cellulose nanocrystals with gradient sulfation degrees. *Nanoscale* 6:5384.
23. Fall, A. B., Burman, A., Wågberg, L. (2014) Cellulosic nanofibrils from eucalyptus, acacia and pine fibers. *Nord Pulp Pap J* 29(1):176–184.
24. Isogai, T., Saito, T., Isogai, A. (2011b) Wood cellulose nanofibrils prepared by TEMPO electro-mediated oxidation. *Cellulose* 18:421–431.
25. Yang, H., Chen, D., van de Ven, T. G. M. (2015) Preparation and characterization of sterically stabilized nanocrystalline cellulose obtained by periodate oxidation of cellulose fibers. *Cellulose* 22:1743–1752.
26. Gehrmayer, V., Potthast, A., Sixta, H. (2012) Reactivity of dissolving pulps modified by TEMPO-mediated oxidation. *Cellulose* 19:1125–1134.
27. Hiraoki, R., Fukuzumi, H., Ono, Y., Saito, T., Isogai, A. (2014) SEC-MALLS analysis of TEMPO-oxidized celluloses using methylation of carboxyl groups. *Cellulose* 21:167–176.
28. Hiraoki, R., Ono, Y., Saito, T., Isogai, A. (2015) Molecular mass and molecular-mass distribution of TEMPO-oxidized celluloses and TEMPO-oxidized cellulose nanofibrils. *Biomacromolecules* 16:675–681.
29. Yackel, E. C., Kenyon, W. O. (1942) The oxidation of cellulose by nitrogen dioxide. *J Am Chem Soc* 64:121–127.
30. Mercer, C., Bolker, H. I. (1906) Keto groups in cellulose and mannan oxidized by dinitrogen tetroxide. *Carbohydr Res* 14:109–113.
31. Unruh, C. C., Kenyon, W. O. (1942) Investigation of the properties of cellulose oxidized by nitrogen dioxide. *J Am Chem Soc* 64:127–131.
32. Bertocci, C., Konowicz, P., Signore, S., Zanetti, F., Flaibani, A., Paoletti, S., Crascenzi, V. (1995) Synthesis and characterization of polyglucuronan. *Carbohydr Polym* 27:295–297.
33. Painter, T. J. (1977) Preparation and periodate oxidation of C-6-Oxycellulose: Conformational interpretation of hemiacetal stability. *Carbohydr Res* 5:95–103.

34. Stilwell, R. L., Marks, M. G., Saferstein, L., Wiseman, D. M. (1998) 15. Oxidized cellulose: Chemistry, processing and medical applications. *Handbook of Biodegradable polym* 7:291.
35. Rosevaere, W. E., Spaulding, D. W. (1955) Effect of swelling and supermolecular structure on reaction of cellulose with nitrogen dioxide. *Ind Eng Chem* 47:2172–2175.
36. Zimnitsky, D. S., Yurkshtovich, T. L., Bychovsky, P. M. (2004) Sunthesis and characterization of oxidized cellulose. *J Polym Sci Part: A* 42:4785–4791.
37. Camy, S., Montanari, S., Rattaz, A., Vignon, M., Condoret, J.-S. (2009) Oxidation of cellulose in pressurized carbon dioxide. *J Supercrit Fluid* 51:188–196.
38. Heinze, T., Klemm, D., Schnabelrauch, M., Nehls, I. (1993) Properties and following reaction of homogeneously oxidized cellulose. In: Kennedy, J. F., Williams, G. O., Williams, P. A. (Eds.), *Cellulosics: Chemical, Biochemical and Material Aspects*, Ellis Harwood, New York.
39. Besemer, A. C., de Nooy, A. E. J., van Bekkum, H. (1998) Cellulose derivatives, Chapter 5, *ACS Symp Ser* 668:73–82.
40. Painter, T. J. (1985) New glucuronoglucans obtained by oxidation of amylose at position 6. *Carbohydr Res* 140:61–68.
41. de Nooy, A. E. J., Pagliaro, M., van Bekkum, H., Besemer, A. C. (1997) Autocatalytic oxidation of primary hydroxyl functions in glucans with nitrogen oxides. *Carbohydr Res* 304:117–123.
42. Kumar, V., Yang, T. (2002) HNO_3/H_3PO_4-$NaNO_2$ mediated oxidation of cellulose—Preparation and characterization of bioabsorbable oxidized celluloses in high yields and with different levels of oxidation. *Carbohydr Polym* 48:404–412.
43. Son, W. K., Youk, J. H., Park, W. H. (2004) Preparation of ultrafine oxidized cellulose mats via electrospinning. *Biomacromolecules* 5:197–201.
44. Belfort, A. M., Wortz, R. B. (1966) Colloidal oxycellulose by nitrogen dioxide treatment of level-off degree of polymerization cellulose. *Ind Eng Chem Prod Res Dev* 5:41–46.
45. Gert, E. V., Torgashov, V. I., Zubets, O. V., Kaputskii, F. N. (2005) Preparation and properties of enterosorbents based on carboxylated microcrystalline cellulose. *Cellulose* 12:517–526.
46. Peng, S., Zheng, Y., Wu, J., Wu, Y., Ma, Y., Song, W., Xi, T. (2012) Preparation and characterization of degradable oxidized bacterial cellulose reacted with nitrogen dioxide. *Polym Bull* 68:415–423.
47. Gert, E. V., Torgashov, V. I., Zubets, O. V., Kaputskii, F. N. (2006) Combination of oxidative and hydrolytic functions of nitric acid in production of enterosorbents based on carboxylated microcrystalline cellulose. *Russ J Appl Chem* 79:1896–1901.
48. de Nooy, A. E. J., Besemer, A. C., van Bekkum, H. (1995b) Selective oxidation of primary alcohols mediated by nitroxyl radical in aqueous solution. Kinetics and Mechanism. *Tetrahedron* 51:8023–8032.
49. Semmelhack, M. F., Chou, C. S., Cortes, D. A. J. (1983) Nitroxyl-mediated electrooxidation of alcohols to aldehydes and ketones. *Am Chem Soc* 105:4492–4494.
50. Anelli, P. L., Biffi, C., Montanari, F., Quici, S. (1986) Fast and selective oxidation of primary alcohols to aldehydes or to carboxylic acids and of secondary alcohols to ketones mediated by oxoammonium salts under two-phase conditions. *J. Org Chem* 52:2559–2562.
51. Davis, N. J., Flitsch, S. L. (1993) Selective oxidation of monosaccharide derivatives to uronic acids. *Tetrahedron Lett* 34:1181–1184.
52. de Nooy, A. E. J., Besemer, A. C., van Bekkum, H. (1994) Highly selective tempo mediated oxidation of primary alcohol groups in polysaccharides. *Recl Trav Ch Pays-Ba* 113:165–166.
53. de Nooy, A. E. J., Besemer, A. C., van Bekkum, H. (1995a) Highly selective nitroxyl radical-mediated oxidation of primary alcohol groups in water-soluble glucans. *Carbohydr Res* 269:89–98.

54. Chang, P. S., Robyt, J. F. (1996) Oxidation of primary alcohol groups of naturally occurring polysaccharides with 2,2,6,6-tetramethyl-1-piperidine oxoammonium ion. *J Carbohydr Chem* 15:819–830.
55. Isogai, A., Kato, Y. (1998) Preparation of polyuronic acid from cellulose by TEMPO-mediated oxidation. *Cellulose* 5:153–164.
56. Saito, T., Nishiyama, Y., Putaux, J.-L., Vignon, M., Isogai, A. (2006) Homogeneous suspensions of individualized microfibrils from TEMPO-catalyzed oxidation of native cellulose. *Biomacromolecules* 7:1687–1691.
57. Okita, Y., Saito, T., Isogai, A. (2009) TEMPO-mediated oxidation of softwood thermo-mechanical pulp. *Holzforschung* 63(5):529–535.
58. Isogai, T., Yanagisawa, M., Isogai, A. (2009) Degrees of polymerization (DP) and DP distribution of cellouronic acids prepared from alkali-treated celluloses and ball-milled native celluloses by TEMPO-mediated oxidation. *Cellulose* 16:117–127.
59. Bragd, P. L., van Bekkum, H., Besemer, A. C. (2004) Oxidation of two major disaccharides: Sucrose and isomaltulose. *Top Catal* 27:49–66.
60. Saito, T., Yanagisawa, M., Isogai, A. (2005) TEMPO-mediated oxidation of native cellulose: SEC-MALLS analysis of water-soluble and -insoluble fractions in the oxidized products. *Cellulose* 12:305–315.
61. Saito, T., Kimura, S., Nishiyama, Y., Isogai, A. (2007) Cellulose nanofibers prepared by TEMPO-mediated oxidation of native cellulose. *Biomacromolecules* 8:2485–2491.
62. Rodionova, G., Saito, T., Lenes, M., Eriksen, Ø., Gregersen, Ø., Kuramae, R., Isogai, A. (2013) TEMPO-Mediated oxidation of norway spruce and eucalyptus pulps: Preparation and characterization of nanofibers and nanofiber dispersions. *J Polym Environ* 21:207–214.
63. Kuramae, R., Saito, T., Isogai, A. (2014) TEMPO-oxidized cellulose nanofibrils prepared from various plant holocelluloses. *Reactive & Functional Polymers* 85:126–133.
64. Shimizu, M., Saito, T., Nishiyama, T., Iwamoto, S., Yano, H., Isogai, A., Endo, T. (2016) Fast and robust nanocellulose width estimation using turbidimetry. *Macromol Rapid Comm* 37:1581–1586.
65. Nechyporchuk, O., Belgacem, M. N., Pignon, F. (2015) Concentration effect of TEMPO-oxidized nanofibrillated cellulose aqueous suspensions on the flow instabilities and small-angle X-ray scattering structural characterization. *Cellulose* 22:2197–2210.
66. Tanaka, R., Saito, T., Hondo, H., Isogai, A. (2015) Influence of flexibility and dimensions of nanocelluloses on the flow properties of their aqueous dispersions. *Biomacomolecules* 16:2127–2131.
67. Tanaka, R., Saito, T., Ishii, D., Isogai, A. (2014) Determination of nanocellulo fibril length by shear viscosity measurement. *Cellulose* 21:1581–1589.
68. Saito, T., Kuramae, R., Wohlert. J., Berglund, L. A. (2013) An ultrastrong nanofibrillar biomaterial: The strength of single cellulose nanofibrils revealed via sonication-induced fragmentation. *Biomacromolecules* 14:248–253.
69. Iwamoto, S., Kai, W., Isogai, A., Iwata, T. (2009) Elastic modulus of single cellulose microfibril from tunicate measured by atomic froce microscopy. *Biomacromolecules* 10:2571–2576.
70. Shinoda, R., Saito, T., Okita, Y., Isogai, A. (2012) Relationship between length and degree of polymerization of TEMPO-oxidized cellulose nanofibrils. *Biomacromolecules* 13:842–849.
71. Milanovic, J., Schiehser, S., Milanovic, P., Potthast, A., Kostic, M. (2013) Molecular weight distribution and functional group profiles of TEMPO-oxidized lyocell fibers. *Carbohydr Polym* 15:444–450.
72. Brodin, F. W., Theliander, H. (2013) High temperature TEMPO oxidation in an heterogeneous reaction system: An investigation of reaction kinetics, pulp properties, and disintegration behavior. *Bioresources* 8:5925–5946.

73. Brochette-Lemoine, S., Joannard, D., Descotes, G., Bouchu, A., Queneau, Y. J. (1999) Sonocatalysis of the TEMPO-mediated oxidation of glucosides. *Mol Catal A: Chemical* 150:31–36.
74. Saito, T., Isogai, A. (2004) TEMPO-mediated oxidation of native cellulose. The effect of oxidation conditions on chemical and crustal structures of the water-insoluble fractions. *Biomacromolecules* 5:1983–1989.
75. Hänninen, T., Orelma, H., Laine, J. (2015) TEMPO oxidized cellulose thin films analysed by QCM-D and AFM. *Cellulose* 22:165–171.
76. Okita, Y., Saito, T., Isogai, A. (2010) Entire surface oxidation of various cellulose microfibrils by TEMPO-mediated oxidation. *Biomacromolecules* 11:1696–1700.
77. Carlsson, D. O., Lindh, J., Strømme, M., Mihrayan, A. (2015) Susceptibility of Iα- and Iβ-dominated cellulose to TEMPO-Mediated oxidation. *Biomacromolecules* 16: 1643–1649.
78. da Silva Perez, D., Montanari, S., Vignon, M. R. (2004) TEMPO-mediated oxidation of cellulose III. *Biomacromolecules* 4:1417–1425.
79. Hirota, M., Furihata, K., Saito, T., Kawada, T., Isogai, A. (2010) Glucose/glucuronic acid alternating co-polysaccharides prepared from TEMPO-Oxidized native celluloses by surface peeling. *Angew Chem Int Ed* 49:7670–7672.
80. Pääkkönen, T., Bertinetto, C., Pönni, R., Tummala, G. K., Nuopponen, M., Vuorinen, T. (2015) Rate-limiting steps in bromide-free TEMPO-mediated oxidation of cellulose—Quantification of the N-Oxoammonium cation by iodometric titration and UV-vis spectroscopy. *Appl Catal A-Gen* 505:532–538.
81. Bragd, P. L., Besemer, A. C., van Bekkum, H. (2001) TEMPO-derivatives as catalysts in the oxidation of primary alcohol groups in carbohydrates. *J Mol Catal A: Chemical* 170:35–42.
82. Hirota, M., Tamura, N., Saito, T., Isogai, A. (2009) Oxidation of regenerated cellulose with $NaClO_2$ catalyzed by TEMPO and NaClO under acid-neutral conditions. *Carbohydr Polym* 78:330–335.
83. Iwamoto, S., Kai, W., Isogai, T., Saito, T., Isogai, A., Iwata, T. (2010) Comparison study of TEMPO-analogous compounds on oxidation efficiency of wood cellulose for preparation of cellulose nanofibrils. *Polym Degrad Stab* 95:1394–1398.
84. Patel, I., Opietnik, M., Böhmdorfer, S., Becker, M., Potthast, A., Saito, T., Isogai, A., Rosenau, T. (2010) Side reactions of 4-acetamido-TEMPO as the catalyst in cellulose oxidation systems. *Holzforschung* 64:549–554.
85. Saito, T., Hirota, M., Tamura, N., Isogai, A. (2010) Oxidation of bleached wood pulp by TEMPO/NaClO/$NaClO_2$ system: Effect of the oxidation conditions on carboxylate content and degree of polymerization. *J Wood Sci* 56:227–232.
86. Tanaka, R., Saito, T., Isogai, A. (2012) Cellulose nanofibrils prepared from softwood cellulose by TEMPO/NaClO/$NaClO_2$ systems in water at pH 4.8 or 6.8. *Int J Biol Macromol* 51:228–234.
87. Takaichi, S., Saito, T., Tanaka, R., Isogai, A. (2014a) Improvement of nanodispersibility of oven-dried TEMPO-oxidized celluloses in water. *Cellulose* 21:4093–4103.
88. Pönni, R., Pääkkönen, T., Nuopponen, M., Pere, J., Vuorinen, T. (2014) Alkali treatment of birch kraft pulp to enhance its TEMPO catalyzed oxidation with hypochlorite. *Cellulose* 21:2859–2869.
89. Pääkkönen, T., Dimic-Misic, K., Orelma, H., Pönni, R., Vuorinen, T., Maloney, T. (2016) Effect of xylan in hardwood pulp on the reaction rate of TEMPO-mediated oxidation and the rheology of the final nanofibrillated cellulose gel. *Cellulose* 23:277–293.
90. Lai, C., Zhang, S., Sheng, L., Liao, S., Xi, T., Zhang, Z. (2013) TEMPO-mediated oxidation of bacterial cellulose in a bromide-free system. *Colloid Polym Sci* 291:2985–2992.
91. Yum, L., Lin, J., Tian, F., Li, X., Bian, F., Wang, J. (2014) Cellulose nanofibrils generated from jute fibers with tunable polymorphs and crystallinity. *J Mater Chem A* 2:6402.

92. Hirota, M., Tamura, N., Saito, T., Isogai, A. (2012) Cellulose II nanoelements prepared from fully mercerized, partially mercerized and regenerated celluloses by 4-acetamido-TEMPO/NaClO/$NaClO_2$ oxidation. *Cellulose* 19:435–442.
93. Isogai, T., Saito, T., Isogai, A. (2010) TEMPO electromediated oxidation of some polysaccharides including regenerated cellulose fiber. *Biomacromolecules* 11:1593–1599.
94. Isogai, A., Fukuzumi, H., Saito, T. (2011a) TEMPO-oxidized cellulose nanofibers. *Nanoscale* 3(71):71–85.
95. Carlsson, D. O., Lindh, J., Nyholm, L., Strømme, M., Mihranyan, A. (2014) Cooxidant-free TEMPO-mediated oxidation of highly crystalline nanocellulose in water. *RCS Advances* 4:52289.
96. Marzorati, M., Danieli, B., Haltrich, D., Riva, S. (2005) Selective laccase-mediated oxidation of sugars derivatives. *Green Chem* 7:310–315.
97. Patel, I., Ludwig, R., Mueangtoom, K., Haltrich, D., Rosenau, T., Potthast, A. (2009) Comparing soluble *Trametes pubescens* laccase and cross-linked enzyme crystals (CLECs) for enzymatic modification of cellulose. *Holzforschung* 63:715–720.
98. Patel, I., Ludwig, R., Haltrich, D., Rosenau, T., Potthast, A. (2011) Studies of the chemoenzymatic modification of cellulosic pulps by the laccase-TEMPO system. *Holzforschung* 65:475–481.
99. Xu, S., Song, Z., Qian, X., Shen, J. (2013) Introducing carboxyl and aldehyde groups to softwood-derived cellulosic fibers by laccase/TEMPO-catalyzed oxidation. *Cellulose* 20:2371–2378.
100. Jaušovec, D., Vogrinčič, R., Kokol, V. (2015) Introduction of aldehyde versus carboxylic groups to cellulose nanofibers using laccase/TEMPO mediated oxidation. *Carbohyd Polym* 116:74–85.
101. Jiang, J., Ye, W., Liu, L., Wang, Z., Fan, Y., Saito, T., Isogai, A. (2017) Cellulose nanofibers prepared using the TEMPO/Laccase/O2 system. *Biomacromolecules* 18:288–294.
102. Aracri, E., Valls, C., Vidal, T. (2012) Paper strength improvement by oxidative modification of sisal cellulose fibers with laccase-TEMPO system: Influence of the process variables. *Carbohyd Polym* 88:830–837.
103. Takaichi S., Isogai A. (2013) Oxidation of wood cellulose using 2-azaadamantane N-oxyl (AZADO) or 1-methyl-AZADO catalyst in NaBr/NaClO system. *Cellulose* 20:1979–1988.
104. Takaichi, S., Hiraoki, R., Inamochi, T., Isogai, A. (2014b) One-step preparation of 2,3,6-tricarboxy cellulose. *Carbohyd Polym* 110:499–504.
105. Coseri, S., Nistor, G., Fras, L., Strnad, S., Harabagiu, V., Simionescu, B. C. (2009) Mild and selective oxidation of cellulose fibers in the presence of N-Hydroxyphthalimide. *Biomacromolecules* 10:2294–2299.
106. Coseri, S., Biliuta, G. (2012) Bromide-free oxidizing system for carboxylic moiety formation in cellulose chain. *Carbohydr Polym* 90:1415–1419.
107. Biliuta, G., Fras, L., Strnad, S., Harabagiu, V., Coseri, S. J. (2010) Oxidation of cellulose fibers mediated by nonpersistent nitroxyl radicals. *Polym Sci Part A* 48:4790–4799.
108. Dobromir, M., Biliuta, G., Luca, D., Aflori, M., Harabagiu, V., Coseri, S. (2011) XPS study of the ion-exchange capacity of the native and surface oxidized viscose fibers. *Colloid Surface A* 381:106–110.
109. Biliuta, G., Fras, L., Drobota, M., Persin, Z., Kreze, T., Stana-Kleinschek, K., Ribitsch, V., Harabagiu, V., Coseri, S. (2013) Comparison study of TEMPO and phthalimide-N-oxyl (PINO) radicals on oxidation efficiency toward cellulose. *Carbohydr Polym* 91:502–507.
110. Rafiee, M., Karimi, B., Alizadeh, S. (2014) Mechanistic study of the electrocatalytic oxidation of alcohols by TEMPO and NHPI. *Chem Electro Chem* 1:455–462.
111. Zitko, V., Bishop, C. T. (1966) Oxidation of polysaccharides by lead tetraacetate in dimethyl sulfoxide. *Can J Chemistry* 44:1749–1756.

112. Abdel-Akher, M., Hamilton, J. K., Montgomery, R., Smith, F. J. (1952) A new procedure for the determination of the fine structure of polysaccharides. *Am Chem Soc* 74:4970–4971.
113. Maekawa, E., Koshijima, T. (1984) Properties of 2,3-Dicarboxy cellulose with various metallic ions. *J Appl Polym Sci* 29:2289–2297.
114. Maekawa, E. (1991) Analysis of oxidized moiety of partially periodate-oxidized cellulose by NMR spectroscopy. *J Appl Polym Sci* 43:417–422.
115. Kristiansen, K. A., Potthast, A., Christensen, B. E. (2010) Periodate oxidation of polysaccharides for modification of chemical and physical properties. *Carbohydr Res* 345:1264–1271.
116. Varma, A. J., Kulkarni, M. P. (2002) Oxidation of cellulose under controlled conditions. *Polym Deg Stabil* 77:25–27.
117. Sirviö, J., Hyväkkö, U., Liimatainen, H., Niinimäki, J., Hormi, O. (2011) Periodate oxidation of cellulose at elevated temperatures using metal salts as cellulose activators. *Carbohydr Polym* 83:1293–1297.
118. Potthast, A., Kostic, M., Schiehser, S., Kosma, P., Rosenau, T. (2007) Studies on oxidative modifications of cellulose in the periodate systam: Molecular weight distribution and carbonyl profiles. *Holzforschung* 61:662–667.
119. Potthast, A., Kostic, M., Schiehser, S., Kosma, P., Rosenau, T. (2009a) Studies on oxidative modifications of cellulose in the periodate system: Molecular weight distribution and carbonyl group profiles. *Holzforschung* 61:662–667.
120. Kim, U.-J., Kuga, S., Wada, M., Okano, T., Kondo, T. (2000) Periodate oxidation of crystalline cellulose. *Biomacromolecules* 1:488–492.
121. Calvini, P., Gorassini, A., Luciano, G., Franceschi, E. (2006) FTIR and WAXS analysis of periodate oxycellulose: Evidence for a cluster mechanism of oxidation. *Vib Spectrosc* 40:177–183.
122. Varma, A. J., Chavan, V. B., Rajmohanan, P. R., Ganapathy, S. (1997) Some observations on the high-resolution solid-state CP-MAS 13C-NMR spectra of periodate-oxidized cellulose. *Polym Degrad Stabil* 58:257–260.
123. Potthast, A., Schiehser, S., Rosenau, T., Kostic, M. (2009b) Oxidative modifications of cellulose in the periodate system—Reduction and beta-elimination reactions. *Cellulose* 63:12–17.
124. Kim, U.-J., Wada, M., Kuga, S. (2004) Solubilization of dialdehyde cellulose by hot water. *Carbohydr Polym* 56:7–10.
125. Chen, D., van de Ven, T. G. M. (2016) Morphological changes of sterically stabilized nanocrystalline cellulose after periodate oxidation. *Cellulose* 23:1051–1059.
126. Yang, H., Tejado, A., Alam, N., Antal, M., van de Ven, T. G. M. (2012) Films prepared from electrosterically stabilized nanocrystalline cellulose. *Langmuir* 28:7834–7842.
127. Yang, H., Alam, M. N., van de Ven, T. G. M. (2013) Highly charged nanocrystalline cellulose and dicarboxylated cellulose from periodate and chlorite oxidized cellulose fibers. *Cellulose* 20:1865–1875.
128. Liimatainen, H., Visanko, M., Sirviö, J. A., Hormi, O. E. O., Niinimäki, J. (2012) Enhancement of the nanofibrillation of wood cellulose through sequential periodate-chlorite oxidation. *Biomacromolecules* 13:1592–1597.
129. Miyawaki, S., Katsukawa, S., Abe, H., Iijima, Y., Isogai, A. (2012) Processes for producing cellulose nanofibers. US8287692 B2.
130. Tsuji, S., Nakayama, T., Miyawaki, S. (2015) Cellulose nanofibers. US 20150267070 A1.
131. Vuorinen, T., Pääkkönen, T., Nuopponen, M. (2015) Method for catalytic oxidation of cellulose and method for making a cellulose product. WO2012168562 A1.
132. Kajanto, I., Nuopponen, M. (2016a) Method and apparatus for controlling the catalytic oxidation of cellulose. WO2016097489 A1.

133. Kajanto, I., Nuopponen, M. (2016b) Process for producing a nanofibrillar cellulose hydrogel. WO2016193548 A1.
134. Miyawaki, S., Katsukawa, S., Abe, H., Iijima, Y., Isogai, A. (2017) Method for recovery/reuse of n-oxyl compound. CA 2755338 C.
135. Nippon Paper Group. (2016) *Nippon Paper Industries Builds the World's Largest Cellulose Nanofiber Large-Scale Production Facility at the Ishinomaki Mill.* http://www.nipponpapergroup.com/english/news/year/2016/news160518003399.html (Accessed April 14, 2017).
136. Nippon.com. (2016) *The Promise of Cellulose Nanofibers.* http://www.nippon.com/en/genre/economy/l00151/ (Accessed April 14, 2017).
137. Nippon Paper Group. (2015) *Launch of World's First Commercial Products Made of Functional Cellulose Nanofibers.* http://www.nipponpapergroup.com/english/news/year/2015/news150916003182.html (Accessed April 14, 2017).

5 Crystallinity and Thermal Stability of Nanocellulose

Alba Santmartí and Koon-Yang Lee

CONTENTS

5.1 Introduction 67
5.2 The Different Crystal Structures of Cellulose 69
5.2.1 X-Ray Diffraction Pattern of Nanocellulose 72
5.2.2 Differences in the Crystallinity of Various Types of Nanocellulose 74
5.3 Thermal Stability of Cellulose 77
5.3.1 Thermal Degradation Pathway of Cellulose 77
5.3.2 Differences in the Thermal Stability of Various Types of Nanocellulose 78
5.4 Conclusion 81
Acknowledgements 82
References 82

5.1 INTRODUCTION

Cellulose is the most abundant organic homopolymer on earth, and it has been used for centuries in various applications in multiple industries including textiles, construction and automotive industry [1]. However, it was not until 1838 that Anselme Payen isolated the components of different plant-based materials and obtained a substance represented by the chemical formula $C_6H_{10}O_5$ that he named cellulose [2]. Cellulose is a linear macromolecule consisting of two D-anhydroglucose unit linked together by $\beta(1 \rightarrow 4)$ glycosidic bonds (Figure 5.1). The degree of polymerisation (DP) of cellulose depends its origin. Cellulose with a DP of 10,000 has been observed in some species of algae such as *Valonia* [3], whereas the cellulose in ground wood pulp was found to possess a DP between 270 and 760 [4].

Individual cellulose chain molecules are assembled into elementary cellulose fibrils of 3–4 nm wide and several micrometers in length via hydrogen bonds between the hydroxyl groups of the anhydroglucose repeating units [5]. These elementary cellulose fibrils contain both ordered (crystalline) and disordered (amorphous) domains. The elementary fibrils assemble via lateral hydrogen bonds with neighbouring elementary fibrils, forming nanostructured fibre bundles known as cellulose microfibrils (Figure 5.2) [6–8]. These microfibrils are also more commonly known as nanocellulose (fibres) in the literature.

Nanocellulose can be obtained by two approaches: (1) top–down or (2) bottom–up. In the top–down approach, woody biomass, such as wood pulp, is disintegrated into

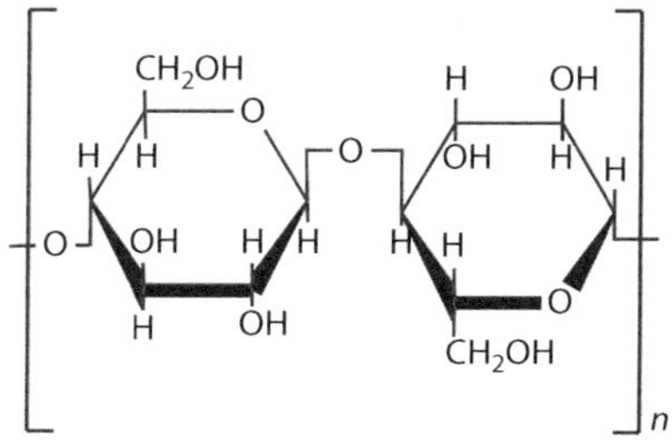

FIGURE 5.1 The chemical structure of cellulose where *n* represents the number of repeating units.

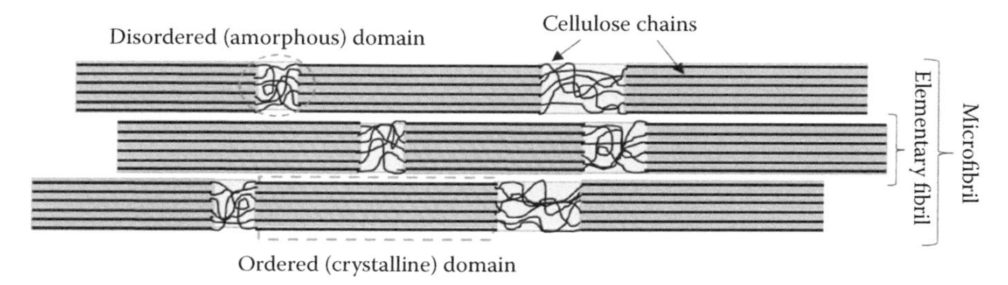

FIGURE 5.2 Schematic of a cellulose microfibril structure showing both amorphous and crystalline regions.

nanocellulose. The first step in the top–down approach to obtain nanocellulose is the pre-treatment of woody biomass to partially or completely eliminate non-cellulosic compounds, such as lignin and pectin [9]. This is often achieved by the use of bleaching and pulping processes to produce wood pulp [10]. The wood pulp is then fed into a high-pressure homogeniser [11,12] and/or stone grinders [13]. The high-shear fibrillation process converts the micrometre-scale wood pulp fibres into nanocellulose (herein termed cellulose nanofibres* or CNF) with fibre diameters between 20 and 100 nm, and several micrometres in length (Figure 5.3a).

In the bottom–up approach, nanocellulose is produced by the fermentation of low molecular weight sugars by cellulose-producing bacteria, such as from the *Acetobacter* species, later renamed as *Komagataeibacter* [15]. Nanocellulose synthesised by bacteria, more commonly known as bacterial cellulose (BC), is excreted by bacteria directly as nanofibres that make up the pellicle (a thick biofilm) in the culture medium. BC possesses a fibre diameter of ~50 nm and a length of several micrometers (Figure 5.3b). Hot aqueous NaOH solution (~0.1 M) is often used to remove any remaining microorganisms and soluble polysaccharides [16,17]. It should be noted that BC is pure cellulose without any impurities such as hemicellulose, lignin and pectin, whereas significant amounts of hemicellulose are still present in CNF [18].

The application of nanocellulose in a range of applications has received significant attention in both academia and industry. This interest [14] stems from the

* The terms nanofibrillated cellulose (NFC) or microfibrillated cellulose (MFC) are also often used in the literature.

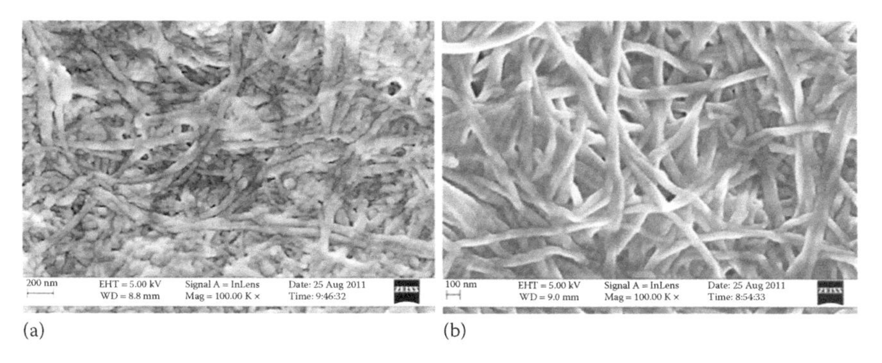

FIGURE 5.3 Scanning electron micrographs of (a) CNF and (b) BC. (Reprinted with permission from Lee, K.Y. et al., *ACS Appl. Mater. Interfaces*, 4, 4078–4086, 2012. Copyright 2012 American Chemical Society.)

fact that nanocellulose combines the physical and chemical properties of cellulose, such as hydrophilicity and chemical reactivity, with other properties such as high mechanical performance, which is estimated to be approximately 100–160 GPa for a single fibre [19–21]. In this chapter, we summarise the crystallinity and thermal degradation behaviour of nanocellulose obtained from various sources.

5.2 THE DIFFERENT CRYSTAL STRUCTURES OF CELLULOSE

Cellulose exhibits a complex intra- and intermolecular hydrogen-bonding network, which gives nanocellulose its outstanding mechanical properties. Different hydrogen-bonding configurations change the packing and molecular orientation of the cellulose chains, leading to the formation of different crystalline structures or polymorphs. Cellulose can have six different polymorphs (I, II, III_I, III_{II}, IV_I and IV_{II}), depending on the source of cellulose, cellulose extraction method and treatments [22–24]. Each polymorph has different unit cell lattice parameters and different cellulose chain packing configurations (Table 5.1).

Cellulose I is the naturally occurring crystalline structure of cellulose. It consists of a mixture of two sub-polymorphs: cellulose Iα and cellulose Iβ [31]. The proportions of

TABLE 5.1
Lattice Parameters of Different Crystalline Structures (Polymorphs) of Cellulose

Polymorph	a (nm)	b (nm)	c (nm)	α (°)	β (°)	γ (°)	Chain Configuration	Reference
Iα	0.672	0.596	1.040	118.08	114.80	80.38	Parallel	[25]
Iβ	0.778	0.820	1.038	90.00	90.00	96.50	Parallel	[26]
II	0.808	0.914	1.039	90.00	90.00	117.00	Antiparallel	[27]
III_I	0.445	0.785	1.051	90.00	90.00	105.10	Parallel	[28]
III_{II}	0.445	0.764	1.036	90.00	90.00	106.96	Antiparallel	[29]
IV_I	0.803	0.813	1.034	90.00	90.00	90.00	Parallel	[30]
IV_{II}	0.799	0.810	1.034	90.00	90.00	90.00	Antiparallel	[30]

TABLE 5.2
Crystallinity Degree and Relative Percentages of Cellulose Iα and Iβ for Different Sources of Cellulose

Cellulose Source	Iα (%)	Iβ (%)	Method	Crystallinity (%)	Method	Reference
Algae						
Glaucocystis	88	12	^{13}C NMR	–	–	[32]
Valonia	64–65	36–35	^{13}C NMR	–	–	[33,35]
Bacterial Cellulose						
BC (*Acetobacter xylinum*)	65	35	^{13}C NMR	–	–	[35]
BC (*Acetobacter xylinum*)	55.3	44.7	FT-IR	49.5	Raman	[36]
BC (*Acetobacter xylinum*)	61–73	27–39	^{13}C NMR	72–80	^{13}C NMR	[37]
Higher Plants						
Cotton	25	75	^{13}C NMR	–	–	[35]
CNF (lemon peel)	42	58	FT-IR	27	FT-IR	[38]
CNF (lemon peel)	51	49	^{13}C NMR	31	^{13}C NMR	[38]
CNF (maize bran)	43	57	FT-IR	10	FT-IR	[38]
CNF (maize bran)	48	52	^{13}C NMR	29	^{13}C NMR	[38]

cellulose Iα and cellulose Iβ within the cellulose vary, depending on the source of cellulose (Table 5.2). Cellulose Iα is the dominant cellulose structure in algal–bacterial cellulose, such as *Glaucocystis* and *Valonia* [32,33]. In cell wall of higher plants, cellulose Iβ is the dominant sub-polymorph [34]. Nishiyama et al. [25,26] determined both the crystal and molecular structure of cellulose Iα and cellulose Iβ using synchrotron and neutron diffraction data. Cellulose Iα structure contains one cellulose chain in a triclinic unit cell, whereas cellulose Iβ unit cell contains two chains and has a monoclinic structure.

Although both crystal structures have a parallel configuration (i.e. all the cellulose chains are arranged such that the β(1→4) glycosidic bonds point in the same direction), the main difference between cellulose Iα and cellulose Iβ is the relative displacement of cellulose sheets. Cellulose Iα exhibits a relative displacement of +c/4, whereas cellulose Iβ has a relative displacement that alternate between +c/4 and –c/4 (Figure 5.4) [22]. This contrast causes a difference in the relative occupancy of the two structures. The relative occupancy was measured by replacing all hydrogen atoms forming hydrogen bonds with deuterium atoms and by determining their positions using neutron diffraction [26]. Cellulose Iβ is densely packed with a relative occupancy between 70% and 80%, whereas only ~55% of the hydrogen atoms of cellulose Iα are involved in forming hydrogen bonds. As a result, the hydrogen bonds in cellulose Iβ are distributed over a region of better geometry than those in cellulose Iα [25,39].

If native cellulose is dissolved and regenerated [41] or mercerised [42], the more thermodynamically stable cellulose II with a monoclinic structure will be obtained. Cellulose regeneration consists of dissolving native cellulose in solvents such as carbon disulphide, concentrated inorganic salts and molten salt hydrates followed by recrystallisation in a coagulation (non-solvent) bath. Mercerisation, on the other hand, involves treating native cellulose with swelling agents such as sodium

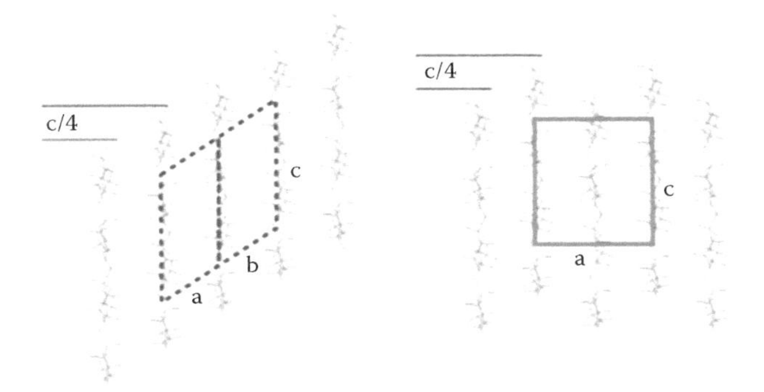

FIGURE 5.4 Schematic of the unit cells for the triclinic cellulose Iα structure (dashed line) and the monoclinic cellulose Iβ structure (solid line). The letters a, b and c stand for the unit cell lattice parameters. The relative displacement of the cellulose chains for Iα is +c/4 and for Iβ is alternating +c/4 and −c/4. (Adapted from Sugiyama, J. et al., *Macromolecules*, 24, 4168–4175, 1991; Moon, R.J. et al., *Chem. Soc. Rev.*, 40, 3941–3994, 2011. Reproduced by permission of The Royal Society of Chemistry.)

hydroxide or concentrated nitric acid solutions (65%) [23]. In contrast to cellulose I, cellulose II has an antiparallel configuration of the cellulose chains that are arranged in a 3D hydrogen-bonded network, whereas cellulose I has a 2D structure formed by hydrogen-bonded layers on top of each other (Figure 5.5) [43].

When cellulose is treated with chemicals containing amine groups such as ammonia or ethylenediamine, its crystalline structure is modified into cellulose III. Cellulose III has two different sub-polymorphs: III_I and III_{II}. The main difference between the two sub-polymorphs is the orientation of the cellulose chains. The type of sub-polymorph obtained depends on the initial crystalline structure of the cellulose source. If the cellulose treated with amines consists of naturally occurring cellulose, cellulose I (both Iα or Iβ) will be transformed into cellulose III_I and will maintain the parallel chain configuration [28]. Instead, if cellulose II is treated with amines, it will be transformed into III_{II} and the cellulose chains will have an antiparallel packing [29].

If cellulose III is heated in water, glycerol or formamide to temperatures up to 180°C, the crystalline structure of cellulose will be transformed into cellulose IV [44].

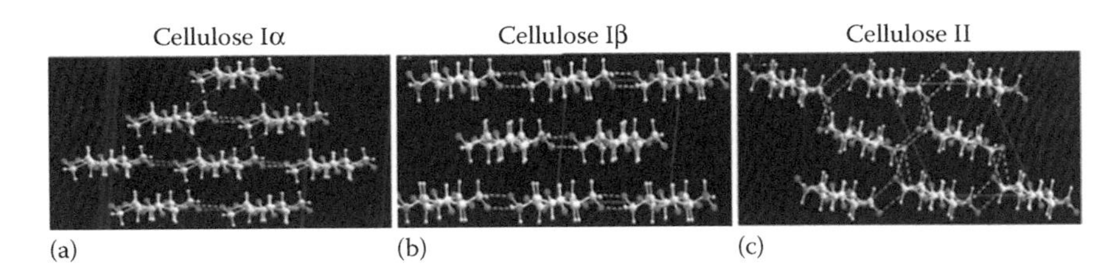

FIGURE 5.5 Projections of the crystal structures of (a) cellulose Iα, (b) cellulose Iβ and (c) cellulose II down the chain axes directions. C, O and H atoms are represented as gray, light gray and white balls, respectively. Covalent and hydrogen bonds are represented as full and dashed sticks, respectively. (Reprinted with permission from Wada, M. et al., *Macromolecules*, 37, 8548–8555, 2004. Copyright 2004 American Chemical Society.)

Cellulose IV also exhibits two sub-polymorphs (IV_I and IV_{II}) and its formation also depends on the cellulose source. If cellulose III_I is heated, the crystalline structure obtained will be IV_I and the chain configuration will be parallel. However, if it is cellulose III_{II} instead, the final crystalline structure will be cellulose IV_{II} with an anti-parallel chain packing [30].

5.2.1 X-Ray Diffraction Pattern of Nanocellulose

X-ray diffraction is a widely used technique to analyse the crystal structure of cellulosic samples. Figure 5.6 shows the idealised X-ray powder diffraction patterns

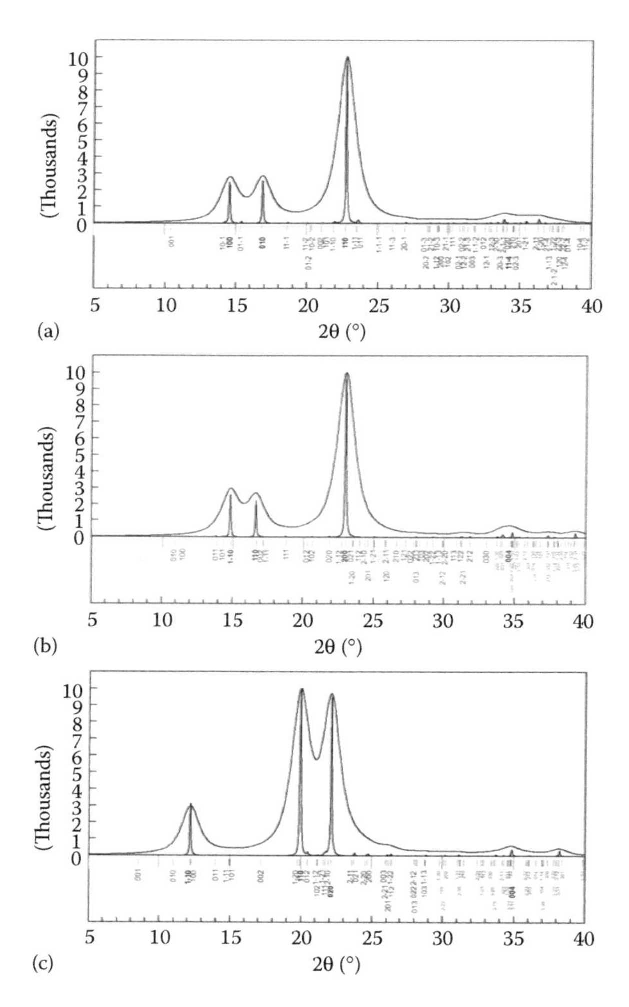

FIGURE 5.6 Simulated (a) Iα, (b) Iβ and (c) II X-ray diffraction patterns with crystallites having preferred orientation along the fibre axis. (With kind permission from Springer Science+Business Media: *Cellulose*, Idealized powder diffraction patterns for cellulose polymorphs, 21, 2010, 885–896, French, A.)

TABLE 5.3
Bragg's Angles and Miller Indices for the Main XRD Peaks for Different Cellulose Polymorphs

Cellulose Iα		Cellulose Iβ		Cellulose II	
Bragg's Angle (°)	**Miller Index**	**Bragg's Angle (°)**	**Miller Index**	**Bragg's Angle (°)**	**Miller Index**
14.5[a]	100	14.9	$1\bar{1}0$	12.2	$1\bar{1}0$
16.9[a]	010	16.7	110	20.0[a]	110
22.9[a]	110	23.0[a]	200	22.1	020
34.0[a]	$11\bar{4}$	34.5[a]	004	34.5[a]	004

Source: French, A., *Cellulose*, 21, 885–896, 2014.
[a] Bragg's angle values have been estimated from graph.

for cellulose Iα, cellulose Iβ and cellulose II [45]. The X-ray diffraction patterns of cellulose Iα and cellulose Iβ overlap, making it extremely difficult to differentiate them by using X-ray powder diffraction. Although cellulose Iα and cellulose Iβ exhibit diffraction peaks at about the same Bragg's angles, the Miller indices are different (Table 5.3). However, if the samples are analysed with electron diffraction, the crystal structure of the two different cellulose I polymorphs can be resolved [25,26].

Cellulose structure is not purely crystalline due to the presence of disordered (amorphous) domains. The relative amount of ordered (crystalline) and disordered (amorphous) domains in cellulose can be described by the crystallinity index (CI) of cellulose. There are several techniques to determine the degree of crystallinity of cellulose such as X-ray diffraction (XRD), solid-state carbon-13 nuclear magnetic resonance (^{13}C NMR), Fourier transform-Infrared (FT-IR) spectroscopy and Raman spectroscopy [46]. The measured crystallinity percentage values vary, depending on the technique utilised, but the method developed by Segal and co-workers is the most used method due to its simplicity [46]. The Segal method [47] is a semi-empirical method derived from cotton cellulose samples that uses X-ray diffraction spectra of cellulose to calculate its CI:

$$CI(\%) = \frac{I_{200} - I_{am}}{I_{200}} \times 100 \tag{5.1}$$

where:

I_{200} is the height of the highest diffraction peak of the (200) lattice and corresponds to the amount of crystalline material

I_{am} is the height of the minimum intensity of the major peaks and matches to the amorphous content of the sample

The CI corresponds to the difference of the height of these two peaks divided by the height of the highest peak [47,48]. The proportion of crystalline and amorphous domains and, therefore, the crystallinity index varies, depending on the source and the treatment the cellulose has undergone. It is worth mentioning that in addition to the Segal's method, Rietveld refinement method can also be used to fit the full X-ray diffraction pattern of cellulose to obtain the crystallinity of cellulose.

5.2.2 Differences in the Crystallinity of Various Types of Nanocellulose

Lee et al. [14] compared the crystallinity of BC and CNF by measuring the area under the curves of the diffraction pattern and obtained a crystallinity of 72% for BC, whereas it was only 41% for CNF. An exemplary XRD pattern of BC and CNF is shown in Figure 5.7. BC exhibited the typical diffraction peaks of cellulose I at 14°, 16°, 22.5° and 34° [34], whereas CNF exhibited a diffractive pattern with only two broad peaks at 15° and 22.5°. The low-measured crystallinity index of CNF and the lower definition of the XRD data compared to BC are due to the presence of hemicellulose in CNF, which was found to be approximately 30%.

Although BC was found to possess a high degree of crystallinity, the culture conditions can change the crystallinity of BC. Watanabe et al. [37] found that BC produced in agitated culture was less crystalline and had a lower content of cellulose Iα than BC grown in static culture. The effect of the shear stress in agitated cultures interferes in the cellulose crystallisation, promoting the formation of smaller crystallites with higher cellulose Iβ content (39% cellulose Iβ instead of 27%). Hirai et al. [49] also found that the cultivation temperature of BC was an important parameter that could affect the crystal structure of BC. BC synthesised at 4°C possessed cellulose II structure, whereas BC synthesised at 28°C possessed a cellulose I structure (Figure 5.8). This was postulated to be due to the cellulose-producing bacteria

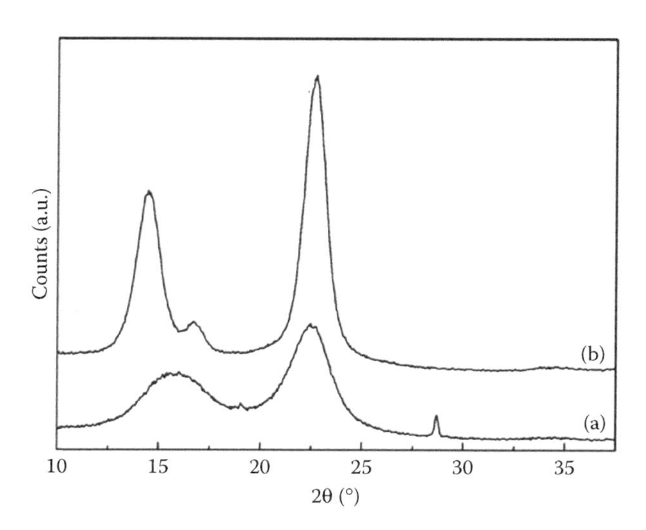

FIGURE 5.7 X-ray diffraction pattern of (a) CNF and (b) BC. (Reprinted with permission from Lee, K.Y. et al., *ACS Appl. Mater. Interfaces*, 4, 4078–4086, 2012. Copyright 2012 American Chemical Society.)

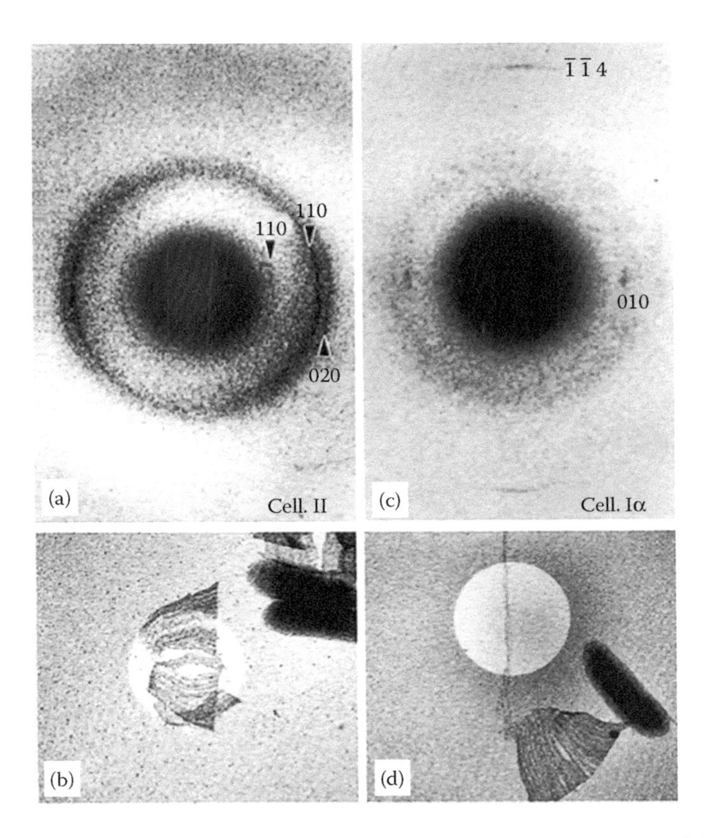

FIGURE 5.8 (a) Selected-area electron diffraction pattern of the band material produced at 4°C. The Miller indices of cellulose II are indicated, (b) morphology corresponding to (a), (c) selected-area electron diffraction pattern of the ribbon assembly produced during incubation for 7 min at 28°C; the crystal plane (010) of cellulose Iα is indicated and (d) morphology corresponding to (c). (With kind permission from Springer Science+Business Media: *Cellulose*, Communication: Culture conditions producing structure entities composed of Cellulose I and II in bacterial cellulose, 4, 1997, 239–245, Hirai, A. et al.)

(*Acetobacter xylinum*) that rotate around its longitudinal axis during the biosynthesis of cellulose ribbons (crystalline structure I) at room temperature but at lower temperature, this rotational movement was hampered, leading to a change in the crystalline structure of the synthesised BC.

The differences in crystallinity between CNF and cellulose nanocrystals (CNCs) have also been compared [50]. CNCs are produced by treating nanocellulose with strong acids, such as sulphuric, phosphoric or hydrochloric acid, to isolate the crystalline domains from native cellulose [51]. The disordered regions of cellulose chains are easily degraded by the acid molecules, whereas the crystalline parts of cellulose are more resistant to the acid hydrolysis and are left intact [23]. As a result, rod-like

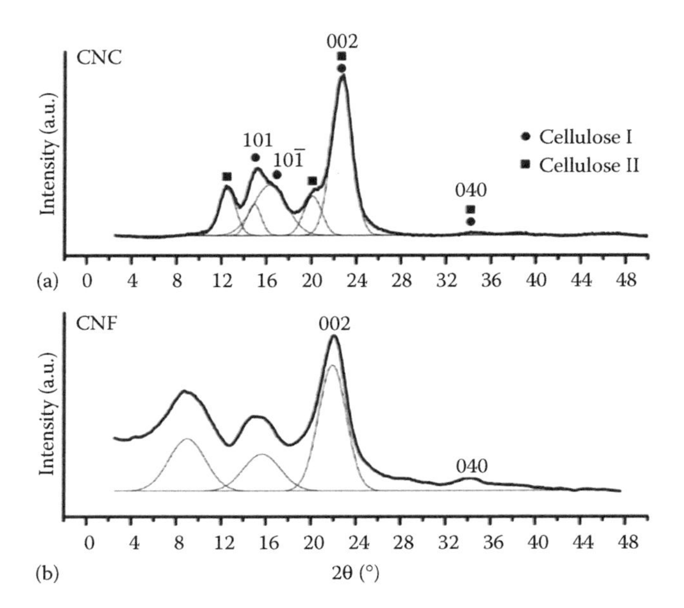

FIGURE 5.9 X-ray diffraction pattern of (a) CNC and (b) CNF. (Reprinted with permission from Xu, X. et al., *ACS Appl. Mater. Interfaces*, 5, 2999–3009, 2013. Copyright 2013 American Chemical Society.)

CNCs are obtained. Xu et al. [50] found that CNC possessed a crystallinity index of 81.0%, whereas CNF possessed a crystallinity index of 64.4% when the crystallinity was calculated using the semi-emperical Segal's method. The difference was even greater when the XRD pattern was deconvoluted. Crystallinity indices of 95% for CNC and 39% for CNF were obtained. The XRD diffraction patterns for CNC exhibit peaks of both cellulose I and cellulose II polymorphs. The peaks in Figure 5.9 at 15.1°, 17.5°, 22.7° and 34.3° correspond to cellulose I, whereas the peaks at 12.5°, 20.1°, 22.7° and 34.0° correspond to cellulose II. When cellulose undergoes an acid hydrolysis treatment, the impurities and most of the amorphous regions of the cellulose chains are removed and only the crystalline parts remain. This acid treatment also converts part of the crystalline structure from cellulose I to cellulose II. In the case of CNF, when cellulose is subjected to intensive mechanical treatment, the cellulose crystals can deform leading to lower crystallinity values although the cellulose still exhibits a crystalline structure of cellulose I.

Peng et al. [52] also investigated the crystallinity differences between CNC and CNF using different drying methods. The CNF samples prepared by spray drying had the highest crystallinity index due to high temperatures involved during the drying process. When cellulose is subjected to heat and humidity treatments, its amorphous regions recrystallise in more ordered and crystalline structures [53,54]. In the case of CNC, different drying processes varied both the crystallinity indexes and the proportion of cellulose I and II of the samples.

5.3 THERMAL STABILITY OF CELLULOSE

Cellulosic materials have been widely used in many applications such as paper, cloth fabric, construction materials and many others. In all these applications, the thermal degradation behaviour of cellulose is key to assess the performance of the material. For this reason, the thermal decomposition of cellulose has been studied extensively [55–59]. In nitrogen atmosphere, the thermal degradation behaviour of cellulose is defined by a straightforward, single-step, irreversible reaction (Figure 5.10a). However, cellulose degradation in air exhibits a two-step thermal degradation behaviour (Figure 5.10b). Unlike pure hydrocarbons, cellulose pyrolysis mechanism is not governed by a universal rate law [60]. It is broadly accepted that the pyrolysis of cellulose consists of a series of complex chemical reactions that are strongly influenced by the physical factors, such as temperature, heating time, the type of atmosphere used during measurements, and chemical factors, such as the degree of crystallinity of cellulose, the presence of impurities and the type of cellulose sample [59,61].

5.3.1 Thermal Degradation Pathway of Cellulose

The thermal degradation of cellulose, both in air and in nitrogen, is believed to consist of three main steps (see the Broido–Shafizadeh model shown in Figure 5.11)

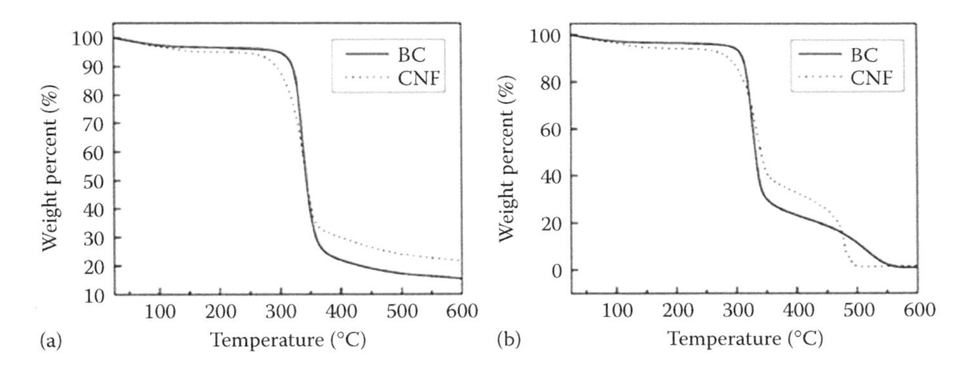

FIGURE 5.10 Thermal degradation behaviour of CNF and BC in (a) nitrogen and (b) air, respectively. (Reprinted with permission from Lee, K.Y. et al., *ACS Appl. Mater. Interfaces*, 4, 4078–4086, 2012. Copyright 2012 American Chemical Society.)

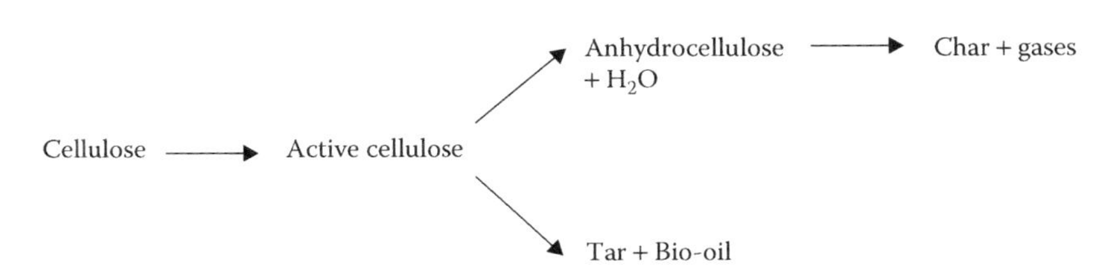

FIGURE 5.11 Broido–Shafizadeh mechanism for cellulose pyrolysis. (Adapted from Mamleev, V. et al., *J. Anal. Appl. Pyrolysis*, 80, 151–165, 2007.)

[62–64]. The first step, also known as initiation step in the cellulose pyrolysis reaction scheme, is the 'cellulose activation'. This initial process is linked to the scission of glycosidic bonds, leading to a reduction in the DP of cellulose without losing any mass. However, some authors [56,61] consider 'active cellulose' as 'anhydrocellulose' although only 'anhydrocellulose' experiences a mass loss due to its dehydration during the water removal. These species of lower molecular weight contain a hydroxyl group or levoglucosan at the end of the chain.

Once the cellulose molecule has been activated, two propagation reactions compete against each other during cellulose pyrolysis. On the one hand, active cellulose can dehydrate to give place to 'anhydrocellulose' that will contribute to the formation of solid char and low molecular gases such as CO_2, CO and H_2O. This process consists of the partial cross-linking of cellulose molecules, resulting in the formation of char. On the other hand, active cellulose can decompose to tars, which consist of a mixture of levoglucosan and other bio-oils such as hydroxyacetaldehyde, hydroxyacetone and furfural [55]. These products are obtained by the depolymerisation of the cellulose chains and are more flammable than solid char. If oxygen is present during the thermal decomposition of cellulose, the combustion of these flammable bio-oils will generate extra energy and heat, promoting the thermal degradation of cellulose. If a cellulose sample exhibits a high degree of cross-linking, the main products of its pyrolysis will be low molecular gases and char, instead of heavier molecules of tar [56]. Although both reactions happen during the cellulose pyrolysis process, catalytic dehydration reaction is favoured at low temperatures (below 100°C to 140°C), whereas the formation of bio-oils is more dominant at higher temperatures [59].

The determination of the activation energy of cellulose pyrolysis is also controversial in the research community. Antal and Várhegyi [65] reported that the thermal degradation of cellulose consists of 'an endothermic reaction governed by a first-order rate law with a high activation energy (ca. 238 kJ/mol)'. However, Milosavljevic and Suuberg [66] found that at temperatures below 327°C and slow heating rates (~1 K/min), the activation energy for cellulose thermal decomposition was about 218 kJ/mol. At temperatures above 327°C and higher heating rates (~60 K/min), the activation energy was about only 140 kJ/mol. These findings are consistent with the Broido–Shafizadeh model in which the competitive reactions of cellulose thermal degradation (Figure 5.11) are dependent on the heating rate and the temperature [59]. The dehydration reaction path that leads to the formation of char and low molecular weight gases has a low activation energy, whereas the depolymerisation reaction path that leads to the production of tar has a high activation energy. Activation energy values for cellulose thermal degradation ranging between 69 kJ/mol to 516.3 kJ/mol have been obtained by various authors (Table 5.4) [67–71]. This variation in the values is due to the difference in the heating rates, the calculation methods employed and the type of cellulose examined in these studies.

5.3.2 Differences in the Thermal Stability of Various Types of Nanocellulose

Nanocellulose is often regarded as a potential candidate to produce strong natural fibre-reinforced thermoplastic composites. Therefore, thermal stability of nanocellulose is

TABLE 5.4
Activation Energy for Cellulose Thermal Degradation for Different Cellulose Sources and Different Methods

Cellulose Source	Method of Analysis	Conversion Rate	Temperature Range (°C)	Ea (kJ/mol)	Reference
Bacterial Cellulose					
BC (*Acetobacter xylinum*)	Coats–Redfern	–	138.1–261.3	85.7	[67]
BC (*Acetobacter xylinum*)	Coats–Redfern	–	280.6–362.3	219.9	[67]
BC (*Acetobacter xylinum*)	Coats–Redfern	–	375.8–419.0	516.3	[67]
BC (*Acetobacter xylinum*)	Coats–Redfern	–	196.1–399.1	79.7	[68]
BC (*Acetobacter xylinum*)	Coats–Redfern	–	399.1–599.2	212.4	[68]
Higher Plants					
Cotton	Coats–Redfern	–	301.1–370.7	245.7	[68]
Cotton	Coats–Redfern	–	447.8–515.2	363.6	[68]
Viscose wood pulp	Coats–Redfern	–	272.1–366.1	238.4	[68]
Viscose wood pulp	Coats–Redfern	–	371.7–509.6	95.9	[68]
Bleached sulphite fibres (*Pinus taeda*)	F–W–O	0.2	–	178.0	[69][a]
Bleached sulphite fibres (*Pinus taeda*)	F–W–O	0.7	–	145.0	[69][a]
Bleached kraft fibres (*Eucalyptus grandis*)	F–W–O	0.2	–	210.0	[69][a]
Bleached kraft fibres (*Eucalyptus grandis*)	F–W–O	0.7	–	167.0	[69][a]
Cellulose Nanocrystals					
CNC (corn stalk)	F–W–O	0.2	–	285.2	[70]
CNC (corn stalk)	F–W–O	0.7	–	343.0	[70]
CNC (BC)	F–W–O	0.25	–	69.0	[71][a]
CNC (BC)	F–W–O	0.7	–	118.0	[71][a]

Note: F–W–O stands for the Flynn–Wall–Osawa method.
[a] Values estimated from a graph.

one of the major properties to be considered. However, the thermal behaviour of nanocellulose varies greatly, depending on the treatment and source of the raw material (Table 5.5). One of the main parameters affecting the thermal degradation behaviour of cellulose is the degree of crystallinity. Many authors have reported that cellulosic materials with a high degree of crystallinity possessed higher thermal stability [14,18,72]. Mostashari and Moafi [73] observed that cellulose thermal degradation starts on the amorphous regions and propagates to its more crystalline domains. These observations agree with the findings of Poletto et al. [74] in which wood cellulose with predominantly amorphous structure is less resistant to heat and high temperatures than those with a compact and ordered crystalline structure.

TABLE 5.5
Onset Degradation Temperature, Maximum Mass Loss Rate Temperature and Carbon Yield for Different Cellulose Samples

Cellulose Source	Onset Degradation Temperature (°C)	Maximum Mass Loss Rate Temperature (°C)	Carbon Yield (%)	Reference
Higher Plants				
Bleached eucalyptus pulp	150–210	300–340	2.0–6.2	[75]
CNF (powdered cellulose)	184–207	328–363	11.5–18.1	[52]
CNF (powdered cellulose)	250–260	310–350	27.0–40.0	[76]
CNF (birch kraft pulp)	247	345[a]	23.0[a]	[14]
CNF (birch kraft pulp)	275–280	320[a]	17.0	[77]
CNC				
CNC (wood pulp)	205–206	253–311	18.4–31.9	[52]
CNC (corn stalk)	230–248	259–269	16.1–22.0	[70]
CNC (BC, acid hydrolysis)	125	184	21.0[a]	[71]
CNC (BC, enzyme hydrolysis)	259	379	30.0[a]	[71]
BC				
BC (*Acetobacter xylinum*)	345[a]	355[a]	2.0[a]	[78]
BC (*Acetobacter xylinum*)	294	345[a]	15.0[a]	[14]
BC (*Acetobacter xylinum*)	320	340[a]	17.0	[77]
BC (*Gluconacetobacter hansenii*)	280	373	10.0[a]	[79]

[a] Values estimated from a graph.

The crystal structure of nanocellulose also influences the thermal degradation of the cellulose samples. Yue [80] compared the thermal stability of neat cotton fibres and mercerised cotton fibres. It was found in this study that cellulose II (mercerised cellulose) is more thermally stable than cellulose I (raw cotton). Similar results were obtained by Peng et al. [52] whereby they observed that the final char residue for CNC (mixture of cellulose I and II) was 31.8% but for CNF (cellulose I), it was only 15.9%. Nonetheless, CNCs have lower onset degradation temperatures than BC and CNF. Roman and Winter [81] attributed these differences to the presence of sulphate groups in hydrolysed CNCs. Sulphate groups enhance dehydration reactions and the water released catalyses cellulose thermal degradation, thereby lowering both the onset temperatures and activation energies of cellulose pyrolysis. However, sulphate groups also act as flame retardants, increasing the amount of charred residue at high temperatures.

The production of CNC via acid hydrolysis leads to the cleavage of the cellulose chains, decreasing their degree of polymerisation. The breakdown of the $\beta(1\rightarrow4)$ glycosidic bonds of the cellulose chains is linked to the production of reducing ends or terminal ends of cellulose. Agustin et al. [78] attributed the decline of the onset degradation temperature of CNC to lower degree of polymerisation and increased number of reducing ends. Each reducing end contains an anomeric carbon with a hemiacetal hydroxyl group that acts as reactive site and starting point during cellulose pyrolysis [82].

In the case of BC and CNF, Lee et al. [14] compared BC and CNF nanopapers and found that even though their thermal degradation behaviour is very similar, CNF has a lower onset degradation temperature due to its lower degree of crystallinity compared to BC. However, the higher residue weight of CNF might not be related to the crystallinity of the samples. CNF nanopaper is much less porous than the BC nanopaper, and previous research has shown that more compact cellulose structures lead to higher carbon yields [52,83].

The presence of impurities might also play a role in determining the thermal degradation of CNF. Although lignin is often completely removed from CNF, hemicellulose contents of up to 30% can be found in CNF samples [84,85]. Although hemicellulose and cellulose have similar structures, it has been found that hemicellulose is easier to degrade [86]. The main weight loss step during thermal decomposition of cellulose occurs between 315°C and 400°C, whereas hemicellulose decomposition takes place at lower temperatures (between 220°C and 315°C) [87]. This could explain why the onset degradation temperature of CNF is shifted to at lower temperatures than for BC, which consists basically of pure cellulose [14].

5.4 CONCLUSION

Cellulose is one of the most studied natural polymers in the world. Although its chemical formula is well known, cellulose structure and its properties will vary depending on its source and the treatment it has been subjected to. Two types of cellulose can be distinguished in the nanoscale (cellulose fibrils with a diameter of less than 100 nm): CNF and BC. When nanocellulose is obtained from plant-based materials, the presence of impurities such as hemicellulose lowers the degree of crystallinity of the samples. BC, on the other hand, has higher degree of crystallinity.

Among other characteristics, crystallinity has a crucial effect on the physical, mechanical and chemical properties of cellulose. Many authors have studied the role of crystallinity on cellulose reactivity and found that the more amorphous the starting material is, the easier it is to degrade [88,89]. This also applies to thermal degradation. When cellulose has a high crystallinity, its chains are more ordered and densely packed, and the cellulose possess higher thermal stability. If the structure of cellulose is predominantly amorphous, the chains are more accessible and the cellulose will be less thermally stable during cellulose pyrolysis. However, other parameters apart from crystallinity also affect the thermal stability of cellulose. The degree of polymerisation and the drying method employed also change the thermal degradation behaviour of the cellulosic materials.

In summary, cellulose is a complex and versatile material due to its unique chemical structure. The hydrogen bonding between the numerous hydroxyl groups present in the cellulose chains allows infinite different configurations of the cellulose fibrils network. The hydrogen bonding governs the stability and conformation of the cellulose crystalline structure. The degree of crystallinity and the crystalline polymorphs of cellulose have been extensively studied for many years [90]. Further study needs to be done on this field to learn more about how the cellulose structure affects cellulose properties such as its thermal stability.

ACKNOWLEDGEMENTS

The authors would like to thank the UK Engineering and Physical Sciences (EP/N026489/1) and Imperial College London for funding Alba Santmartí (AS).

REFERENCES

1. J. Müssig, *Industrial Applications of Natural Fibres: Structure, Properties and Technical Applications*, John Wiley & Sons, UK, 2010.
2. A. Payen, Memoire sur la composition du tissu propre des plantes et du ligneux, *Compt. Rend.* 7 (1838) 1052–1056.
3. K. Blomqvist, S. Djerbi, H. Aspeborg, T. Teeri, Cellulose Biosynthesis in Forest Trees, in: M. Brown, and I. Saxena, *Cellulose: Molecular and Structural Biology: Selected Articles on the Synthesis, Structure, and Applications of Cellulose*, Springer, Austin, TX, 2007.
4. K. Vizárová, S. Kirschnerová, F. Kačík, A. Briškárová, S. Šutý, S. Katuščák, Relationship between the decrease of degree of polymerisation of cellulose and the loss of groundwood pulp paper mechanical properties during accelerated ageing, *Chem. Pap.* 66 (2012) 1124–1129.
5. A. Frey-Wyssling, K. Mühlethaler, Die elementarfibrillen der cellulose, *Makromol. Chem.* 62 (1963). doi:10.1002/macp.1963.020620103.
6. L. J. Gibson, The hierarchical structure and mechanics of plant materials, *J. R. Soc. Interface.* 9 (2012) 2749–2766. doi:10.1098/rsif.2012.0341.
7. I. Siró, D. Plackett, Microfibrillated cellulose and new nanocomposite materials: A review, *Cellulose.* 17 (2010). doi:10.1007/s10570-010-9405-y.
8. H. Meier, Chemical and morphological aspects of the fine structure of wood, *Pure Appl. Chem.* 5 (1962). doi:10.1351/pac196205010037.
9. Q. Li, S. McGinnis, C. Sydnor, A. Wong, S. Renneckar, Nanocellulose life cycle assessment, *ACS Sustain. Chem. Eng.* 1 (2013) 919–928. doi:10.1021/sc4000225.
10. M. A. Hubbe, O. J. Rojas, L. A. Lucia, M. Sain, Cellulosic Nanocomposites: A Review, *BioResources.* 3 (2008) 929–980. doi:10.15376/biores.3.3.929-980.
11. A. F. Turbak, F. W. Snyder, K. R. Sandberg, Microfibrillated cellulose, a new cellulose product: Properties, uses and commercial potential, *J. Appl. Polym. Sci.* 37 (1983) 815–827.
12. F. W. Herrick, R. L. Casebier, J. K. Hamilton, K. R. Sandberg, Microfibrillated cellulose: Morphology and accessibility, *J. Appl. Polym. Sci. Appl. Polym. Symp.* 37 (1983) 797–813.
13. T. Taniguchi, K. Okamura, New films produced from microfibrillated natural fibres, *Polym. Int.* 47 (1998) 291–294.
14. K. Y. Lee, T. Tammelin, K. Schulfter, H. Kiiskinen, J. Samela, A. Bismarck, High performance cellulose nanocomposites: Comparing the reinforcing ability of bacterial cellulose and nanofibrillated cellulose, *ACS Appl. Mater. Interfaces.* 4 (2012) 4078–4086. doi:10.1021/am300852a.
15. J. Brown, XLIII.—On an acetic ferment which forms cellulose, *J. Chem. Soc., Trans.* 49 (1886) 432–439.

16. J. J. Blaker, K.-Y. Lee, A. Bismarck, Hierarchical composites made entirely from renewable resources, *J. Biobased Mater. Bioenergy.* 5 (2011) 1–16. doi:10.1166/jbmb.2011.1113.
17. K. Y. Lee, J. J. Blaker, A. Bismarck, Surface functionalisation of bacterial cellulose as the route to produce green polylactide nanocomposites with improved properties, *Compos. Sci. Technol.* 69 (2009) 2724–2733. doi:10.1016/j.compscitech.2009.08.016.
18. M. Jonoobi, J. Harun, A. Shakeri, M. Misra, K. Oksmand, Chemical composition, crystallinity, and thermal degradation of bleached and unbleached kenaf bast (*Hibiscus cannabinus*) pulp and nanofibers, *BioResources.* 4 (2009) 626–639. doi:10.15376/biores.4.2.626-639.
19. S. J. Eichhorn, G. R. Davies, Modelling the crystalline deformation of native and regenerated cellulose, *Cellulose.* 13 (2006) 291–307. doi:10.1007/s10570-006-9046-3.
20. M. Matsuo, C. Sawatari, Y. Iwai, F. Ozaki, Effect of orientation distribution and crystallinity on the measurement by X-ray diffraction of the crystal lattice moduli of cellulose I and II, *Macromolecules.* 23 (1990) 3266–3275.
21. Y.-C. Hsieh, H. Yano, M. Nogi, S. J. Eichhorn, An estimation of the Young's modulus of bacterial cellulose filaments, *Cellulose.* 15 (2008) 507–513. doi:10.1007/s10570-008-9206-8.
22. R. J. Moon, A. Martini, J. Nairn, J. Simonsen, J. Youngblood, Cellulose nanomaterials review: Structure, properties and nanocomposites, *Chem. Soc. Rev.* 40 (2011) 3941–3994. doi:10.1039/c0cs00108b.
23. Y. Habibi, L. A. Lucia, O. J. Rojas, Cellulose nanocrystals: Chemistry, self-assembly, and applications, *Chem. Rev.* 110 (2010) 3479–3500. doi:10.1021/cr900339w.
24. A. C. O'Sullivan, Cellulose: The structure slowly unravels, *Cellulose.* 4 (1997) 173–207.
25. Y. Nishiyama, J. Sugiyama, H. Chanzy, P. Langan, Crystal structure and hydrogen bonding system in cellulose Iα from synchrotron X-ray and neutron fiber diffraction, *J. Am. Chem. Soc.* 125 (2003) 14300–14306.
26. Y. Nishiyama, P. Langan, H. Chanzy, Crystal structure and hydrogen-bonding system in cellulose Iβ from synchrotron X-ray and neutron fiber diffraction yoshiharu, *J. Am. Chem. Soc.* 124 (2002) 9074–9082.
27. J. A. Kaduk, T. N. Blanton, An improved structural model for cellulose II, *Powder Diffr.* 28 (2013) 194–199. doi:10.1017/S0885715613000092.
28. M. Wada, H. Chanzy, Y. Nishiyama, P. Langan, Cellulose IIII crystal structure and hydrogen bonding by synchrotron X-ray and neutron fiber diffraction, *Macromolecules.* 37 (2004) 8548–8555.
29. M. Wada, L. Heux, Y. Nishiyama, P. Langan, X-ray crystallographic, scanning microprobe X-ray diffraction, and cross-polarized/magic angle spinning (13)C nuclear magnetic resonance studies of the structure of cellulose III(II), *Biomacromolecules.* 10 (2009) 302–309. doi:10.1021/bm8010227.
30. E. Gardiner, A. Sarko, Packing analysis of carbohydrates and polysaccharides. 16. The crystal structures of celluloses IVI and IVII, *Can. J. Chem.* 63 (1985) 173–180.
31. R. Atalla, D. Vanderhart, Studies on the structure of cellulose using Raman spectroscopy and solid state 13C NMR, Cellulose and Wood: Chemistry and Technology, in: C. Schuerch (Ed.) *Proceedings of the 10th cellulose conference*, New York, John Wiley & Sons, 1989, 169–187.
32. T. Imai, J. Sugiyama, T. Itoh, F. Horii, Almost pure Iα cellulose in the cell wall of glaucocystis, *J. Struct. Biol.* 127 (1999) 248–257. doi:10.1006/jsbi.1999.4160.
33. H. Yamamoto, F. Horii, CPMAS carbon-13 NMR analysis of the crystal transformation induced for Valonia cellulose by annealing at high temperatures, (2002). http://pubs.acs.org/doi/abs/10.1021/ma00058a020#.WLaxntxeH5w.mendeley (Accessed March 1, 2017).
34. M. Wada, T. Okano, J. Sugiyama, Allomorphs of native crystalline cellulose I evaluated by two equatorial d-spacings, *J. Wood Sci.* 47 (2001) 124–128. doi:10.1007/BF00780560.

35. D. L. VanderHart, R. H. Atalla, Studies of microstructure in native celluloses using solid-state carbon-13 NMR, *Macromolecules.* 17 (1984) 1465–1472. doi:10.1021/ma00138a009.
36. M. Szymańska-Chargot, J. Cybulska, A. Zdunek, sensing the structural differences in cellulose from apple and bacterial cell wall materials by raman and FT-IR spectroscopy, *Sensors (Basel).* 11 (2011) 5543–5560. doi:10.3390/s110605543.
37. K. Watanabe, M. Tabuchi, Y. Morinaga, F. Yoshinaga, Structural features and properties of bacterial cellulose produced in agitated culture, *Cellulose.* 5 (1998) 187–200. doi:10.1023/A:1009272904582.
38. C. Rondeau-Mouro, B. Bouchet, B. Pontoire, P. Robert, J. Mazoyer, A. Buléon, Structural features and potential texturising properties of lemon and maize cellulose microfibrils, *Carbohydr. Polym.* 53 (2003) 241–252. doi:10.1016/S0144-8617(03)00069-9.
39. V. I. Kovalenko, Crystalline cellulose: Structure and hydrogen bonds, *Russ. Chem. Rev.* 79 (2010) 231–142.
40. J. Sugiyama, R. Vuong, H. Chanzy, Electron diffraction study on the two crystalline phases occurring in native cellulose from an algal cell wall, *Macromolecules.* 24 (1991) 4168–4175.
41. C. Woodings, A brief history of regenerated cellulosic fibres, in: C. Woodings, *Regenerated Cellulose Fibres,* Cambridge: Woodhead Publishing Limited. 2001.
42. F. J. Kolpak, M. Weih, J. Blackwell, Mercerization of cellulose: 1. Determination of the structure of Mercerized cotton, *Polymer (Guildf).* 19 (1978) 123–131. doi:10.1016/0032-3861(78)90027-7.
43. L. M. J. Kroon-Batenburg, J. Kroon, The crystal and molecular structures of cellulose I and II, *Glycoconj. J.* 14 (1997) 677–690.
44. S. H. Zeronian, H.-S. Ryu, Properties of cotton fibers containing the cellulose IV crystal structure, *J. Appl. Polym. Sci.* 33 (1987) 2587–2604. doi:10.1002/app.1987.070330725.
45. A. French, Idealized powder diffraction patterns for cellulose polymorphs, *Cellulose.* 21 (2014) 885–896.
46. S. Park, J. Baker, M. Himmel, P. Parilla, D. Johnson, Cellulose crystallinity index: Measurement techniques and their impact on interpreting cellulase performance, *Biotechnol Biofuels.* 3 (2010) 10.
47. C. M. Conrad, L. Segal, J. J. Creely, A. E. Martin Jr, An empirical method for estimating the degree of crystallinity of native cellulose using the X-ray diffractometer, *Text. Res. J.* 29 (1962) 786–794.
48. A. D. French, M. S. Cintrón, Cellulose polymorphy, crystallite size, and the segal crystallinity index, *Cellulose.* 20 (2013) 583–588. doi:10.1007/s10570-012-9833-y.
49. A. Hirai, M. Tsuji, F. Horii, Communication: Culture conditions producing structure entities composed of Cellulose I and II in bacterial cellulose, *Cellulose.* 4 (1997) 239–245. doi:10.1023/A:1018439907396.
50. X. Xu, F. Liu, L. Jiang, J. Y. Zhu, D. Haagenson, D. P. Wiesenborn, Cellulose nanocrystals versus cellulose nanofibrils: A comparative study on their microstructures and effects as polymer reinforcing agents, *ACS Appl. Mater. Interfaces.* 5 (2013) 2999–3009. doi:10.1021/am302624t.
51. J. George, S. N. Sabapathi, Cellulose nanocrystals: Synthesis, functional properties, and applications, *Nanotechnol. Sci. Appl.* 8 (2015) 45. doi:10.2147/NSA.S64386.
52. Y. Peng, D. J. Gardner, Y. Han, A. Kiziltas, Z. Cai, M. A. Tshabalala, Influence of drying method on the material properties of nanocellulose I: Thermostability and crystallinity, *Cellulose.* 20 (2013) 2379–2392. doi:10.1007/s10570-013-0019-z.
53. H. Hatakeyama, T. Hatakeyama, Structural change of amorphous cellulose by water- and heat-treatment, *Makromol. Chem.* 182 (1981) 1655–1668.
54. S. Yildiz, E. Gümüşkaya, The effects of thermal modification on crystalline structure of cellulose in soft and hardwood, *Build. Environ.* 42 (2007) 62–67. doi:10.1016/j.buildenv.2005.07.009.

55. D. K. Shen, S. Gu, The mechanism for thermal decomposition of cellulose and its main products, *Bioresour. Technol.* 100 (2009) 6496–6504. doi:10.1016/j.biortech.2009.06.095.
56. V. Mamleev, S. Bourbigot, J. Yvon, Kinetic analysis of the thermal decomposition of cellulose: The main step of mass loss, *J. Anal. Appl. Pyrolysis.* 80 (2007) 151–165. doi:10.1016/j.jaap.2007.01.013.
57. W. Jin, K. Singh, J. Zondlo, Pyrolysis kinetics of physical components of wood and wood-polymers using isoconversion method, *Agriculture.* 3 (2013) 12–32.
58. P. K. Chatterjee, Thermogravimetric analysis of cellulose, *J. Polym. Sci.* 6 (1968) 3217–3233.
59. P. K. Chatterjee, R. F. Schwenker, Jr, Instrumental methods in the study of oxidation, degradation, and pyrolysis of cellulose, in: *Instrumensts Analysis of Cotton Cellulose and Modified Cotton Cellulose*, 1972: pp. 275–338.
60. K. K. Kuo, *Principles of Combustion*, John Wiley & Sons, Hoboken, NJ, 2005.
61. G. Várhegyi, E. Jakab, Is the Broido-Shafizadeh model for cellulose pyrolysis true? *Energy & Fuels.* 8 (1994) 1345–1352.
62. A. Broido, M. A. Nelson, Char yield on pyrolysis of cellulose, *Combust. Flame.* 24 (1975) 263–268.
63. F. Shafizadeh, Introduction to pyrolysis of biomass. Review, *J. Anal. Appl. Pyrolysis.* 3 (1982) 283–305.
64. F. Shafizadeh, G. W. Bradbury, Thermal degradation of cellulose in air and nitrogen at low temperatures, *J. Appl. Pol. Sci.* 23 (1979) 1431–1442.
65. M. J. Antal, G. Varhegyi, Cellulose pyrolysis kinetics: The current state of knowledge, *Ind. Eng. Chem. Res.* 34 (1995) 703–717.
66. I. Milosavljevic, E. M. Suuberg, Cellulose thermal decomposition kinetics: Global mass loss kinetics, *Ind. Eng. Chem. Res.* 34 (1995) 1081–1091.
67. A. H. Basta, H. El-Saied, Performance of improved bacterial cellulose application in the production of functional paper, *J. Appl. Microbiol.* 107 (2009) 2098–2107. doi:10.1111/j.1365-2672.2009.04467.x.
68. H. El-Saied, A. I. El-Diwany, A. H. Basta, N. A. Atwa, D. E. El-Ghwas, Production and characterization of economical bacterial cellulose, *BioResources.* 3 (2008) 1196–1217. doi:10.15376/biores.3.4.1196-1217.
69. M. Poletto, V. Pistor, A. J. Zattera, Structural characteristics and thermal properties of native cellulose, in: T. v. de Ven and L. Godbout (Eds.) *Cellulose—Fundamental Aspects*, InTech: Croatia, 2013.
70. S. Huang, L. Zhou, M.-C. Li, Q. Wu, D. Zhou, Cellulose nanocrystals (CNCs) from corn stalk: Activation energy analysis, *Materials (Basel).* 10 (2017) 80. doi:10.3390/ma10010080.
71. J. George, K. V. Ramana, A. S. Bawa, Siddaramaiah, bacterial cellulose nanocrystals exhibiting high thermal stability and their polymer nanocomposites, *Int. J. Biol. Macromol.* 48 (2011) 50–57. doi:10.1016/j.ijbiomac.2010.09.013.
72. M. Poletto, H. L. Ornaghi Jr, A. J. Zattera, Native cellulose: Structure, characterization and thermal properties, *Materials (Basel).* 7 (2014) 6105–6119. doi:10.3390/ma7096105.
73. S. M. Mostashari, H. F. Moafi, Thermogravimetric analysis of a cellulosic fabric incorporated with ammonium iron (II)-sulfate hexahydrate as a flame-retardant, *J. Ind. Text.* 37 (2007) 31–42.
74. M. Poletto, A. J. Zattera, M. M. C. Forte, R. M. C. Santana, Thermal decomposition of wood: Influence of wood components and cellulose crystallite size, *Bioresour. Technol.* 109 (2012) 148–153. doi:10.1016/j.biortech.2011.11.122.
75. M. E. Calahorra, M. Cortazar, J. I. Eguiazabal, Thermogravimetric analysis of cellulose: Effect of the molecular weight on thermal decomposition, *J. Appl. Polym. Sci.* 37 (1989) 3305–3314.

76. N. Quiévy, N. Jacquet, M. Sclavons, C. Deroanne, M. Paquot, J. Devaux, Influence of homogenization and drying on the thermal stability of microfibrillated cellulose, *Polym. Degrad. Stab.* 95 (2010) 306–314. doi:10.1016/j.polymdegradstab.2009.11.020.
77. A. Mautner, J. Lucenius, M. Österberg, A. Bismarck, Multi-layer nanopaper based composites, *Cellulose.* (2017) 1–15. doi:10.1007/s10570-017-1220-2.
78. M. B. Agustin, F. Nakatsubo, H. Yano, The thermal stability of nanocellulose and its acetates with different degree of polymerization, *Cellulose.* (2015). doi:10.1007/s10570-015-0813-x.
79. H. G. O. Barud, H. D. S. Barud, M. Cavicchioli, T. S. Do Amaral, O. B. De Oliverira Jr, D. M. Santos, A. L. D. O. A. Petersen et al., Preparation and characterization of a bacterial cellulose/silk fibroin sponge scaffold for tissue regeneration, *Carbohydr. Polym.* 128 (2015) 41–51. doi:10.1016/j.carbpol.2015.04.007.
80. Y. Yue, A comparative study of cellulose I and II fibers and nanocrystals, Louisiana State University, Baton Rouge, LA, 2011.
81. M. Roman, W. T. Winter, Effect of sulfate groups from sulfuric acid hydrolysis on the thermal degradation behavior of bacterial cellulose, *Biomacromolecules.* 5 (2004) 1671–1677. doi:10.1021/bm034519+.
82. S. Matsuoka, H. Kawamoto, S. Saka, Thermal glycosylation and degradation reactions occurring at the reducing ends of cellulose during low-temperature pyrolysis, *Carbohydr. Res.* 346 (2011) 272–279. doi:10.1016/j.carres.2010.10.018.
83. P. Rämänen, P. Penttilä, K. Svedström, S. Maunu, R. Serimaa, The effect of drying method on the properties and nanoscale structure of cellulose whiskers, *Cellulose.* 19 (2012) 901–912.
84. K.-Y. Lee, Y. Aitomäki, L. A. Berglund, K. Oksman, A. Bismarck, On the use of nanocellulose as reinforcement in polymer matrix composites, *Compos. Sci. Technol.* 105 (2014) 15–27. doi:10.1016/j.compscitech.2014.08.032.
85. S. Arola, J.-M. Malho, P. Laaksonen, M. Lille, M. B. Linder, The role of hemicellulose in nanofibrillated cellulose networks, *Soft Matter.* 9 (2013). doi:10.1039/c2sm26932e.
86. M. V. Ramiah, Thermogravimetric and differential thermal analysis of cellulose, hemicellulose, and lignin, *J. Appl. Polym. Sci.* 14 (1970) 1323–1337. doi:10.1002/app.1970.070140518.
87. H. Yang, R. Yan, H. Chen, D. H. Lee, C. Zheng, Characteristics of hemicellulose, cellulose and lignin pyrolysis, *Fuel.* 86 (2007) 1781–1788. doi:10.1016/j.fuel.2006.12.013.
88. M. Möller, F. Harnischab, U. Schröder, Hydrothermal liquefaction of cellulose in subcritical water—The role of crystallinity on the cellulose reactivity, *RSC Adv.* 3 (2013) 11035–11044.
89. M. Hall, P. Bansal, J. H. Lee, M. J. Realff, A. S. Bommarius, Cellulose crystallinity—A key predictor of the enzymatic hydrolysis rate, *FEBS J.* 277 (2010) 1571–1582. doi:10.1111/j.1742-4658.2010.07585.x.
90. A. O'Sullivan, Cellulose: The structure slowly unravel, *Cellulose.* 4 (1997) 173–207.

6 Crucial Interfacial Features of Nanocellulose Materials

Andreas Mautner, Minna Hakalahti, Ville Rissanen and Tekla Tammelin

CONTENTS

6.1 Short Introduction to Different Grades of Cellulosic Nanomaterials............. 87
6.2 Dimensions and the Specific Surface Area of Nanoscaled Cellulosic Materials............. 88
6.2.1 Specific Surface Area of Nanoscaled Cellulosic Materials............. 91
6.3 Chemical Composition of the Nanocellulosic Building Blocks............. 96
6.4 Surface Charge of the Nanocellulosic Building Blocks............. 98
6.4.1 Characterisation of Surface Charge and Mobility of Nanocellulosic Materials: Zeta Potential............. 100
6.5 Hydrophilic Nanomaterial with Amphiphilic Character............. 104
6.6 Reactivity and Surface Modification of Nanocellulosic Materials............. 106
6.7 Water Interactions of Nanocellulosic Materials............. 108
6.7.1 Water Vapour Sorption Analysed Using Surface-Sensitive Methods..... 111
6.7.2 Swelling Behaviour Analysed Using Surface-Sensitive Methods.... 111
6.8 Case Examples: Importance of Nanocellulose Interfaces in Nanocellulose Applications............. 113
6.8.1 Adsorption of Contaminants onto Cellulose Nanomaterials............. 113
6.8.1.1 Mechanism of Contaminant Adsorption on Nanocelluloses............. 113
6.8.2 Application of Interfacial Properties of Nanocellulose in Membranes............. 116
6.8.3 Nanocellulose as Emulsion Stabilisers............. 117
References............. 118

6.1 SHORT INTRODUCTION TO DIFFERENT GRADES OF CELLULOSIC NANOMATERIALS

This chapter addresses the behaviour and performance of cellulosic nanomaterials and how these are attributed to the interfacial features of this particular bionanomaterial. Cellulosic nanomaterials can be roughly categorised with respect to their size and overall dimensions as well as with respect to the surface charge and chemical

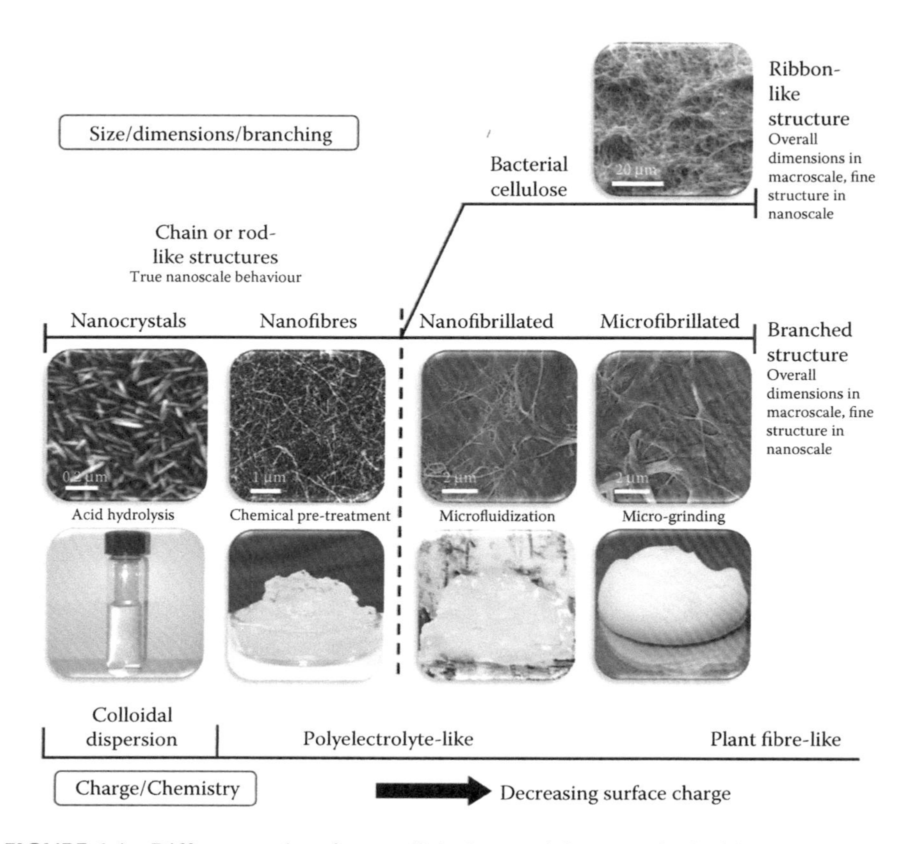

FIGURE 6.1 Different grades of nanocellulosic materials categorised with respect to size, dimensions and branching as well as surface charge and chemistry.

composition (Figure 6.1). These features mainly define the structure–function interdependencies and structure–property relationships of the different nanocellulose grades, and they can be considered as the most relevant characteristics to be quantified in order to achieve the full performance in a desired application area. In fact, in order to significantly improve and develop functional nanomaterial structures using lignocellulose-derived building blocks, the understanding of the interfacial interactions becomes highly relevant. In addition, the reader is referred to several thorough review papers [1–3] and a recent text book [4] in which different nanocellulose grades, their manufacturing, characteristic features and exploitation in various applications are comprehensively treated.

6.2 DIMENSIONS AND THE SPECIFIC SURFACE AREA OF NANOSCALED CELLULOSIC MATERIALS

Fibril dimensions can be thoroughly characterised with microscopic methods, see a short review by Kangas et al. [5] and a textbook chapter by Moon et al. [6] Optical microscopy gives an overview on fibrillated materials and is often utilised in order to follow the degradation of fibre-like material during extensive mechanical disintegration. Scanning electron microscopy (SEM) is able to give information from the millimetre scale down to the nanometre scale, whereas transmission electron microscopy (TEM) and atomic force microscopy (AFM) are capable of revealing the true nanoscale dimensions. Another measure of high importance, which is directly linked to the nanoscaled dimensions, is the specific surface area (SSA). Here, a separate section is dedicated to the determination of the SSA in general, and in addition, a collection of values and references for different nanocellulosic materials is presented.

A generic definition of cellulosic nanomaterials, following the IUPAC definition of nanomaterials, is that at least one of the external dimensions is in the size range of 1–100 nm. The main types of the nanocellulose materials, fulfilling this definition, can be divided into three different categories: cellulose nanocrystals (CNCs), cellulose nano- and microfibrils (CNFs and CMFs) and bacterial cellulose (BC). CNC and CNF/CMF grades are manufactured via the so-called top–down approach, which involves the degradation and disintegration of the plant cell wall structure by chemical/enzymatic and mechanical means. In turn, BC can be obtained by the bottom–up process. Thereby, certain types of bacteria, for example *Gluconobacter xylinum* [7], extrude cellulose fibrils that were synthesised from sugar educts, as first observed by Brown in 1886 [8]. Thus, this type of nanocellulose was discovered almost 100 years before the production of cellulose nanomaterials utilising the top–down process was established. Due to the production of the cellulose fibrils from sugar medium, opposite to plant-derived CNF, BC pellicles consist almost exclusively of cellulose without by-compounds such as lignin, hemicelluloses or pectin [9]. Accordingly, BC fibrils can reach very high degrees of crystallinity [10], which is responsible for very high Young's modulus and strength [11]. BC fibrils can be easily cultured but are also already commercially available. They constitute a certain fraction (~1 wt.%) of a food product, nata de coco, popular in Southeast Asia. Harvesting of BC fibrils via an extraction process directly yields nanofibrils [12]. Typically, BCs have diameters in the range between 20 and 100 nm with lengths on the micrometre scale [1].

The preparation of CNCs involves acid hydrolysis of cellulose fibres with strong acids, for example, H_2SO_4 or HCl. The glycosidic bonds of the less ordered domains are efficiently cleaved, leading to a release of individual nanocrystals. The dimensions of these twisted, rod-like nanocrystals are dependent on the cellulose source and hydrolysis conditions. Basically, the length of the crystals varies within 100–1000 nm, whereas the width is only in the range of a few nanometres up to tens of nanometres resulting in aspect ratio values of approximately 10–80.

Fibrillated cellulose grades (CNF and CMF) are manufactured using intensive mechanical disintegration methods often combined with chemical and/or enzymatic pre-treatments to facilitate the deliberation of the fibrils from the plant fibres [13,14]. The outcome of the final nanofibril morphology greatly varies, depending on the mechanical fibrillation protocol in relation to the level of energy consumption and the selection of pre-treatment procedures. As shown in Figure 6.1, CMFs are more coarse compared to CNFs, and for example CNF grades are needed to fabricate smooth and even film structures for the film preparation purposes. The most efficient pre-treatment concept involving 2,2,6,6-tetramethylpiperidine-1-oxyl radical (TEMPO)-mediated oxidation of cellulosic fibres yields the liberation of individual and almost monodisperse nanofibres after rather low energy-intensive mechanical disintegration [15]. The width of individual TEMPO CNF can be as low as 3–5 nm, whereas the length is on the micrometre scale.

Typically, CNF and CMF grades, which are manufactured by means of mechanical disintegration without oxidation pre-treatment, possibly assisted only with enzymatic pre-treatments, are more polydisperse with broad size distributions. Moreover, the assembly of such materials is branched resulting in a structure in which the overall dimensions are microscale, whereas the fine structure is nanoscale. Conventionally, all CNF and CMF grades form strong hydrogels with very low solid content (1–3 wt.%) due to the large surface area, high aspect ratio, nanoscale fine structure and high swelling ability. A recently developed technique to produce CNF in high consistency (20% to 40%) by exploiting specific tailored cellulase enzymes generates a new type of CNF grade, which is based on combined action of the enzymes and fibre–fibre friction during gentle mixing at low water content (Figure 6.2) [16]. Through a peeling type of action, a paste-like fibril network is formed with high yield (typically about 90%). As no chemicals are involved in the treatment, the basic properties of the starting cellulose material, for example, charge and crystallinity, remain unchanged. Based on SEM and AFM imaging, the lateral width of the fibrils is typically between 15 and 200 nm.

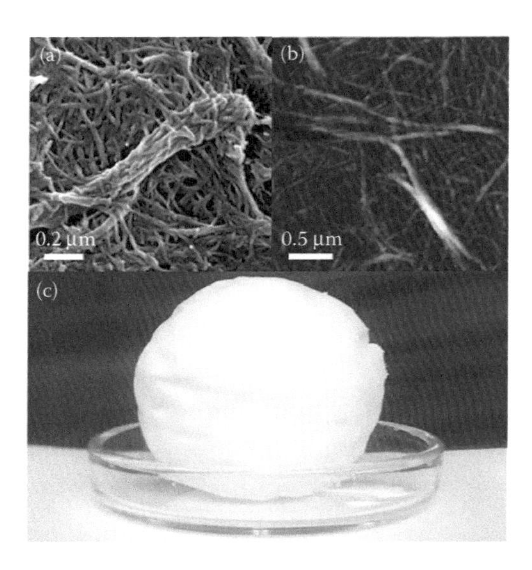

FIGURE 6.2 A new high-consistency grade of cellulose nanofibrils (HefCel) produced by combining enzymatic actions and fibre–fibre friction during mixing. (a) SEM image, (b) AFM height image that reveals the HefCel fine structure and (c) photograph of HefCel that shows the appearance of the fibril paste.

6.2.1 Specific Surface Area of Nanoscaled Cellulosic Materials

The SSA is a measure of great importance for the characterisation of materials, in particular nanosized materials, such as cellulose nanomaterials. It is defined as surface area per unit mass of a material (m^2/g). The determination of the SSA is most commonly based on the physico-chemical phenomenon of adsorption of gases on solid surfaces. Thereby, gas molecules, usually an inert gas such as nitrogen or a noble gas, are adsorbed on the sample of interest and adsorption isotherms established from which the SSA and also the pore size of materials could be analysed. It is important to note that the measurement principle is based on physisorption and not chemisorption. Physisorption is the process through which gas molecules are attached to a solid surface but not chemically bonded. Hence, the energy of adsorption is rather low, comparable to liquefaction, due to weak interaction between adsorbate and adsorbent via van der Waals forces. Furthermore, the process is completely reversible, which is important for the determination of pore sizes, and multilayer adsorption is 'allowed', which is not in chemisorption. For a more detailed account on how these gas adsorption models are derived, the reader is referred to specialised literature on which the following discussion is based [17].

In order to classify the adsorption behaviour of gases on solids, five types of adsorption isotherms were established by Brunauer [18]. Adsorption isotherms show the amount of gas molecules adsorbed as a function of the partial pressure of the gas. Hence, the higher the partial pressure, the higher the amount of molecules adsorbed on the solid. The amount of molecules adsorbed can be displayed in terms of surface coverage, that is, θ is the ratio of the number of adsorption sites filled to the number of available adsorption sites. Based on the principal types of adsorption isotherms, the surface area can be characterised at the point where monolayer coverage (θ_m) is reached. The easiest mathematical treatment of adsorption measurements is by application of the Langmuir isotherm [19]. It is based on several assumptions that limit their applicability. It requires that all adsorption sites are equivalent, the surface containing the adsorbing sites is uniform and homogeneous, that is, a perfectly flat plane with no corrugations, the heat of adsorption does not vary with coverage and only monolayer adsorption takes place whereby each site can hold at most one molecule of the adsorbate. Furthermore, no interaction between adsorbed molecules on adjacent sites is present and the adsorbing gas adsorbs into an immobile state. Even though these stringent limitations exist, it is quite often a useful tool to describe practical adsorption experiments. For the case of adsorption of gases, in a linear form, it can be written as follows:

$$\frac{p}{V} = \frac{p}{V_{\text{mono}}} + \frac{1}{K_L V_{\text{mono}}} \quad (6.1)$$

where:

p is the partial pressure
V is the volume of the gas adsorbed
V_{mono} is the volume of the gas at full monolayer coverage
K_L is the Langmuir constant taking into account the adsorption enthalpy and the temperature

By measuring pressure and gas volume during an adsorption experiment, it is possible to figure out the monolayer capacity of the gas and hence the surface area.

In practical terms, the adsorption isotherm after Brunauer–Emmet–Teller (BET) is more commonly applied. It is based on similar assumption as the Langmuir isotherm but allows for dealing with (incomplete) multilayer adsorption. Thus, under certain circumstances, that is, at low pressures and hence only monolayer adsorption, this isotherm reduces to the Langmuir isotherm. In other words, the BET theory is an extension of the Langmuir treatment to allow for multilayer adsorption on non-porous solid surfaces. It is given by

$$\frac{1}{V\left(\frac{p_0}{p}-1\right)} = \frac{1}{BV_{\mathrm{mono}}} + \frac{B-1}{BV_{\mathrm{mono}}}\frac{p}{p_0} \tag{6.2}$$

where:

p_0 is the saturation vapour pressure

B (see Equation 6.3) is a system-depending constant that account for temperature as well as adsorption (ΔH_{ads}) and condensation (ΔH_{cond}) enthalpies:

$$B = \mathrm{e}^{\frac{\Delta H_{\mathrm{ads}} - \Delta H_{\mathrm{cond}}}{RT}} \tag{6.3}$$

The BET isotherm can be derived by balancing the rates of evaporation and condensation for the various adsorbed molecular layers, based on the simplifying supposition that a characteristic heat of adsorption applies to the first monolayer, whereas the heat of liquefaction of the vapour in question applies to adsorption in the second and subsequent molecular layers. Just as the Langmuir isotherm, the application of the BET adsorption isotherm is based on a number of assumptions and hence exhibits certain limitations, such as failure to account for surface roughness and to account for adsorbate–adsorbate interactions. Accordingly, it can be considered to be an accurate account of adsorption only within a restricted relative pressure range ($0.05 < (p/p_0) < 0.35$).

Being aware of these limitations, this equation can be used to determine the surface area by first identifying V_{mono} from a $1/V((p_0/p)-1)$ versus p/p_0 plot (Figure 6.3).

From V_{mono}, the surface area (A_s) can then be calculated:

$$A_s = A_m N_{\mathrm{mono}} = A_m \frac{V_{\mathrm{mono}}}{V_{\mathrm{T,p}}} N_A \tag{6.4}$$

where:

A_m is the cross-sectional area of a single gas molecule (m^2) (for N_2: $A_m = 16.2 \times 10^{-20}\,m^2$, Ar: $A_m = 16.6 \times 10^{-20}\,m^2$, Kr: $A_m = 21.0 \times 10^{-20}\,m^2$)

N_{mono} is the number of adsorbate molecules that is required to cover the surface with a single monolayer

N_A is the Avogadro's constant

$V_{\mathrm{T,p}}$ is the molar volume of the adsorbed gas (L/mol)

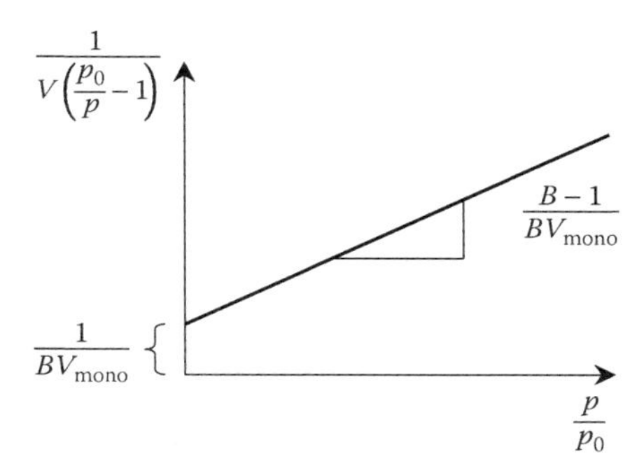

FIGURE 6.3 Determination of V_{mono} from a BET experiment.

By adsorption experiments, not only the surface area can be determined but also the pore size and pore size distributions. The analysis of the pore size is thus based on the differences of the condensation (and vapour) pressure depending on the pore radius, as described by the Kelvin Equation 6.5:

$$\ln\left(\frac{p}{p_0}\right) = \frac{2\gamma V_m}{RTr_p} \tag{6.5}$$

where:

p/p_0 is the relative pressure
γ is the interfacial tension
V_m is the molar gaseous volume
r_p is the radius of the pore

In principle, the pore size can be calculated based on various empirical models by establishing the volume of adsorbed gas for a certain relative pressure, which corresponds to a certain pore size.

In the case of fibrous materials of high aspect ratio (i.e. the ratio of length to diameter), the SSA is directly correlated to the thickness of the fibres/fibrils: the lower the thickness, the higher the SSA. In the case of cellulose fibrous materials, the relationship between SSA and thickness (fibril diameter) is depicted in Figure 6.4.

It is apparent that down to a fibril diameter of about 50 nm, the influence of the fibril diameter onto the SSA is only a matter of minor significance. However, below 50 nm, and particularly below 20 nm, the SSA increases significantly. It is thus of utmost importance to be aware of the fibril dimensions in this size regime as the consequences following the impact onto the SSA are tremendous. For example, the modification of surface –OH groups and its success directly depend on how many surface groups are available for modification, which in turn is dependent on the SSA and thus majorly on the fibril diameter. Furthermore,

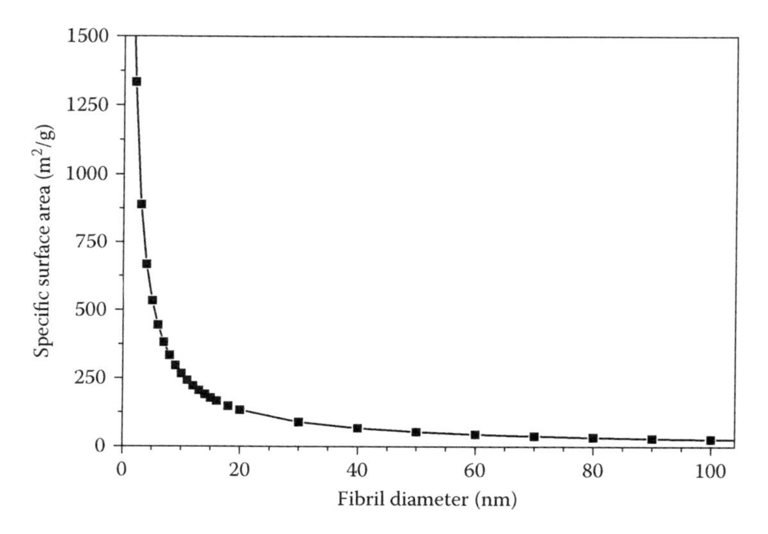

FIGURE 6.4 Correlation between the specific surface area and the fibril diameter for fibrous cellulose nanomaterials of high aspect ratio.

the adsorption efficiency of materials onto nanomaterials is greatly affected by the available SSA and the emulsifying efficiency of nanomaterials applied as emulsifiers.

Even though there is a great number of literatures available that deals with the use of cellulose nanomaterials in various applications – most of which are particularly dependent on the dimensions of the nanocellulose – the SSA is frequently a neglected affair and not assessed. Often the SSA is only calculated from measured dimensions, that is, fibril diameters and lengths. Nonetheless, there are certain publications available, which give measured numbers for cellulose nanomaterials. These values and references are summarised in Table 6.1 together with the determination method, surface charge (where available) and the corresponding reference.

From these reported values, some conclusions can be drawn. First of all, these measurements confirm the theory that the SSA is primarily dependent on the fibril diameter, that is, TEMPO-oxidised CNFs that commonly possess the lowest fibril diameters also show the highest SSA among nanocellulose. CNC, which usually also possess very low diameters, exhibit high SSA as well. Furthermore, the degree of oxidation in TEMPO-oxidised materials plays a significant role: a higher degree of oxidation commonly results in lower fibril diameter and hence higher SSA. The form of the sample also influences the measured SSA to a great extent. Usually a high SSA is established for aerogels and freeze-dried samples, whereas papers possess only a relatively low SSA. However, a high SSA was reported for papers that were prepared via a supercritical drying process.

TABLE 6.1
Reported Specific Surface Areas and Surface Charges (Where Available), Determination Method and Corresponding References

Material	Method/Remarks	Surface Area (m^2/g)	Surface Charge (mmol/g)	SSA Determination Method	References
CNF		50–70		Theory	[20]
CNC		250		Theory	[3]
TEMPO		600	0.70–1.00	Theory	[21]
MFC		<0.5		BET	[22]
Cellulose	*Cladophora* algae	90.5			
Cellulose	*Cladophora* algae	89.9			
Cellulose	*Cladophora* algae	102		BET	[23]
BC	f-d	37		BET	[24]
BC		2.7			
BC	Aerogel scCO_2	200		BET	[25]
BC	f-d	55		BET	[26]
BC	Paper	7			
CNF		37			
CNF	Aerogel f-d	20–66		BET	[27]
CNF	Aerogel	11–42		BET	[28]
CNF	Aerogel	11–15		BET	[29]
CNF		84		Congo red adsorption	[30]
CNF	Sludge	112			
CNF		146–219		Calculated from AFM	[31]
CNC		138–226			
CNC		131		Calculated from AFM	[32]
CNC	Phosphorylated	66			
CNF	Phosphorylated	26			
CNF	Paper	5.8		BET	[33]
CNC	Paper	7.6			
CNC	Tunicin	150–170		–	[34]
CNC		150–250		–	[35]
CNC	Aerogel scCO_2	216–605		BET	[36]
CNF	Paper scd	304		BET	[37]
TEMPO	Paper scd	482			
CNF	Paper CO_2 evaporation	262			
TEMPO	Paper CO_2 evaporation	415			
CNF	Tert-butanol f-d	117			
TEMPO	Tert-butanol f-d	45			

(Continued)

TABLE 6.1 (*Continued*)
Reported Specific Surface Areas and Surface Charges (Where Available), Determination Method and Corresponding References

Material	Method/Remarks	Surface Area (m^2/g)	Surface Charge (mmol/g)	SSA Determination Method	References
CNF		229	0.10	BET	[38]
TEMPO		330	0.26		
TEMPO		267	0.86		
TEMPO		303	1.21		
TEMPO		345	1.50		
CNF	Aerogel 1 step fibrill.	153		BET	[39]
CNF	Aerogel 6 step fibrill.	249			
TEMPO	Aerogel 1 step fibrill.	254			
TEMPO	Aerogel 6 step fibrill.	284			
TEMPO		188	1.46	BET	[40]
TEMPO		134	0.6	Calculated from	[41]
TEMPO		215	1.5	SEM	

f-d, freeze-dried; $scCO_2$, supercritical CO_2; scd, supercritical drying; step fibrill, fibrillation steps.

6.3 CHEMICAL COMPOSITION OF THE NANOCELLULOSIC BUILDING BLOCKS

In contrast to BC with high chemical purity of cellulose, lignocellulosic biomass-derived nanocellulosic building blocks rarely exist as pure cellulose. Plant-derived CNF/CMF grades contain hemicelluloses, lignin and extractives, depending on the pulp fibre source and processing. Bleached kraft pulps from softwood and hardwood can contain more than 20 wt.% of hemicelluloses (softwood pulps contain glucomannan and xylan, whereas hardwood pulps mainly contain xylan) [42]. Especially, the anionically charged glucuronoxylan located on the nanofibril surface has a decisive role when dealing with interfacial interactions and physical properties of nanocellulosic materials [43–46]. The accessible and loosely bound anionically charged xylan has been shown to efficiently stabilise the CNF dispersion via electrosteric interactions (Figure 6.5) [46]. Here, electrostatic contribution is derived from the repulsion between charged acidic groups, whereas the steric contribution is derived from the mobile xylan chains that are located on the fibril surface. Removal of xylan deteriorates the dispersion stability and significantly alters the rheological behaviour and affects the swelling behaviour of the individual fibrils. Hemicellulose-rich CNFs have shown to be tolerant towards changes

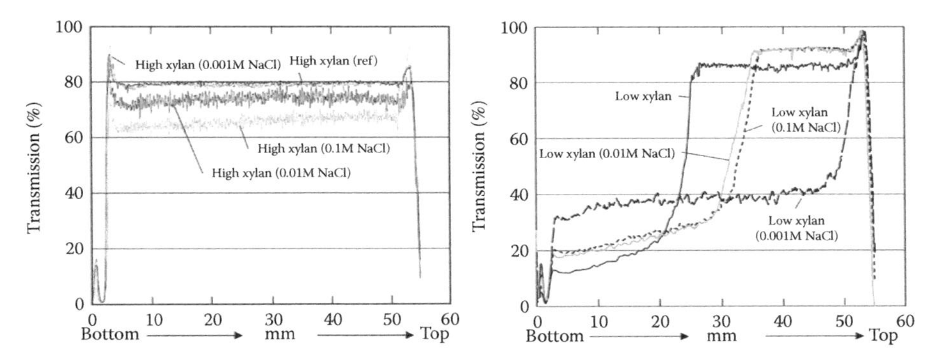

FIGURE 6.5 Light transmission profiles determined by Turbiscan of high xylan content CNF (~23 wt.% of glucuronoxylan) and low xylan content CNF (~13 wt.% of glucuronoxylan) at different concentrations of NaCl at pH ~6.5. CNF is prepared from bleached birch kraft pulp via mechanical disintegration. Samples are scanned from bottom to top. $t = 2$ h. (From Tenhunen, T.-M. et al., *React Funct Polym*, 85, 157–166, 2014.)

in ionic strength and pH, and the fibrils were in a swollen state even at high salt concentration and at low pH levels [47].

Although research on bleached pulp grades as CNF/CMF raw materials has gained extensive attention, lignin containing CNFs possesses several qualities relevant for specific applications in which polarity and hydrophilic/hydrophobic balance adjustments are required. Residual lignin located on the fibre surface radically changes the interfacial properties and physical performance of the fibrillated materials. For example, Lahtinen et al. [48], Spence et al. [49,50] and Hoeger et al. [51] compared different wood pulps with the aim to clarify the role and impact of lignin on the fibrillation and quality of the produced nanofibrils. These investigations consistently showed that unbleached kraft pulps with low residual lignin content (<10 wt.%) were suitable raw materials for CNF production. Several reasons for the improved fibrillation performance were given such as simultaneous higher hemicellulose content (higher surface charge) when compared to bleached pulps and the amorphous nature of lignin. An interesting contributor to the enhanced fibrillation tendency of unbleached kraft pulp, as suggested by Solala et al. [52], is the formation of mechano-radicals during intensive refining, and that lignin has a radical scavenging ability. They suggested that in the absence of lignin, recombination reactions might take place between cellulose radicals that may result in unwanted cross-linking reactions. Therefore, this may lead to the formation of more rigid structures that can be expected to interfere with the fibrillation action.

Lahtinen et al. [48], Spence et al. [49,50] and Hoeger et al. [51] also concluded that the fibrillation of the mechanical pulps with considerably higher lignin content (total lignin content as high as 30 wt.%) produced heterogeneous materials with stiff fibre fragments and flake-like particles with lower SSA and lower apparent viscosity levels when compared to kraft pulps. As stated in the paper by Hoeger et al. [51], lignin has a significant role in protecting the cell wall. Therefore, the cell wall breakdown in order to achieve a sufficient level of fibrillation is less severe compared to the pulps with lower lignin content.

CNF with residual lignin can be considered as an advantageous material when lowered water absorbency, less polar surfaces and high thermal stability are required. Rojo et al. [53] investigated various surface specific properties of sulphonated residual lignin-containing CNF (product from SO_2–ethanol–water (SEW) pulping process). They showed that the lower surface energy of the lignin-containing CNF films contributed to the lower wettability of the films. This particular lignin was able to smoothen and densify the film structure due to hot pressing, and thus the water penetration through the film structure was efficiently prevented.

Nair and Yan [54] reported high thermal stability of high lignin content nanofibrils produced from alkali-treated pine bark: Thermal degradation of high residual lignin-containing nanofibrils (21 wt.% of lignin) started at 306°C, whereas the degradation temperature for the same but delignified fibrils was at mere 278°C. The water uptake ability of lignin-containing fibrils was somewhat lower compared to their delignified counterparts although the water absorption was still rather extensive despite of the high lignin content. This result is in agreement with the findings of Spence et al. [49] who showed that the presence of lignin does not solely decrease the water absorption ability of the CNF films, but the fine structure of the film (porosity–density) has a substantial impact on the water absorption and water vapour transmission rates.

6.4 SURFACE CHARGE OF THE NANOCELLULOSIC BUILDING BLOCKS

The surface charge of pulp and all cellulosic nanomaterials has inevitably the largest and most evident impact on their behaviour and interfacial interactions. The pulp charge (anionic charge) is generally analysed by conductometric titration with the standard titration method (SCAN CM 65:02) and as described by Katz et al. [55] The charge state is not altered, at least not substantially, due to the mechanical fibrillation process. The surface charge of individual nanoscaled fibrils and colloidal nanocellulosic particles can be determined using mobility measurements (zeta potential [ζ-potential]). A brief theoretical background and some examples on surface charge and mobility of nanocellulosic materials are given in a separate section.

It is well known that anionic charges derived from anionic polysaccharides, especially from glucuronoxylan, increase the fibre swelling ability and facilitate pulp beatability [56]. Similarly, higher amounts of hemicelluloses assist fibrillation and improve the efficiency of the production of CNF/CMF materials. Fibre cell wall delamination with decreased energy demand can be further improved by introducing anionic charges by chemically pre-treating the pulp fibres by means of, for example, TEMPO-mediated oxidation [15], carboxymethylation [57] or by the more recently reported route with periodate–chlorite oxidation [58]. Here the routes are discussed only briefly, and the reader is referred to the numerous papers published by the groups from the University of Tokyo, Japan, the Royal Institute of Technology (KTH)/Inventia, Sweden and the University of Oulu, Finland. Briefly, all pre-treatments introduce carboxyl groups on the fibre surface of which TEMPO-oxidation is highly selective converting only primary hydroxyl groups of cellulose to carboxyl groups. During TEMPO-oxidation and carboxymethylation, the glucopyranose ring of cellulose remains intact, whereas during the periodate-oxidation the hydroxyl groups at positions 2 and 3 are oxidised to aldehydes concurrently breaking the carbon–carbon bond. The surface charge of the anionic fibrils is considerably higher compared to the charge of fibrils from native fibre sources. Surface charge values determined by conductometric titration as high as 1.5 mmol/g for TEMPO-oxidised nanofibrils, ~0.7 mmol/g for periodate–chlorite-oxidised nanofibrils and 0.5 mmol/g for carboxymethylated nanofibrils have been reported, whereas, for comparison, the charge of the xylan containing birch kraft pulp, containing approximately 23 wt.% of glucuronoxylan, is ~40 mmol/g [46].

Fall et al. [59] studied the effect of surface charge on the colloidal stability of the carboxymethylated CNF dispersions, and they explained the stability to be a result of the electrostatic repulsion due to deprotonation of the carboxyl groups located on the fibril surface. They showed that the aggregation tendency at low pH is due to increased proton concentration in the bulk and in close proximity to the fibril surface. Protonation of the carboxyl groups leads to reduction in repulsive forces, thus inducing fibril aggregation. At high salt concentration, counterions occupy the fibril surface, and with increased ion concentration, a decrease in the surface charge is induced leading to aggregation of the fibrils. They also concluded that the unmodified soft wood-derived CNFs were highly susceptible to electrolyte and pH-induced aggregation.

On the other hand, a high amount of hemicelluloses on the fibril surfaces has been shown to enhance the stability of the CNF dispersions and increase the tolerance towards high salt concentrations and low pH values. Improved stability and tolerance when exposed to different levels of ionic strength and/or pH were explained by changes in conformation of the low charge, but mobile, hemicellulose layer located on the CNF surface (CNF prepared from hemicellulose rich pulp). High amounts of hemicellulose on the fibril surface seemed to be a source for steric interaction even in a more compact conformation, and severe agglomeration and sedimentation could be avoided even at high NaCl concentrations (100 mM; see also Figure 6.5).

A few examples on pulp pre-treatments involving quarterization in order to prepare CNFs in aqueous environment with cationic functionality exist [60–62]. Reactions using 2,3-epoxypropyl trimethylammonium chloride resulted in a charge density of 0.35 meq/g (degree of substitution [DS] 0.08 mol/mol anhydroglucose unit [AGU]), whereas reactions using trimethyl ammoniumchloride gave charge densities in the range of 0.59–2.31 mmol/g (DS 0.1–0.37) determined using N-analysis and conductometric titration, respectively. Common features for both cationised CNF grades were very low lateral dimensions (fibril width in the range of 2–3 nm measured in dry state) and behaviour, which is typical for polyelectrolyte gels (pH and ionic strength-dependent behaviour with high adsorption capacity) [60,61]. The reaction route introduced by Liimatainen et al. [62] encompasses periodate oxidation and a further cationisation step including the reaction with Girard's reagent T ((2-hydrazinyl-2-oxoethyl)trimethylazanium chloride). Here, fibril widths of 10–50 nm were achieved with charge densities ranging between 0.2 and 1 meq/g, as determined by polyelectrolyte titration. Regarding the crystalline structure, cellulose I was retained in all reaction conditions although peak broadening in the X-Ray diffraction (XRD) analysis suggests that the periodate oxidation plus cationisation does not selectively occur on the fibril surface only.

The surface charge and the stability of CNCs are greatly dependent on the acid utilised during the hydrolysis. Crystals achieved using HCl generate neutral colloids with poor stability due to the scarcity of charges or other sources for repulsive surface interactions. In turn, the aqueous dispersions of sulphuric acid hydrolysed CNCs are highly stable, thanks to the strong electrostatic repulsion between the anionic sulphate ester groups on the crystal surface. On average, every fourth glucose unit on the CNC surface carries the sulphate group (DSs = ~0.25) giving a charge density of ~0.36 mmol/g, as analysed with conductometric titration [63–65].

6.4.1 Characterisation of Surface Charge and Mobility of Nanocellulosic Materials: Zeta Potential

Surface charge and mobility are key parameters of colloidal systems and thus also for cellulose nanomaterials. These two parameters can be used to describe the four electrokinetic phenomena. They are differentiated into two main processes: (1) an externally imposed electromotive force (EMF) results in motion and (2) an externally imposed motion produces the EMF. Process (1) can be further divided into two phenomena: (a) electrophoresis (the solid moves, the liquid is stationary) and (b) electro-osmosis (the liquid moves, the solid is stationary). For process (2), the two corresponding phenomena are streaming potential (the liquid is forced to move past the solid) and sedimentation potential (solids are forced to move through the liquid). In order to characterise the charge and mobility of colloidal systems, electrophoresis and the streaming potential are of significance. Both can be utilised to determine the ζ-potential of dispersed media. For a detailed account on surface charge and mobility, the reader is guided to dedicated literature, which constitutes the base of the following discussion [17].

The ζ-potential is the electrokinetic potential in the interfacial double layer at the location of the slipping plane (Figure 6.6) relative to a point in the bulk fluid away from the interface. It constitutes the potential difference between the dispersion medium and the stationary layer of fluid attached to the surface of the dispersed particle. The ζ-potential is caused by the net electrical charge contained within the region bounded by the slipping plane and depends on the location of that plane. It is widely used for the quantification of the magnitude of the charge. The ζ-potential is not equal to the Stern potential Ψ_δ or the electric surface potential in the double layers, which are defined at different locations. Most importantly, the ζ-potential is often the only available path for characterisation of double-layer properties.

For the stability of colloidal dispersions, the ζ-potential is a key indicator. It indicates the degree of electrostatic repulsion between adjacent, similarly charged particles in a dispersion. Generally, high ζ-potential indicates stable dispersions, that is, the dispersion will resist aggregation. In other words, colloids with high (negative or positive) ζ-potential are electrically stabilised. On the contrary, low ζ-potential indicates that attractive forces may exceed the repulsion of like particles, and the dispersion may break and coagulate or flocculate. Factors that affect the ζ-potential are dependent on the type of solid and the solution conditions. For the former, the presence of ionic groups, hydrophilicity/hydrophobicity and the swellability in water are important.

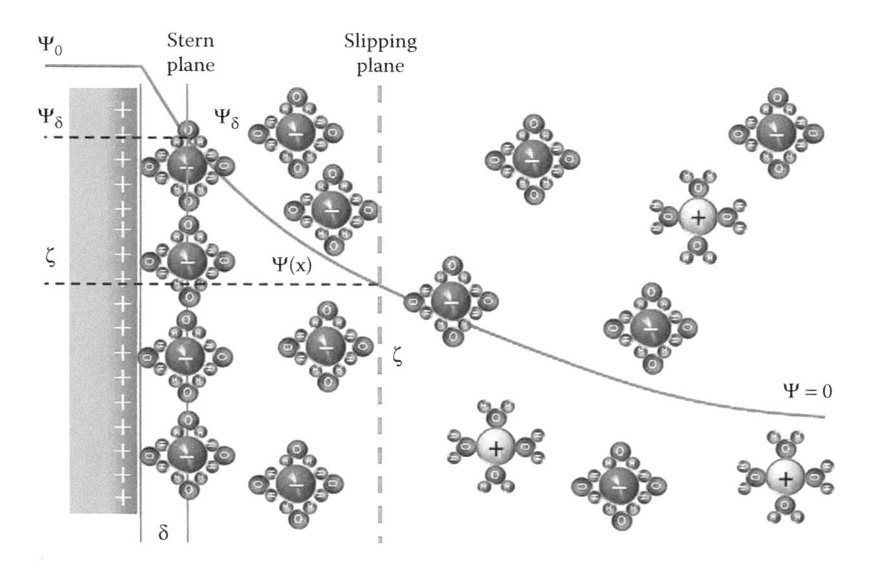

FIGURE 6.6 Exemplary position of the slipping plane and ζ-potential, accordingly. Ψ_0, surface potential; Ψ_δ, Stern potential; $\Psi(x)$, potential function in dependence of the distance to the surface; δ, position of the Stern plane.

Solution conditions influencing the ζ-potential are temperature, pH and electrolyte concentration. Thus, without information about the specifying measurement conditions, the quote of a ζ-potential value is meaningless. Most often, the ζ-potential is determined as a function of pH. The presence of pH-depending functional groups, such as carboxyl or amine groups, is an important feature of the surface charge of materials. Carboxyl groups are deprotonised at higher pH, resulting in higher negative ζ-potential. Contrary to this, the carboxyl groups are protonised at low pH, resulting in lower negative ζ-potential. At some point, usually corresponding to the pK_a value, the ζ-potential is zero. This pH value is called the isoelectric point (IEP). At the IEP, the potential energy of repulsion between particles is minimal, thus constituting the optimal conditions for coagulation. This is of particular interest in case that coagulation is aimed for [66].

Even though the ζ-potential is only an indirect measure of the surface charge, in practical terms, it is of great importance for it gives direct information about the stability of dispersions. Although the surface charge is of theoretical importance, the ζ-potential is influenced by counterions attached to the charged surface and adsorbed charged particles. Hence, the ζ-potential is what particles in dispersion effectively 'see'/'feel' of each other. Accordingly, it is a direct measure for the stability of dispersions. In general, dispersion can be considered to be stable outside the critical ζ-potential range, which is between −10 and +10 mV.

The ζ-potential is mainly examined with two methods, whereby electrophoresis is primarily used to analyse the mobility of particles in dispersion in dependence of the surface charge, that is, ζ-potential, and with the streaming potential it is possible to determine the surface charge of flat surfaces by studying the behaviour of an electrolyte solution being forced to flow over it. By electrophoresis, the direction and velocity (i.e. electrophoretic mobility U_E) of particles in dispersions can be analysed. The velocity of a particle in an electric field given by the Henry equation (see Equation 6.6) is dependent on the strength of the electric field, the dielectric constant of the liquid ε, the viscosity of the liquid η, the ζ-potential ζ and the Henry function $f(\kappa a)$ with κ being the Debye Hückel parameter and a the particle radius.

$$U_E = \frac{2\varepsilon\zeta f(\kappa a)}{3\eta} \tag{6.6}$$

By determination of the streaming potential, it is possible to characterise the surface charge of films, papers, and so on, via application of the Helmholtz–Smoluchowski Equation 6.7:

$$\frac{\Delta U}{\Delta p} = \frac{\varepsilon\varepsilon_0\zeta}{\eta\kappa} \tag{6.7}$$

where:

ε is the dielectric constant of the electrolyte
ε_0 is the permittivity in vacuum
η is the electrolyte viscosity
κ is the conductivity of the electrolyte
ΔU is the induced potential
Δp is the applied pressure

Typical values for the ζ-potential of nanocelluloses from electrophoresis measurements are shown in Figure 6.7 as function of pH in a 1 mM KCl electrolyte [67].

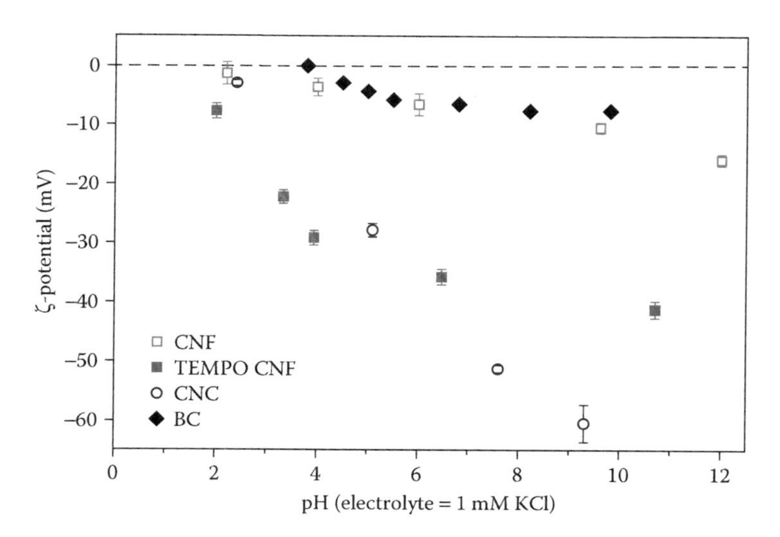

FIGURE 6.7 ζ-potential as function of pH as determined by electrophoresis for BC, unmodified CNF, TEMPO CNF and CNC. (From Mautner, A. et al., *React Funct Polym*, 86, 209–214, 2015.)

For TEMPO CNF and CNC, a negative ζ-potential at high pH, with TEMPO CNF showing a plateau, and a very low IEP were found, which is characteristic for acidic solid surfaces as both materials contain a multitude of carboxylic acid groups. The magnitude of the ζ-potential is dependent on the relative pK_a value and the concentration of acidic groups present on the surface. By decreasing the pH, that is, addition of protons, the ζ-potential decreases as a result of the protonation of the acidic functional groups. The IEP was reached at 1.5 (by extrapolation) and 2.2, respectively, for TEMPO CNF and CNC. For BC and unmodified CNF, principally a similar behaviour was found, but at a different level. Due to the absence, or much lower content, respectively, of carboxyl groups the ζ-potential was around −10 mV at high pH. A decrease of the ζ-potential with decreasing pH could be observed, whereby the IEP was reached at 2.2 and 3.8 for unmodified CNF and BC, respectively.

From streaming potential measurements, typical results streaming potential measurements of nanopapers from unmodified CNF, cationic CNF, phosphorylated CNF and BC are shown in Figure 6.8 [67–69]. From streaming potential measurements, the ζ-potential is slightly more accentuated compared to electrophoresis. For both unmodified CNF and BC, a more negative ζ-potential was found. This is inherently coupled to the differences in the measurement process of nanopapers and single fibrils in suspension. Cationic CNF in which ammonium groups are attached onto the CNF surface show a positive ζ-potential over the whole pH range. At low pH, the ammonium groups present are fully protonated, resulting in a high positive surface charge. CNF modified with phosphoric acid shows a very low negative ζ-potential plateau at high pH, due to the acidic phosphate groups being fully deprotonated. At low pH, in

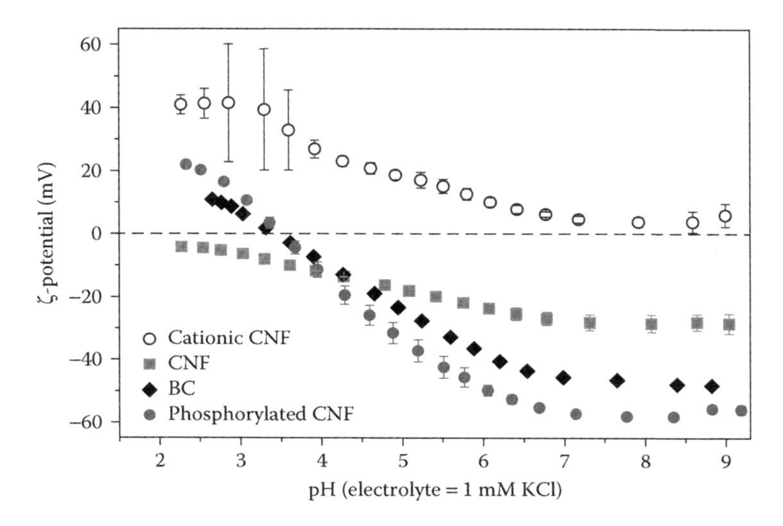

FIGURE 6.8 ζ-potential as function of pH as determined by streaming potential measurements of nanopapers from BC, unmodified CNF, cationic CNF and phosphorylated CNF. (From Mautner, A. et al., *React. Funct. Polym.*, 86, 209–214, 2015; Mautner, A. et al., *Int J Environ Sci Technol.*, 13, 1861, 2016a; Mautner, A. et al., *Environ Sci: Water Res Technol.*, 2, 117–124, 2016b.)

particular below the IEP, which is at 3.5, the hydrogen phosphate groups are protonated, hence giving a positive ζ-potential. Another important aspect is the dependency on the salt concentration of the nanocellulose suspension. Mautner et al. [66] showed that via the concentration of, particularly multivalent, metal ions the ζ-potential can be tuned and thus the coagulation behaviour can be controlled.

6.5 HYDROPHILIC NANOMATERIAL WITH AMPHIPHILIC CHARACTER

The high SSA of nanoscale cellulosic materials results in large amounts of available hydroxyl groups located on the fibril/crystal surface. The cellulose molecule possesses one primary hydroxyl group and two secondary hydroxyl groups per glucose unit that promote the significant hydrophilicity of the material. Static equilibrium water contact angle values as low as ~10° for regenerated cellulose films and clean nanofibrillated cellulose films have been reported [70–72]. The high wettability of the cellulosic films is argued to be due to higher density of the hydroxyl groups when compared to other, even water-soluble hydrophilic polymers such as polyvinyl alcohol (PVA) that give a static water contact angle value of 36° [70,73]. On the other hand, despite the high hydrophilicity, cellulose is not soluble in water. This fact can be at least partly explained by the amphiphilic character of cellulose [74–76] while bearing in mind the role of hydrogen bonding and the rigidity of the cellulose chain also. In addition, cellulosic materials have also been shown to display different surface properties due to the structural anisotropy: different wetting behaviour of the surfaces have been shown to correlate with the orientation of the crystal planes [70,77]. The equatorial plane direction is hydrophilic because hydroxyl groups are protruding this direction, whereas the axial direction of the glucopyranose unit can be considered more hydrophobic because C–H bonds are located in axial positions. Thus, cellulose can be considered as a hydrophilic material with amphiphilic character, and this can be deduced from the following experimental evidence.

As already mentioned, the wetting behaviour of the regenerated cellulose with changing crystallinity index values is dependent on structural anisotropy, that is, different planes possess markedly different polarities. The higher the planar orientation index, the lower the water contact angle value ranging between 11° and 37°, suggesting that cellulose has both hydrophobic and hydrophilic domains [77]. In addition, the same authors showed that the higher the crystallinity index, the higher the wettability and that amorphous cellulose displayed the highest contact angle value of 42°. Similar contact angle values have been reported by Holmberg et al. [71] and Suchy et al. [78] for dry amorphous cellulose thin films. In addition, both groups showed that wetting/swelling of the cellulose surface radically changes the contact angle and surface energy of the cellulose films. The advancing contact angle of the dry substrate values was 45°, whereas the receding contact angle for the wetted cellulose surface was <10° [71]. Similarly, Suchy et al. [78] showed that the swollen and thoroughly dried cellulose film displayed a contact angle decrease of 20 units (47° → 27°) and simultaneously the surface energy increased for the swollen cellulose surface. Likewise, from research involving regenerated cellulosic materials, identical wetting behaviour has been demonstrated as for the films consisting of nanofibrillated cellulose (Figure 6.9) [79].

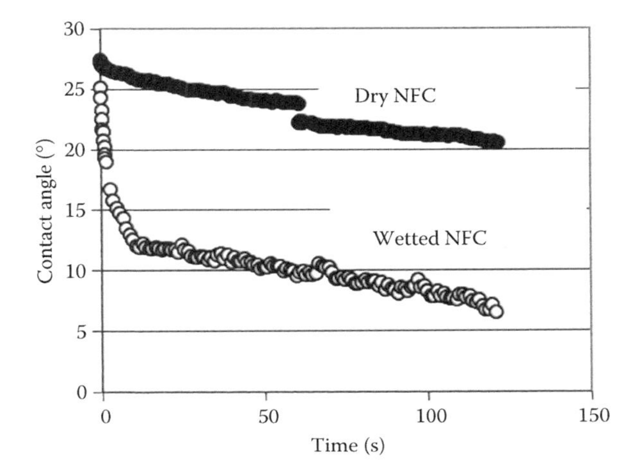

FIGURE 6.9 Changes in water contact angles for spin-coated CNF on silica surfaces covered with a layer of anchoring polymer recorded during 2 min. The dry CNF (Dry NFC) was measured after film preparation and thorough drying. The wetted CNF (Wetted NFC) was measured after exactly the same sample had been thoroughly wetted with water and quickly dried with nitrogen gas. Note that the contact angle for the anchoring polymer is about 15° and does not change by the wetting treatment. (From Johansson, L.-S. et al., *Soft Matter*, 7(22), 10917–10924, 2011.)

These findings indicate that the cellulose molecule is capable of adapting its conformation in order to find the energetically most favourably state, depending on the medium. When exposed to non-polar media (air, organic solvents), cellulose undergoes partly irreversible organisation and/or surface passivation, which involves the accumulation of airborne contaminants. The existence of surface passivation can be substantiated by following the C–C carbon signal determined using X-ray photoelectron spectroscopy. This particular signal gives information about non-cellulosic materials appearing on the cellulose surface. When the nanocellulose surface was exposed to amphiphilic media (dimethyl acetamide, DMAc), the level of the C–C component was even lower compared to the pure reference material (cotton filter paper). This may indicate that the CNF surface can change its conformation by turning the more hydrophobic axial plane towards the amphiphilic solvent, which results in lowered surface energy. Therefore, CNF surfaces do not extensively accumulate carbonaceous contaminants required for surface energy reduction that is taking place during drying. This cannot be explained merely by raw material residuals, such as lignin and extractives, because this behaviour can be evidenced for the same cellulose samples derived from relatively pure sources. In addition, the surface free energy-driven adaptation process is common for hydrophilic mineral surfaces [80].

Jiang and Hsieh [81] investigated the effect of protonation of TEMPO CNF fibrils along with the influence of the drying method on water and toluene absorption capacity of the formed structures. They were able to conclude that slowly air-dried, very dense structures absorbed low amounts of water (0.6–2.6 g/g), whereas the toluene absorption did not occur at all more or less in spite of the surface carboxyl protonation level. In contrast, freeze-dried and highly porous structures were able to absorb rather high amounts of both toluene and water (61–95 g/g of water and 101–145 g/g toluene). This was explained by the ability of CNF to self-assemble into structures with distinct affinity towards polar and non-polar solvents resulting from the availability of both hydrophobic and hydrophilic planes.

Due to the peculiar and intriguing wetting characteristics and ability for the surface adaptation derived from the amphiphilicity, nanoscale cellulosic materials adsorb and assemble at the oil/water interface. All different nanocellulosic materials (CNC, CNF, CMF and even highly hydrophilic TEMPO CNF and TEMPO-oxidised BC) have been shown to efficiently stabilise oil-in-water emulsions via the Pickering mechanism [65,82]. Both the strong network of nanocellulose and assembly at the interface facilitate the stability, and emulsion droplet coalescence can be avoided. Similar tendencies have also been shown in foam systems [83,84].

6.6 REACTIVITY AND SURFACE MODIFICATION OF NANOCELLULOSIC MATERIALS

The high reactivity of cellulosic nanomaterials for modification arises from their high SSA and numerous surface hydroxyl groups available for the chemical reactions. The reaction pathways are mainly chemistry of alcohols including, among others, esterifications, etherifications and oxidations, whereby functional groups, monomers or polymers are covalently attached to the hydroxyl groups of cellulose. Principally, two modes of chemical modifications are existent: grafting reactions and direct chemical modification [85]. Grafting reactions comprise 'grafting-from' (attachment of monomers to the polymer backbone and subsequent chain reaction with further monomer molecules to grow the polymer chain) and 'grafting-to' (attachment of a polymer chain to another polymer chain), whereby with regards to modification of cellulose, the grafting-from mechanism is more important. Grafting-from cellulose can be established via various routes, for example, chemically initiated grafting, photo-grafting, high-energy radiation grafting, click chemistry and so on. Furthermore, by application of direct chemical modification methods, monomers or a polymer-chain are not attached to the hydroxyl groups, but the hydroxyl groups are directly modified, for example, by esterification, etherification, silylation, phosphorylation, oxidation or alkaline treatment. For more details on the chemical modification of the nanocellulosic materials, the reader is referred to the excellent reviews by Habibi [86] and Eyley and Thielemans [87].

When considering cellulose nanomaterials in their native form (not as soluble polymeric substances), researchers have primarily been interested in surface modification of crystals/fibrils in order to preserve the crystalline structure of cellulose I and hence their high SSA and nanoscale structure [79,88]. However, this approach brings limitations to the reactivity and reaction efficiency because the hydroxyl accessibility is dependent on its crystallinity: less ordered and amorphous celluloses can be functionalised more readily than the crystalline cellulose structures, as shown by Tasker et al [89].

The surface reactivity of cellulosic nanomaterials has also been linked to the cellulose surface adaptation and its tendency to minimise the surface free energy [79]. Surface passivation, involving accumulation of contaminants, hinders chemical modifications as parts of the hydroxyl groups are no longer available for further reactions. This was demonstrated by comparing the reaction efficiency of a simple silylation reaction in a cellulose compatible solvent (DMAc) and in non-polar toluene. Severe CNF aggregation took place in toluene and the presence of aliphatic compounds, as detected with X-Ray photoelectron spectroscopy (XPS) by tracking the C–C carbon signal, was substantial. Silylation conducted in toluene resulted in a significantly lower degree of substitution compared to silylation in DMAc giving values of 0.03 and 0.9, respectively.

Similarly, acetone and toluene were compared as reaction media for esterification reactions [90]. Although higher grafting efficiency was achieved nominally in toluene, the fibrillar nanostructure was completely deteriorated due to aggregation and the inherent film formation tendency was destroyed. In general, it can be concluded that the optimal reaction conditions are often a compromise so that they are favourable for both the nanoscale materials and the reagents, that is, the nanoscale structure of the cellulosic nanomaterials can be retained during the modification with the feasible consumption of chemicals and sufficient reagent solubility simultaneously avoiding unwanted side reactions.

In order to improve the reaction efficiency of cellulose nanomaterials, with the aim to increase the OH-group accessibility and to avoid the agglomeration tendency, several approaches have been suggested. Labet and Thielemans [91] showed the importance of an extensive purification step by Soxhlet extraction with ethanol to remove harmful impurities adsorbed on the surface of the nanocrystalline cellulose during acid hydrolysis. They also demonstrated that CNC purification is vital for reproducible modifications carried out on the crystal surfaces.

The passivation of the nanocellulose surfaces can be avoided by using appropriate solvent exchange procedures with cellulose compatible solvents (e.g. DMAc, acetone or methanol) before drying. However, the seminal difficulty with chemical modification of cellulose fibrils lies in their hydrophilicity: water is their natural medium, and tedious solvent exchange procedures to non-polar solvents are required to enable the introduction of new organic functionalities on the surface of individual fibrils. Even with cellulose compatible solvents in which the high SSA of CNF can be retained, the removal of reagents and side-products can be challenging requiring, for example, prolonged dialysis and other conventional purification procedures. The CNF dispersion-related challenges could be eluded by first preparing a water/solvent stable structure such as a film of CNF and by subsequently performing interfacial chemical reaction directly onto the surface of the macroscopic structure. By following this approach, surface-only modification is achieved and the purification of the product is easy involving only simple rinsing. Successful modifications on the assembled films have been reported using gas-phase reactions [92,93] and chemical reactions taking place in cellulose compatible solvents [94,95]. In addition, this modification strategy has shown drastic improvements in the application-driven performance. As demonstrated by Hakalahti et al., [94] only a 8% coverage of polymeric thermoresponsive N-isopropylacrylamide (NIPAM) grafted to the CNF film surface resulted in a clear temperature switchable effect in water permeability experiments.

The reaction efficiency of the already assembled structures can be improved by activating the CNF surface using a commonly applied cleaning method based on UV

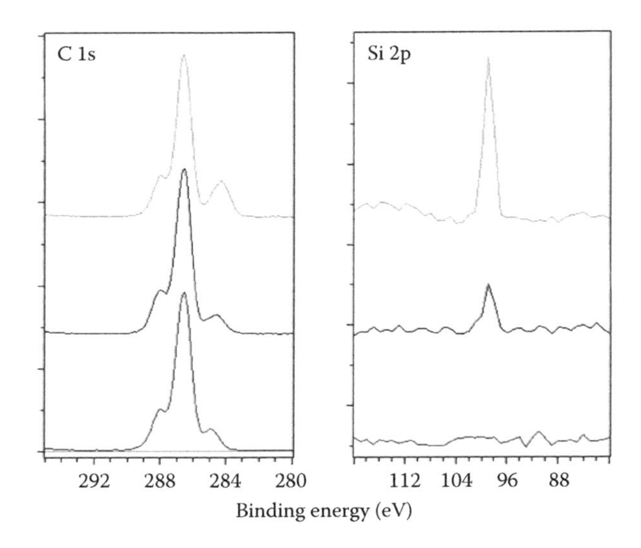

FIGURE 6.10 XPS C 1s and Si 2p high-resolution regions of untreated film (bottom spectrum), untreated and silylated film (middle spectrum) and UV/O_3 treated and silylated film (upper spectrum). (From Österberg, M. et al., *Cellulose*, 20(3), 983–990, 2013.)

radiation and ozone [72]. UV/O_3 treatment removes organic contaminants from the surface by cleaving bonds such as C–C, C–H and O–H into volatile organics [96]. UV/O_3 treatment itself has been shown to increase the hydrophilicity and to drastically improve the reaction efficiency. Degree of surface substitution calculated based on the relative amount of silicon on the CNF film surface was 0.07 for unactivated and 0.26 for UV/O_3-treated CNF film, corresponding to a fourfold increase in reaction efficiency due to the UV/O_3 treatment (Figure 6.10).

6.7 WATER INTERACTIONS OF NANOCELLULOSIC MATERIALS

All cellulosic (nano) materials interact strongly with water and water vapour due to their high hydrophilicity and hygroscopicity – a feature that sets them apart from many other nanoscaled materials that may have other comparable properties, such as high aspect ratios and mechanical properties [3,97]. Binding of water onto cellulosic materials has been extensively studied throughout the past century [98–101] due to its practical relevance for, for example, papermaking and food preservation. However, a vast majority of the published studies have been conducted using bulk methods, mainly the dynamic vapour sorption [102–104]. Alongside bulk methods, the surface-sensitive approach has gained considerable ground in this realm, as several surface-sensitive techniques, for example, surface plasmon resonance (SPR) [105], quartz crystal microbalance with dissipation (QCM-D) [106] and spectroscopic ellipsometry (SE) [107] have emerged, and the benefits related to their use with respect to complex biomaterials have become clear. The surface-sensitive techniques offer precise, versatile, and non-invasive ways to quantify the water-binding behaviour directly at the solid–gas [108–110] or at solid–liquid interface [99,111–113]. In this section, the aim is to briefly summarise the principles of the most relevant surface-sensitive tools and to give an overview of selected recent findings with respect to insights into the intrinsic hydrophilic character and water interactions of cellulosic nanomaterials.

Supported ultrathin films – the cornerstone of the surface-sensitive approach – are very thin (generally less than 100 nm) and have smooth surfaces with well-defined chemical compositions and morphologies [114]. Considerable efforts are put into the characterisation of the ultrathin film samples, as any changes due to water vapour uptake are always proportional to the mass and dimensions of the film in its original state. A versatile range of ultrathin (nano)cellulosic film samples can be prepared by several established deposition techniques, such as Langmuir–Blodgett [115] deposition, layer-by-layer assembly [116] and spin coating [117,118] and combinations thereof. Compared to macroscale materials, ultrathin films coupled with surface-sensitive analytical methods have unquestionable advantages with regard to sensitivity, providing explicit information about material interfaces.

The QCM-D is used for in situ monitoring of changes in mass and viscoelastic properties of ultrathin (cellulosic) film samples interacting with water or water vapour [106]. For QCM-D, the ultrathin film sample is deposited on a piezoelectric quartz crystal sensor surface, which is sandwiched between two electrodes. The sample is mounted in a measurement chamber and a pulsed electric field is applied, causing the sensor to oscillate at a specific fundamental frequency (f) and its overtones. When mass is added to the sample, for example, through water (vapour) sorption, the frequency of the oscillation decreases (Figure 6.11a). The mass change can be deduced with nanogram precision from the Sauerbrey relationship [119,120]:

$$\Delta m = C \Delta f n^{-1} \tag{6.8}$$

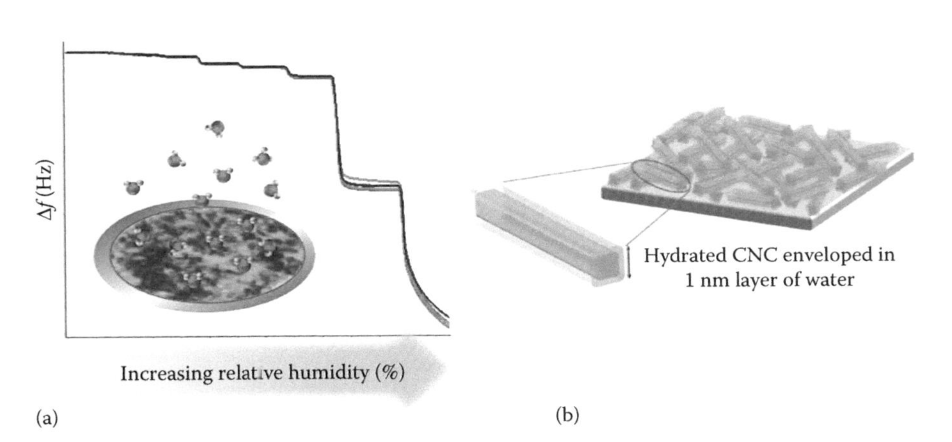

FIGURE 6.11 (a) Changes in frequency (with different overtones) detected by QCM-D during water vapour adsorption measurements onto an ultrathin cellulose nanofibril film (schematic shown in the inset) and (b) Schematic illustration of hydrated cellulose nanocrystals. (Adapted from Niinivaara, E. et al., *Langmuir*, 31(44), 12170–12176, 2015.)

where:

C is the sensitivity constant of the sensor crystal

n is the overtone number

Simultaneously, when the voltage is cut off, the oscillation gradually decreases and the resonance amplitude attenuates due to frictional losses in the adsorbed ultrathin film layer. The attenuation of the amplitude, that is, dissipation of energy, D, can be used as a measure of viscoelastic properties of the sample [120] and it can be presented as

$$D = \frac{E_{\text{dissipation}}}{2\pi E_{\text{storage}}} \tag{6.9}$$

where:

$E_{\text{dissipation}}$ is the energy dissipated

E_{storage} is the total energy stored during one oscillation cycle in the oscillator

SPR is an optical method based on a physical phenomenon called surface plasmon resonance occurring at metal–dielectric interfaces. It is used for detecting thickness and refractive index of ultrathin films. A beam of monochromatic p-polarised light passes through a prism and is reflected off from the backside of a sensor chip into a detector. At the angle of total internal reflection, electrons in the middle of the film start to resonate due to absorption of light. The resonating electrons are also called surface plasmons, and they are sensitive to their surrounding environment near the metal–dielectric interface. Coupling of photons (light) and plasmons consumes energy, which appears as an intensity loss in the reflected beam. This dip in the intensity can be used to convey information about the surface. The angle of total internal reflection is sensitive to changes in the refractive index (n) outside the sensor surface, which makes it suitable for analysing organic material attached to the SPR sensor surface [105].

SE is an optical surface and thin film measurement technique based on the polarisation of light. It is used for determining optical constants and film thickness with sub-nanometre preciseness. SE tracks the change in the state of polarisation of light (phase and amplitude) upon reflection from a film surface. The monitored quantity is the complex reflectance ratio (ρ), denoted as the ratio of p and s components of the reflected light:

$$\rho = \frac{r_p}{r_s} = \tan(\Psi)e^{i\Delta} \tag{6.10}$$

where:

Ψ is the amplitude ratio upon reflection

Δ is the phase shift upon reflection

Based on Ψ and Δ, physically meaningful values, such as the thickness of the film, can be inferred according to Fresnel equations using ideal layer models, for example, the classical Cauchy model for transparent (nano) materials [107].

6.7.1 Water Vapour Sorftion Analysed Using Surface-Sensitive Methods

Films made from cellulose nanoparticles exhibit distinctive and unique water uptake behaviour in response to changing humidity conditions in air. Using a QCM-D equipped with a humidity chamber, Tenhunen et al. [46] showed the response of CNF thin films due to water vapour uptake upon exposure to different levels of humid air. In accordance with other types of cellulose thin films [109,121,122], instantaneous responses were detected as a response to changes in the relative humidity of the air, signalling the high moisture sensitivity of CNF towards water. Higher water vapour uptake capability was associated with CNF with high content of xylan (glucuronoxylan) in comparison to CNF with low xylan content – a difference that was not apparent in dynamic vapour sorption (DVS) measurements. Hence, the authors concluded that the significance of xylan diminishes on macroscale, and the water vapour uptake is dominated by the fibrillar network [46].

Water vapour uptake into CNC thin films was studied by Niinivaara et al. [110] using the complementary combination of QCM-D and SE equipped with humidity chambers. Impregnable to water, CNCs provided an ideal platform for studying the hydration of cellulosic materials and the underlying causes. Through quantitative analysis of the QCM-D and SE data, it was shown that at the point of hydration (97 RH%) each individual CNC became enveloped by a 1 nm layer of water vapour caused by the adsorption of two to four monolayers of water, as shown in Figure 6.11b. In a subsequent paper, Niinivaara et al. [123] studied two-layer systems consisting of amorphous cellulose and CNC in varying ratios to mimic the humidity response of a plant cell wall. Based on the quantitative results, they concluded that the swelling of the system is strongly dependent on the thickness of the amorphous overlayer (more swelling when the amorphous layer is thinner) rather than exclusively on the ratio of amorphous layer to crystalline layer thickness.

The combination of SE and QCM-D was also utilised by Hakalahti et al. [124] to reveal underlying mechanisms of water vapour sorption into TEMPO CNF. The Langmuir – Flory–Huggins – clustering model was found to describe the sorption data accurately over the entire relative humidity range. The model suggests the existence of distinct zones of dominant water vapour sorption mechanisms into cellulosic thin films: specific adsorption below 10% RH, association of the Flory–Huggins population of water molecules at 10%–75% RH and clustering above 75% RH.

Recently, the water vapour uptake capacity of cellulose nanoparticle films was enhanced by using electrophoretic deposition (EPD) for preparation in place of established cellulosic thin film fabrication methods. QCM-D measurements revealed almost twofold water uptake capacity for a TEMPO CNF thin film deposited using EPD in comparison to spin-coated TEMPO CNF thin film at high relative humidity air (97% RH). The difference was attributed to higher roughness and porosity of the EPD-deposited film [125].

6.7.2 Swelling Behaviour Analysed Using Surface-Sensitive Methods

Aulin et al. [112] used QCM-D to study swelling properties of thin films of cellulose nanoparticle films, clarifying the effect of crystalline ordering and mesostructure of the thin film on the water uptake in liquid. The highest degree of swelling (48%) was detected for amorphous film made from cellulose dissolved in Li/DMAc (crystalline ordering 14.7%). However, contradicting the hypothesis, CNCs (crystalline ordering 87%) showed the second highest degree of swelling (24%), exceeding the swelling capacity of less ordered films, for example, films deposited using the Langmuir–Schäfer technique (crystalline ordering 63%) with a degree of swelling of 7%. Hence, the swelling behaviour of cellulosic films is determined by a balance between the molecular structure of the cellulose and the mesostructure of the films [112].

Swelling and water uptake properties have been shown to depend on the charge of the cellulosic material, electrolyte concentration and pH. A number of studies [60,113,121,126] have explored this topic, as the causal relationship between these interconnected attributes and the swelling is not straightforward. Using QCM-D, Ahola et al. [126] demonstrated that in general, higher charge leads to a higher degree of swelling. Electrolyte (NaCl) addition increased the water uptake of thin films with low charge (enzymatically pre-treated fibrils) and thin films with high negative charge (carboxymethylated) fibrils, although the effect was stronger for the more charged film. According to the Donnan effect, the electrolytes increased the pH inside the film, dissociating the carboxyl groups and making the film more susceptible towards water. However, on reaching a threshold at high electrolyte concentration (10 mM), reduction of the osmotic pressure leads to a decrease in the water uptake [126].

Olszewska et al. [60] reported pH-dependent swelling of cationic (2,3-epoxypropyl trimethylammonium chloride functionalised) CNF in aqueous media. Regardless of the pH, considerable swelling of the film took place dominantly due to the charged groups in cationic CNF and the very high surface area. At pH 4.5, partial protonation of the carboxyl groups caused the cationic CNF to exhibit a swelling of 303 wt.%, which was considerably higher compared to 137 wt.% at pH 8.0. At pH 4.5, the film behaved as a polyelectrolyte gel, expelling water upon increase of the electrolyte concentration in the range from 1 to 100 mM of NaCl. On the other hand, at pH 8 the film swelling was insensitive to electrolyte concentration due to electrostatic cross-linking between the carboxyls and tertiary amino groups [60].

Reid et al. [99] used SPR to monitor the swelling of CNC thin films in water. The authors suggested that the water uptake of CNC thin films occurs through two distinct steps: rapid penetration of the solvent (water) through the porous structure by capillary action, followed by swelling induced by the solvent wetting the surfaces and continuing to fill and create new space between CNCs. In agreement with Niinivaara et al. [110] and Aulin et al., [112] maximum spacing between CNCs at swollen state was found to be 1.2–1.6 nm corresponding to four to six molecular layers of water. Reid et al. [99] reported that the films of CNC contain 25 wt.% of water at swollen state – a significantly lower value compared to 74 wt.% for CNC and 70–85 wt.% for CNF as reported by Kittle et al. [127] and Kontturi et al. [113], respectively. Kontturi et al. [113] also pointed out that the effect of the charges is relatively moderate compared to the hydration of the vast amount of hydroxyl groups.

Ultrasensitive surface analytics have made it possible to experimentally study fundamental aspects of complex biomaterials with unprecedented precision. The largest benefit of ultrathin films and the surface-sensitive analytics approach with regard to water vapour sorption is related to understanding of the sorption phenomena, interactions of water with biomaterials and the underlying physico-chemical mechanisms at the molecular level. The fundamental and quantitative information can be linked to macroscale properties and ennobled into useful knowledge for advanced future biomaterial applications in the fields of, for example, membranes and sensors.

6.8 CASE EXAMPLES: IMPORTANCE OF NANOCELLULOSE INTERFACES IN NANOCELLULOSE APPLICATIONS

6.8.1 Adsorption of Contaminants onto Cellulose Nanomaterials

One important application of cellulose nanomaterials that exploits their inherent interfacial properties is the adsorption of contaminants, particularly in water treatment operations. Cellulose nanomaterials exhibit very high SSAs, up to several hundred m^2/g, which enable attachment of appropriate molecules to a high extent. Moreover, due to the abundance of anionic groups present, electrostatic interactions between adsorbent and adsorbate can easily be established. Cellulose nanomaterials usually exhibit negative ζ-potential over a wide range of pH; only below the IEP, which lies between pH 1 and 4 for many of these materials, the ζ-potential turns positive [66,67,128]. Hence, above this pH threshold, adsorption of positively charged materials, that is, metal ions and specifically heavy metal ions, which constitute a serious problem for the quality of drinking water, is very prone to adsorption on these materials. Especially, TEMPO-oxidised CNF and CNC exhibit negative surface charge at very low pH, whereby they are of particular interest for adsorption processes of positively charged contaminants.

6.8.1.1 Mechanism of Contaminant Adsorption on Nanocelluloses

The adsorption of contaminants on nanocelluloses is most commonly based on electrostatic interactions between the functional groups of the cellulose materials. These groups can be the inherently present hydroxyl groups, carboxyl or other acid groups attached during the preparation of the material, that is, via TEMPO modification or established during preparation of CNC with strong acids, or purposely introduced functional groups, such as ester groups, phosphate groups or grafted monomers. Whereas particularly cationic contaminants are already prone to adsorption onto pristine nanocellulose, in order to facilitate adsorption of anionic materials, modification with cationic functional groups needs to be performed. Moreover, hydrogen bonding is an important factor in the adsorption of certain classes of contaminants. Thus, generally, various physical forces, such as van der Waals forces (Keesom: interaction between permanent dipoles; Debye: interaction between permanent and induced dipoles; London dispersion force: interaction between induced dipoles), hydrophobicity, hydrogen bonds, polarity and steric interaction, are responsible for the adsorption processes that are taking place. Only in some exceptional cases, chemisorption plays a role. Based on these interactions, various sub-processes can contribute to the complex process of adsorption. For cellulose adsorbent materials, a variety of mechanisms have been reported: complexation, adsorption–complexation on surface and pores, ion exchange, microprecipitation and surface adsorption [85]. One particularly important mechanism thereby is ion exchange, especially in the case of heavy metal ions, which are exchanged against protons and thus attached onto the adsorbent. Co-ordination to electron-rich nucleophiles such as nitrile groups is also an important mechanism.

Crucial parameters for the successful application of adsorbents are pH, temperature, concentration of pollutants, contact time and particle size of the adsorbent, that is, the SSA available for adsorption [129]. In order to systematically describe adsorption processes and grade adsorbent materials, adsorption isotherms are frequently used. The two most often used isotherms are the ones established by Langmuir [130] and Freundlich [131]. As these isotherms are based on many simplifications, a lot of extensions and improvements have been applied to them to enable better fitting of practical data with theory [132]. The basic Langmuir isotherm (Equation 6.11) describes the adsorption process for monolayer adsorption, whereby all adsorption sites are considered to be equivalent, and the surface containing the adsorbing sites is assumed to be homogeneous and uniform, that is, a perfectly flat plane with no corrugations. Furthermore, the enthalpy of adsorption is assumed to be constant for all surface coverages from the first particles adsorbed until the formation of a full monolayer, no interaction between adsorbed molecules on adjacent sites takes place and the adsorbate is adsorbed into an immobile state. Even though these restrictions are very stringent, the Langmuir isotherm is still capable of describing many adsorption processes quite well.

$$q_e = q_m \frac{c_e K_L}{1 + c_e K_L} \tag{6.11}$$

where:

q_e (mg/g) is the equilibrium adsorption capacity at equilibrium ion concentration c_e (mg/L)

q_m (mg/g) is the maximum adsorption amount per unit weight of adsorbent in order to form a complete monolayer coverage on the surface

K_L is the Langmuir constant, which is related to the affinity of binding sites

The second important isotherm that describes adsorption processes is the Freundlich model (Equation 6.12). This model accounts for energetically heterogeneous surfaces on which the adsorbate molecules are interactive and the adsorbate loading increases infinitely with increasing concentration. It further takes into account that the adsorption enthalpy is exponentially dependent on the coverage.

$$q_e = K_F c_e^n \tag{6.12}$$

K_F and n are dimensionless Freundlich parameters, whereby K_F is the adsorption coefficient, which characterises the strength of the adsorption. The exponent n indicates the energetic heterogeneity. In a linear form, the Freundlich model can be expressed as

$$\ln q_e = \ln K_F + n \ln c_e \tag{6.13}$$

By application of this equation, K_F and n can be easily calculated from a linear plot of $\ln q_e$ versus $\ln c_e$.

Mechanistic studies on the adsorption of contaminants on nanocellulose materials have been performed, for example, by Liu et al. [41] for TEMPO-oxidised CNF, by Yu et al. [133] for CNC, by Yang et al. [134] for modified CNF and by Zhu et al. [135] for BC–Fe_3O_4 composites. Whereas non-nanosized cellulose, particularly those modified with various kinds of functional groups, has been employed in water treatment operations for quite some time [136], the use of cellulose nanomaterials has only very recently come into the focal point of research. A pioneering work by Sakairi et al. [137] already in 1998, demonstrating the use of BC for metal-ion adsorption, was the sole example for almost a decade. In terms of CNF from plant material, Ma and co-workers [138] demonstrated the high affinity of nanocellulose nanofibres to UO_2^{2+} ions. Thereafter, the number of literature on this topic has risen significantly. The use of nanocellulose materials in adsorption applications has been reviewed recently by several authors [85,139].

Initially, compared to traditional adsorbent materials, adsorption capacities of unmodified nanocellulose turned out to be only moderate at best. Accordingly, various modification techniques have been employed to improve the adsorption capacity of cellulose nanomaterials. These approaches usually do have in common that for the electrostatic attachment of metal cations, a very high negative surface charge is anticipated to be favourable. Thus, TEMPO modification and attachment of both carboxylic and phosphoric acid groups have resulted in a much higher adsorption capacity [32,38,140]. Furthermore, a vast amount of modifications has been postulated that ensures improved adsorption capacity towards metal ions. Exemplarily, cysteine-functionalised 2,3-dialdehyde cellulose [141], bisphosphonate nanocellulose [142] and phosphorylated nanocellulose [32,143] have recently been successfully introduced.

Apart from targeting only positively charged contaminants, there was the intention to achieve removal of negatively charged contaminants, such as nitrates, fluorides, sulphates or phosphates. As unmodified cellulose nanomaterials have a positive ζ-potential only at very low pH values, it is necessary to modify these materials in a way to realise positive ζ-potential also at and around neutral pH. To achieve this, exemplarily, ammonium groups have been attached to nanocellulose [144]. This resulted in turning the ζ-potential of nanocelluloses positive over the whole range of pH, thus enabling adsorption of negatively charged contaminants.

Cellulose nanomaterials have also been successfully applied in the removal of dyes. Opposed to metal ions or anions, the adsorption of dyes cannot be discussed with regard to electrostatic interactions of charged compounds. Whereas dyes can be of cationic or anionic nature, whose adsorption follows the same principles as mentioned earlier for anions or cations, they can also be uncharged. In this case, the mechanism of dye removal is based on physical entrapment or the dissolution of the dye in the adsorbent. Furthermore, reactive dyes are susceptible to the establishment of covalent bonds with the adsorbent [145]. Just as for metal ions, unmodified nanocellulose is capable of removal of positively charged dyes [146]. This effect could be significantly improved with the introduction of anionic groups, for example, by TEMPO modification or with negatively charged CNCs [147–150]. For negatively charged dyes, functional groups had to be attached that introduced positive surface charge, and hence enable adsorption of negatively charged materials as well [151].

Oil pollution of water sources is another severe problem that can be addressed by adsorbents made from nanocellulose, particularly aerogels. Recent incidents such as the explosion of the oil rig Deepwater Horizon in the Gulf of Mexico have demonstrated the vulnerability of environments towards oil pollution. This issue can be highlighted by the fact that several million litres of water can be made unfit for human consumption with just one litre benzene [152]. Oil sorbent materials can be characterised by high sorption capacity and oil/water selectivity, high porosity, fast oil sorption rate and high floatability (i.e. low density), ideally being of low cost, environmentally friendly and recyclable [153]. Aerogels prepared from cellulose nanomaterials are thus a recent development that matches those requirements to a high degree. However, in order to achieve the necessitated performance, nanocellulose materials need to be modified. In their pristine state, these materials possess intrinsic hydrophilicity, hence constituting exactly the opposite behaviour as desired. Thus, super-hydrophobic materials had to be developed. Furthermore, the establishment of a 3D network via a gelation step is of highest importance in order to enable the formation of the aerogels [153]. Possible pathways to achieving the required aerogel structure plus hydrophobicity can either be based on creation of the aerogel and post-modification, for example, by coating with TiO_2 [154], silylation [155] or grafting of styrene-acrylic monomers [156], or integration of those steps [157]. Oil sorption capacities of several hundred g/g were achieved based on these approaches. For further details on this application, the reader is directed to the review of Liu et al [153].

The examples presented earlier show that in order to apply sustainable, benign, industrially scalable adsorbent materials with a low chemical burden and toxic solvent-free reactions, cellulose nanomaterials can be one of the ideal candidates as base material for water treatment [158]. The performance of these materials has substantially been improved by the application of chemical modifications, enriching the surface of nanocelluloses with functional groups, thus increasing the adsorption capacity. Following this approach further on led to the development of composite and hybrid materials comprising cellulose nanomaterials. This was done to specifically tailor the surface of nanocellulose for selective and enhanced interactions with contaminants. Exemplarily, aerogels from cross-linked hairy nanocrystalline cellulose and modified chitosan have been employed for dye removal [159], CNFs have been modified with dopamine in order to construct an iron conjugate prone to heavy metal adsorption [158], and microgels based on nanocellulose and polyvinylamine for the removal of anionic dyes [160] were developed. Generally, an approach commonly employed is the combination of (modified) nanocellulose with an inorganic compound. For example, CNFs have been grafted with methyl methacrylate and itaconic acid in order to prepare a composite with nanobentonite or magnetite for heavy metal adsorption [161,162]. Another current example is the application of a polyaniline impregnated nanocellulose composite [163]. In the recent review of Olivera et al. [164] the interested reader will find many more examples for this approach.

6.8.2 Application of Interfacial Properties of Nanocellulose in Membranes

The inherent interfacial properties applied in various adsorption applications were also already researched in membrane applications. Nanocellulose of particular affinity towards various kinds of contaminants was processed into membranes and applied in continuous membrane processes. Exemplarily, phosphorylated nanocellulose and nanocrystals were applied in heavy metal adsorption membranes [68,165,166]; nanopapers bearing ammonium groups were efficient for nitrate removal [69], whereas CNCs were used as functional entity in dye-adsorption membranes [147].

In this regard, not only the affinity of nanocellulose towards certain classes of materials was exploited but also the opposite process: the rejection of microorganisms, exemplarily for the development of antimicrobial filters [167], or hydrophobic compounds in order to restrict or even prevent membrane fouling. Membrane fouling [168] is a major problem that limits the applicability of membranes to a great extent, whereby the extent of fouling is determined by the interfacial properties [169]. It is the reduction of membrane flux caused by the formation of a film on the surface or inside the membrane due to the adsorption of hydrophobic compounds [170]. The inherent hydrophilicity of cellulose nanomaterials is thereby an important property [171] to minimise detrimental effects of this process for hydrophilicity that is anticipated to decrease fouling [172].

6.8.3 Nanocellulose as Emulsion Stabilisers

The use of cellulose nanomaterials as emulsion stabiliser is another important development especially regarding environmental issues. The pollution of aquatic environments by surfactants is a problem known already for decades [173]. Hence, the use of a renewable, non-toxic material as alternative to these materials would be highly desirable.

Emulsions are (often colloidal) dispersions of one fluid phase in another, that is, two immiscible liquids, kinetically stabilised by the presence of suitable emulsifiers. A suitable emulsifier, for example, surfactant (short for surface active agent), is characterised by strong adsorption at the interface, lowering the interfacial tension and hence the driving force for phase separation [174]. Commonly, surfactants are lipids, that is, long chain aliphatic hydrocarbon derivatives, which have been applied for centuries, whereas the use of colloidal particles as emulsifiers only dates back to the early twentieth century. It was first described in pioneering works of Ramsden [175] and Pickering [176]; hence, these colloidal particle-stabilised emulsions are commonly known as 'Pickering emulsions' [177]. The use of cellulose as Pickering emulsifier was first suggested by Oza and Frank [178]. They applied microcrystalline cellulose in order to stabilise oil-in-water (o/w) emulsions. The cellulose formed a network around emulsified oil droplets, whereby this structure provides a mechanical barrier at the o/w interface, thus stabilising the emulsion without the necessity for decreasing interfacial tension, as in conventional surfactant-stabilised emulsions. Just as in this first example, the application of cellulose as emulsifier commonly results in establishing o/w emulsions; in order to achieve water-in-oil (w/o) emulsion, the cellulose surface needs to be hydrophobically modified [179,180].

Generally, an o/w emulsion will form preferably when the three-phase contact angle between oil, solid and water is $<90°$. In the case when angle is bigger than 90°, a w/o emulsion will be formed [181,182]. Particles present at the interface are able to form rigid structures that sterically inhibit the coalescence of emulsion droplets [183].

In the case of nanocellulose, it was proposed that a 3D network is established in the continuous phase, thus further retarding the coalescence through entrapment of emulsion droplets [180,184]. Alternatively, CNF acting as emulsion stabilisers can be present as single or dispersed fibrils [185].

Principally, in nanocellulose-stabilised emulsions, the droplet size in the emulsion was found to be dependent on the particle size of the nanocellulose and thus on the number of passes of the cellulose material through the homogeniser during preparation [186]. Moreover, the appearance of the emulsion was observed to be largely dependent on the size and type of nanocellulose material [65]. Emulsions stabilised with unmodified CNF and CNC formed a thick creaming layer with a clear aqueous phase (CNF) below, indicating the formation of droplet clusters, or a turbid phase below (CNC). With TEMPO CNF, two distinct phases were formed within 24 h after emulsification, whereby the turbid phase below the creaming layer indicates that the emulsion phase contains well dispersed droplets that coexist with clusters in the creaming layer. Furthermore, it was found in the same study that nanocelluloses stabilise emulsions against coalescence rather than flocculation.

Surface modification of nanocellulose emulsifiers has not only been important to switch the type of emulsion and enable w/o emulsion [179] but also to enhance the stability of these emulsions. For example, Lee et al. [187] studied the pH-triggered transitional phase behaviour of w/o Pickering emulsions stabilised by hydrophobised BC. They esterified neat BC carboxyl acids of varying length (C2- to C12-) and observed that C6– and C12–BC stabilised emulsions exhibited a pH-triggered reversible transitional phase separation, whereas the C2–BC underwent an irreversible pH-triggered transitional phase separation and inversion. This was attributed to hydrolysis of the ester bonds of C2–BC at high pH. For a more detailed summary of this topic, the reader is referred to the reviews of Salas et al., [185] Dickinson [188] and Lee et al [177]. More recent developments comprise the use of bifunctionalised CNC [189], surfactant-free stabilisation of w/o/w systems [190], and the achievement of highly stable emulsions by application of TEMPO-oxidised BC [82].

REFERENCES

1. Klemm D, Kramer F, Moritz S, Lindström T, Ankerfors M, Gray D, Dorris, A. Nanocelluloses: A new family of nature-based materials. *Angew Chem Int Ed.* 2011;50:5438–5466.
2. Siró I, Plackett D. Microfibrillated cellulose and new nanocomposite materials: A review. *Cellulose.* 2010;17(3):459–494.
3. Eichhorn SJ, Dufresne A, Aranguren M, Marcovich NE, Capadona JR, Rowan SJ, Weder C et al. Review: Current international research into cellulose nanofibres and nanocomposites. *J Mater Sci.* 2010;45(1):1–33.
4. Dufresne A. *Nanocellulose: From Nature to High Performance Tailored Materials.* Walter De Gruyter, Berlin, Germany, 2012/2013.
5. Kangas H, Lahtinen P, Sneck A, Saariaho AM, Laitinen O, Hellen E. Characterization of fibrillated celluloses. A short review and evaluation of characteristics with a combination of methods. *Nord Pulp Pap Res J.* 2014;29(1):129–143.

6. Moon RJ, Pöhler T, Tammelin T. Microscopic characterization of nanofibers and nanocrystals, in: *Handbook of Green Materials, Vol. 1. Bio Based Nanomaterials: Separation, Processes, Characterisation and Properties*, Oksman K, Mathew AP, Bismarck A, Rojas O, Sain M (Eds.), World Scientific Publishing, Singapore, 2014, pp. 159–180.
7. Ferguson A, Khan U, Walsh M, Lee K-Y, Bismarck A, Shaffer MSP, Coleman JN, Bergin SD. Understanding the dispersion and assembly of bacterial cellulose in organic solvents. *Biomacromolecules*. 2016;17(5):1845–1853.
8. Brown AJ. The chemical action of pure cultivations of bacterium aceti. *J Chem Soc*. 1886;49:172–187.
9. Iguchi M, Yamanaka S, Budhiono A. Bacterial cellulose – A masterpiece of nature's art. *J Mater Sci*. 2000;35(2):261–270.
10. Czaja W, Romanovicz D, Brown RM. Structural investigations of microbial cellulose produced in stationary and agitated culture. *Cellulose*. 2004;11(3–4):403–411.
11. Hsieh YC, Yano H, Nogi M, Eichhorn SJ. An estimation of the Young's modulus of bacterial cellulose filaments. *Cellulose*. 2008;15(4):507–513.
12. Lee K-Y, Buldum G, Mantalaris A, Bismarck A. More than meets the eye in bacterial cellulose: Biosynthesis, bioprocessing, and applications in advanced fiber composites. *Macromol Biosci*. 2014;14(1):10–32.
13. Turbak AF, Snyder FW, Sandberg KR. Microfibrillated cellulose: A new cellulose product. Properties, uses and commercial potential. *J Appl Polym Sci Polym Symp*. 1983;37:815–827.
14. Pääkkö M, Ankerfors M, Kosonen H, Nykänen A, Ahola S, Österberg M, Ruokolainen J et al. Enzymatic hydrolysis combined with mechanical shearing and high-pressure homogenization for nanoscale cellulose fibrils and strong gels. *Biomacromolecules*. 2007;8(6):1934–1941.
15. Saito T, Nishiyama Y, Putaux JL, Vignon M, Isogai A. Homogeneous suspensions of individualized microfibrils from TEMPO-catalyzed oxidation of native cellulose. *Biomacromolecules*. 2006;7(6):1687–1691; Saito T, Kimura S, Nishiyama Y, Isogai A. Cellulose nanofibers prepared by TEMPO-mediated oxidation of native cellulose. *Biomacromolecules*. 2007;8(8):2485–2491.
16. Hiltunen J, Kemppainen K, Pere J. Process for producing fibrillated cellulose material. WO2015/092146 A1. Finnish Patent FI126698.
17. Hiemenz PC, Rajagopalan R. *Principles of Colloid and Surface Chemistry*, (3rd Ed.). Marcel Dekker, New York. 1997; Shaw DJ. *Introduction to Colloid and Surface Chemistry*, (3rd Ed.). Butterworth & Co., London. 1983; Berg JC. *An Introduction to Interfaces and Colloids – The Bridge to Nanoscience*. World Scientific Publishing, Singapore, 2012, http://www.worldscientific.com/worldscibooks/10.1142/7579.
18. Brunauer S. *Physical Adsorption of Gases and Vapours*, Oxford University Press, London, UK, 1944.
19. Langmuir I. The adsorption of gases on plane surfaces of glass, mica and platinum. *J Am Chem Soc*. 1918; 40(9):1361–1403.
20. Missoum K, Belgacem MN, Bras J. Nanofibrillated cellulose surface modification: A review. *Materials*. 2013; 6(5):1745–1766.
21. Ma H, Burger C, Hsiao BS, Chu B. Ultrafine polysaccharide nanofibrous membranes for water purification. *Biomacromolecules*. 2011;12(4):970–976; Ultra-fine cellulose nanofibers: New nano-scale materials for water purification. *J Mater Chem*. 2011;21:7507–7510.
22. Mihranyan A, Esmaeili M, Razaq A, Alexeichik D, Lindström T. Influence of the nanocellulose raw material characteristics on the electrochemical and mechanical properties of conductive paper electrodes. *J Mater Sci*. 2012;47(10):4463–4472.

23. Ruan C, Strømme M, Lindh J. A green and simple method for preparation of an efficient palladium adsorbent based on cysteine functionalized 2,3-dialdehyde cellulose. *Cellulose.* 2016;23(4):2627–2638.
24. Kim D-K, Nishiyama Y, Kuga S. Surface acetylation of bacterial cellulose. *Cellulose.* 2002;9(3–4):361–367.
25. Liebner F, Haimer E, Wendland M, Neouze M-A, Schlufter K, Miethe P, Heinze T, Potthast A, Rosenau T. Aerogels from unaltered bacterial cellulose: Application of $scCO_2$ drying for the preparation of shaped, ultra-lightweight cellulosic aerogels. *Macromol. Biosci.* 2010;10(4):349–352.
26. Lee K-Y, Qian H, Tay FH, Blaker JJ, Kazarian SG, Bismarck A. Bacterial cellulose as source for activated nanosized carbon. *J Mater Sci.* 2013;48(1):367–376.
27. Leung ACW, Lam E, Chong J, Hrapovic S, Luong JHT. Reinforced plastics and aerogels by nanocrystalline cellulose. *J Nanopart Res.* 2013;15(5):1636; Pääkkö M, Vapaavuori J, Silvennoinen R, Kosonen H, Ankerfors M, Lindström T, Berglund LA, Ikkala O. Long and entangled native cellulose I nanofibers allow flexible aerogels and hierarchically porous templates for functionalities. *Soft Matter.* 2008;4(12):2492–2499.
28. Cervin NT, Aulin C, Larsson PT, Wågberg L. Ultra porous nanocellulose aerogels as separation medium for mixtures of oil/water liquids. *Cellulose.* 2012;19(2):401–410.
29. Aulin C, Netrval J, Wågberg L, Lindstrom T. Aerogels from nanofibrillated cellulose with tunable oleophobicity. *Soft Matter.* 2010;6(14):3298–3305.
30. Jonoobi M, Mathew AP, Oksman K. Producing low-cost cellulose nanofiber from sludge as new source of raw materials. *Ind Crops Prod.* 2012;40:232–238.
31. Liu P, Sehaqui H, Tingaut P, Wichser A, Oksman K, Mathew AP. Cellulose and chitin nanomaterials for capturing silver ions (Ag+) from water via surface adsorption. *Cellulose.* 2014;21(1):449–461.
32. Liu P, Borrell PF, Bozic M, Kokol V, Oksman K, Mathew AP. Nanocelluloses and their phosphorylated derivatives for selective adsorption of Ag^+, $Cu2^+$ and $Fe3^+$ from industrial effluents. *J Hazard Mat.* 2015;294:177–185.
33. Bayer T, Cunning BV, Selyanchyn R, Nishihara M, Fujikawa S, Sasaki K, Lyth SM. High temperature proton conduction in nanocellulose membranes: Paper fuel cells. *Chem Mater.* 2016;28(13):4805–4814.
34. Angles M, Dufresne A. Plasticized starch/tunicin whiskers nanocomposite materials. 2. Mechanical Behaviour. *Macromolecules.* 2001;34(9):2921–2931; Sturcova A, Davies, GR, Eichhorn SJ. Elastic modulus and stress-transfer properties of tunicate cellulose whiskers. *Biomacromolecules.* 2005;6(2):1055–1061.
35. Hooshmand S, Cho S-W, Skrifvars M, Mathew A, Oksman K. Melt spun cellulose nanocomposite fibres: Comparison of two dispersion techniques. *Plast Rubber Compos.* 2014;43(1):15–24.
36. Heath L, Thielemans W. Cellulose nanowhisker aerogels. *Green Chem.* 2010;12(8): 1448–1453.
37. Sehaqui H, Zhou Q, Ikkala O, Berglund LA. Strong and tough cellulose nanopaper with high specific surface area and porosity. *Biomacromolecules.* 2011;12(10):3638–3644.
38. Sehaqui H, Perez de Larraya U, Liu P, Pfenninger N, Mathew AP, Zimmermann T, Tingaut P. Enhancing adsorption of heavy metal ions onto biobased nanofibers from waste pulp residues for application in wastewater treatment. *Cellulose.* 2014;21(4):2831–2844.
39. Sehaqui H, Zhou Q, Berglund LA. High-porosity aerogels of high specific surface area prepared from nanofibrillated cellulose (NFC). *Compos Sci Technol.* 2011;71(13):1593–1599.
40. Dwivedi AD, Sanandiya ND, Singh JP, Husnain SM, Chae KH, Hwang DS, Chang Y-S, Tuning and characterizing nanocellulose interface for enhanced removal of dual-sorbate (AsV and CrVI) from water matrices. *ACS Sustainable Chem. Eng.* 2017;5(1):518–528.

41. Liu P, Garrido B, Oksman K, Mathew AP. Adsorption isotherms and mechanisms of Cu(II) sorption onto TEMPO-mediated oxidized cellulose nanofibers. *RSC Adv.* 2016, 6(109): 107759–107767.
42. Sjöström E. *Wood Chemistry, Fundamentals and Applications*, (1st Ed.). Academic Press, New York, 1981.
43. Penttilä PA, Várnai A, Pere J, Tammelin T, Salmén L, Siika-aho M, Viikari L, Serimaa R. Xylan as limiting factor in enzymatic hydrolysis of nanocellulose. *Bioresour Technol.* 2013;129:135–141.
44. Uetani K, Yano H. Zeta potential time dependence reveals the swelling dynamics of wood cellulose nanofibrils. *Langmuir.* 2012;28(1):818–827.
45. Arola S, Malho J-M, Laaksonen P, Lille M, Linder M. The role of hemicellulose in nanofibrillated cellulose networks. *Soft Matter.* 2013;9(4):1319–1326.
46. Tenhunen T-M, Peresin MD, Penttilä PA, Pere J, Serimaa R, Tammelin T. Significance of xylan on the stability and water interactions of cellulosic nanofibrils. *React Funct Polym.* 2014;85:157–166.
47. Tanaka R, Saito T, Hänninen T, Ono Y, Hakalahti M, Tammelin T. Viscoelastic properties of core–shell-structured, hemicellulose-rich nanofibrillated cellulose in dispersion and wet-film states. *Biomacromolecules.* 2016;17(6):2104–2111.
48. Lahtinen P, Liukkonen S, Pere J, Sneck A, Kangas H. A comparative study of fibrillated fibers from different mechanical and chemical pulps. *Bioresources.* 2014;9(2):2115–2127.
49. Spence KL, Venditti RA, Rojas OJ, Habibi Y, Pawlak JJ. The effect of chemical composition on microfibrillar cellulose films from wood pulps: Water interactions and physical properties for packaging applications. *Cellulose.* 2010a;17(4): 835–848.
50. Spence KL, Venditti RA, Habibi Y, Rojas OJ, Pawlak JJ. The effect of chemical composition on microfibrillar cellulose films from wood pulps: Mechanical processing and physical properties. *Bioresource Technol.* 2010b;101(15):5961–5968.
51. Hoeger IC, Nair SS, Ragauskas AJ, Deng Y, Rojas OJ, Zhu JY. Mechanical deconstruction of lignocellulose cell walls and their enzymatic saccharification. *Cellulose.* 2013;20(2):807–818.
52. Solala I, Volperts A, Andersone A, Dizhbite T, Mironova-Ulmane N, Vehniäinen A, Pere J, Vuorinen T. Mechanoradical formation and its effects on birch kraft pulp during the preparation of nanofibrillated cellulose with Masuko refining. *Holzforschung.* 2012;66(4):477–483.
53. Rojo E, Peresin MS, Sampson WW, Hoeger IC, Vartiainen J, Laine J, Rojas OJ. Comprehensive elucidation of the effect of residual lignin on the physical, barrier, mechanical and surface properties of nanocellulose films. *Green Chem.* 2015;17(3):1853–1866.
54. Nair SS, Yan N. Effect of high residual lignin on the thermal stability of nanofibrils and its enhanced mechanical performance in aqueous environments. *Cellulose.* 2015;22(5):3137–3150.
55. Katz K, Beatson RP, Scallan AM. The determination of strong and weak acidic groups in sulphite pulps. *Sven Papperstidn.* 1984;87(6):48–53.
56. Centola G, Borruso D. The influence of hemicelluloses on the beatability of pulps. *TAPPI J.* 1967;50(7):344–347.
57. Wågberg L, Decher G, Norgren M, Lindström T, Ankerfors M, Axnäs K. The build-up of polyelectrolyte multilayers of microfibrillated cellulose and cationic polyelectrolytes. *Langmuir.* 2008;24(3):784–795.
58. Liimatainen H, Visanko M, Sirviö JA, Hormi O, Niinimäki J. Enhancement of the nanofibrillation of wood cellulose through sequential periodate-chlorite oxidation. *Biomacromolecules.* 2012;13(5):1592–1597.

59. Fall AB, Lindström SB, Sundman O, Ödberg L, Wågberg L. Colloidal stability of aqueous nanofibrillated cellulose dispersions. *Langmuir.* 2011;27(18):11332–11338.
60. Olszewska A, Eronen P, Johansson LS, Malho JM, Ankerfors M, Lindström T, Ruokolainen J, Laine J, Österberg M. The behaviour of cationic nanofibrillar cellulose in aqueous media. *Cellulose.* 2011;18(5):1213–1226.
61. Pei A, Butchosa N, Berglund LA, Zhou Q. Surface quaternized cellulose nanofibrils with high water absorbency and adsorption capacity for anionic dyes. *Soft Matter.* 2013;9(6):2047–2055.
62. Liimatainen H, Suopajärvi T, Sirviö JA, Hormi O, Niinimäki J. Fabrication of cationic cellulosic nanofibrils through aqueous quaternization pretreatment and their use in colloid aggregation. *Carbohyd Polym.* 2014;103:187–192.
63. Dong XM, Revol JF, Gray DG. Effect of microcrystallite cellulose preparation conditions on the formation of colloid crystals of cellulose. *Cellulose.* 1998;5(1):19–32.
64. Beck S, Méthot M, Bouchard J. General procedure for determining cellulose nanocrystal sulfate half-ester content by conductometric titration. *Cellulose.* 2015;22(1):101–116.
65. Gestranius M, Stenius P, Kontturi E, Sjöblom J, Tammelin T. Phase behaviour and droplet size of oil-in-water Pickering emulsions stabilised with plant-derived nanocellulosic materials. *Colloids Surf A.* 2016;519:60–70.
66. Mautner A, Lee K-Y, Lahtinen P, Hakalahti M, Tammelin T, Li K, Bismarck A. Nanopapers for organic solvent nanofiltration. *Chem Commun.* 2014;50:5778–5781.
67. Mautner A, Lee K-Y, Tammelin T, Mathew AP, Nedoma AJ, Li K, Bismarck A. Cellulose nanopapers as tight aqueous ultra-filtration membranes. *React Funct Polym.* 2015;86:209–214.
68. Mautner A, Maples HA, Kobkeatthawin T, Kokol V, Karim Z, Li K, Bismarck A. Phosphorylated nanocellulose papers for copper adsorption from aqueous solutions. *Int J Environ Sci Technol.* 2016a;13:1861.
69. Mautner A, Maples HA, Sehaqui H, Zimmermann T, Perez de Larraya U, Mathew AP, Lai C-Y, Li K, Bismarck A. Nitrate removal from water using a nanopaper ion-exchanger. *Environ Sci: Water Res Technol.* 2016b;2:117–124.
70. Yamane C. Structure formation of regenerated cellulose from its solution and resultant features of high wettability: A review. *Nord Pulp Pap Res J.* 2015;30(1):78–91.
71. Holmberg M, Berg J, Rasmusson J, Stemme S, Ödberg L, Claesson P. Surface force studies of Langmuir–Blodgett cellulose films. *J Colloid Interf Sci.* 1997;186(2):369–381.
72. Österberg M, Peresin MS, Johansson L-S, Tammelin T. Clean and reactive nanostructured cellulose surface. *Cellulose.* 2013;20(3):983–990.
73. Matsunaga T, Ikada Y. Surface modifications of cellulose and polyvinyl alcohol and determination of the surface density of the hydroxyl group. *Adv Colloid Sri Symp Ser.* 1980;121:391–406.
74. Lindman B, Karlström G, Stigsson L. On the mechanism of dissolution of cellulose. *J Mol Liq.* 2010;156(1):76–81.
75. Medronho B, Lindman B. Competing forces during cellulose dissolution: From solvents to mechanisms. *Curr Opin Colloid Interface Sci.* 2014;19(1):32–40.
76. Medhorno B, Duarte H, Alves L, Antunes F, Romano A, Lindman B. Probing cellulose amphiphilicity. *Nord Pulp Pap Res J.* 2015;30(1):58–66.
77. Yamane C, Aoyagi T, Ago M, Sato K, Okajima K, Takahashi T. Two different surface properties of regenerated cellulose due to structural anisotrophy. *Polym J.* 2006;38(8):819–826.
78. Suchy M, Virtanen J, Kontturi E, Vuorinen T. Impact of drying on wood ultrastructure observed by deuterium exchange and photoacoustic FT-IR spectroscopy. *Biomacromolecules.* 2009;11(2):515–520.
79. Johansson L-S, Tammelin T, Campbell JM, Setälä H, Österberg M. Experimental evidence on medium driven cellulose surface adaptation demonstrated using nanofibrillated cellulose. *Soft Matter.* 2011;7(22):10917–10924.

80. Birch W, Carre A, Mittal KI. Wettability techniques to monitor the cleanliness of surfaces. *Developments in Surface Contamination and Cleaning*. William Andrew, (ed.) R. Kohli and K.L. Mittal, Waltham, San Diego, 2015: pp. 1–32.
81. Jiang F, Hsieh Y-L. Self-assembling of TEMPO oxidized cellulose nanofibrils as affected by protonation of surface carboxyls and drying methods. *ACS Sustainable Chem. Eng*. 2016;4(3):1041–1049.
82. Jia Y, Zhai X, Fu W, Liu Y, Li F, Zhong C. Surfactant-free emulsions stabilized by tempo-oxidized bacterial cellulose. *Carbohydr Polym*. 2016;151:907–915.
83. Cervin NT, Andersson L, Ng JBS, Olin P, Bergström L, Wågberg L. Lightweight and strong cellulose materials made from aqueous foams stabilized by nanofibrillated cellulose. *Biomacromolecules*. 2013;14(2):503–511.
84. Kinnunen K, Hjelt T, Kenttä E, Forsström U. Thin coatings for paper by foam coating. *Tappi J*. 2014;13(7):9–19.
85. Hokkanen S, Bhatnagar A, Sillanpää M. A review on modification methods to cellulose-based adsorbents to improve adsorption capacity. *Water Research*. 2016;91:156–173.
86. Habibi Y. Key advances in the chemical modification of nanocelluloses. *Chem Soc Rev* 2014;43:519–1542. doi:10.1039/C3CS60204D.
87. Eyley S, Thielemans W. Surface modification of cellulose nanocrystals. *Nanoscale*. 2014;6(14):7764–7779.
88. Lee K-Y, Quero F, Blaker JJ, Hill CAS, Eichhorn SJ, Bismarck A. Surface only modification of bacterial cellulose nanofibers with organic acids. *Cellulose*. 2011;18(3):595–605.
89. Tasker S, Badyal JPS, Backson SCE, Richards RW. Hydroxyl accessibility in celluloses. *Polymer*. 1994;35(22):4717–4721.
90. Vuoti S, Talja R, Johansson L-S, Heikkinen H, Tammelin T. Solvent impact on esterification and film formation ability of nanofibrillated cellulose. *Cellulose*. 2013;20(5):2359–2370.
91. Labet M, Thielemans W. Improving the reproducibility of chemical reactions on the surface of cellulose nanocrystals: ROP of ε-caprolactone as a case study. *Cellulose*. 2011;18(3):607–617.
92. Chinga-Carrasco G, Kuznetsova N, Garaeva M, Leirset I, Galiullina G, Kostochko A, Syverud K. Bleached and unbleached MFC nanobarriers: Properties and hydrophobisation with hexamethyldisilazane. *J Nanopart Res*. 2012;14(12):1280–1289.
93. Rodionova G, Lenes M, Eriksen Ø, Gregersen Ø. Surface chemical modification of microfibrillated cellulose: Improvement of barrier properties for packaging applications. *Cellulose*. 2011;18(1):127–134.
94. Hakalahti M, Mautner A, Johansson L-S, Hänninen T, Setälä H, Kontturi E, Bismarck A, Tammelin T. Direct interfacial modification of nanocellulose films for thermoresponsive membrane template. *ACS Appl Mater Interfaces*. 2016;8(5):2923–2927.
95. Peresin MS, Kammiovirta K, Heikkinen H, Johansson L-S, Vartiainen J, Setälä H, Österberg M, Tammelin T. Surface functionalized nanocellulose film with controlled interactions with oxygen and water. *Carbohyd Polym*. 2017;174:309–317.
96. Ye T, McArthur EA, Borguet E. Mechanism of UV photoreactivity of alkylsiloxane self-assembled monolayers. *J Phys Chem B*. 2005;109(20):9927–9938.
97. Lavoine N, Desloges I, Dufresne A, Bras J. Microfibrillated cellulose – Its barrier properties and applications on cellulosic materials: A review. *Carbohydr Polym*. 2012;90(2):735–764.
98. Klemm D, Philipp B, Heinze T, Heinze U, Wagenknecht W. Swelling of cellulose in water, in: *Comprehensive Cellulose Chemistry: Fundamentals and Analytical Methods*, Vol. 1. Wiley-VCH Verlag GmbH & Co. KGaA, Weinheim, Germany, 1998, pp. 45–50.
99. Reid MS, Villalobos M, Cranston ED. Cellulose nanocrystal interactions probed by thin film swelling to predict dispersibility. *Nanoscale*. 2016;8(24):12247–12257.

100. Shresta S, Diaz JA, Ghanbari S, Youngblood JP. Hygroscopic swelling determination of cellulose nanocrystal (CNC) films by polarized light microscopy digital image correlation. *Biomacromolecules*. 2017;18(5):1482–1490.
101. Urquhardt AR, Eckersall N. The moisture relations of cotton. VII – A study of hysteresis. *J Text Inst*. 1930;21(10):499–510.
102. Belbekhouche S, Bras J, Siqueira G, Chappey C, Lebrun L, Khelifi B, Marais S, Dufresne A. Water sorption behaviour and gas barrier properties of cellulose whiskers and microfibrils films. *Carbohydr Polym*. 2011;83(4):1740–1748.
103. Driemeier C, Mendes FM, Oliviera M. Dynamic vapour sorption and thermoporometry to probe water in celluloses. *Cellulose*. 2012;19(4):1051–1063.
104. Henriksson M, Berglund LA. Structure and properties of cellulose nanocomposite films containing melamine formaldehyde. *J Appl Polym Sci*. 2007;106(4):2817–2824.
105. Schasfoort RB, Tudos AJ. *Handbook of Surface Plasmon Resonance*. Royal Society of Chemistry, Cambridge, UK, 2008.
106. Reviakine I, Johannsmann D, Richter RP. Hearing what you cannot see and visualizing what you hear: Interpreting quartz crystal microbalance data from solvated interfaces. *Anal Chem*. 2011;83:8838–8848.
107. Losurdo M, Hingerl K. *Ellipsometry at the Nanoscale*. Springer Science & Business Media. Heidelberg, Berlin, 2013.
108. Li R, Faustini M, Boissière C, Grosso D. Water capillary condensation effect on the photocatalytic activity of porous TiO_2 in air. *J Phys Chem C*. 2014;118(31): 17710–17716.
109. Tammelin T, Abburi R, Gestranius M, Laine C, Setälä H, Österberg M. Correlation between cellulose thin film supramolecular structures and interactions with water. *Soft matter*. 2015;11(21):4273–4282.
110. Niinivaara E, Faustini M, tammelin T, Kontturi E. Water vapour uptake of ultrathin films of biologically derived nanocrystals: Quantitative assessment with quartzc microbalance and spectroscopic ellipsometry. *Langmuir*. 2015;31(44):12170–12176.
111. Tammelin T, Saarinen T, Österberg M, Laine J. Preparation of Langmuir/Blodgett-cellulose surfaces by using horizontal dipping procedure. Application for polyelectrolyte adsorption studies performed with QCM-D. *Cellulose*. 2006;13(5):519–535.
112. Aulin C, Ahola S, Josefsson P, Nishino T, Hirose Y, Österberg M, Wågberg L. Nanoscale cellulose films with different crystallinities and mesostructures – Their surface properties and interaction with water. *Langmuir*. 2009;25(13):7675–7685.
113. Kontturi KS, Kontturi E, Laine J. Specific water uptake of thin films from nanofibrillar cellulose. *J Mat Chem A*. 2013;1(43):13655–13663.
114. Kontturi E, Tammelin T, Österberg M. Cellulose – Model films and the fundamental approach. *Chem Soc Rev*. 2006;35(12):1287–1304.
115. Blodgett KB. Films built by depositing successive monomolecular layers on a solid surface. *J Am Chem Soc*. 1935;7(6):1007–1022.
116. Decher G. Fuzzy nanoassemblies: Toward layered polymeric multicomposites. *Science*. 1997;277(5330):1232–1237.
117. Emslie AG, Bonner FT, Peck LG. Flow of a viscous liquid on a rotating disc. *J Appl Phys*. 1958;29(5):858–862.
118. Meyerhofer D. Characteristics of resist films produced by spinning. *J Appl Phys*. 1978;49(7):3993–3997.
119. Sauerbrey G. The use of quartz oscillators for weighing thin layers and for microweighing. *Z Phys*. 1959;155:206–222.
120. Höök F, Rodahl M, Brzezinski P, Kasemo B. Energy dissipation kinetics for protein and antibody – Antigen adsorption under shear oscillation on a quartz crystal microbalance. *Langmuir*. 1998;14(4):729–734.

121. Fält S, Wågberg L, Vesterlind E-L. Swelling of model films of cellulose having different charge densities and comparison to the swelling behaviour of corresponding fibers. *Langmuir*. 2003;19(19):7895–7903.
122. Rechfeldt F, Tanaka M. Hydration forces in ultrathin films of cellulose. *Langmuir*. 2003;19(5):1467–1473.
123. Niinivaara E, Faustini M, Tammelin T, Kontturi E. Mimicking the humidity response of the plant cell wall by using two-dimensional systems: The critical role of amorphous and crystalline polysaccharides. *Langmuir*. 2016;32(8):2032–2040.
124. Interfacial Mechanisms of Water Vapor Sorption into Cellulose Nanofibril Films as Revealed by Quantitative Models. Biomacromolecules. 2017;18:2951–2958.
125. Wilson B, Yliniemi K, Gestranius M, Putkonen M, Lundström M, Murtomäki L, Karppinen M, Tammelin T, Kontturi E. .
126. Ahola S, Salmi J, Johansson L-S, Laine J, Österberg M. Model films from native cellulose nanofibrils. Preparation, swelling, and surface interactions. *Biomacromolecules*. 2008;9(4):1273–1282.
127. Kittle JD, Du X, Jiang F, Qian C, Heinze T, Roman M, Esker AR. Equilibrium water contents of cellulose films determined via solvent exchange and quartz crystal microbalance with dissipation monitoring. *Biomacromolecules*. 2011;12(8):2881–2887.
128. Lee KY, Tammelin T, Schulfter K, Kiiskinen H, Samela J, Bismarck A. High performance cellulose nanocomposites: Comparing the reinforcing ability of bacterial cellulose and nanofibrillated cellulose. *ACS Appl. Mater. Interfaces*. 2012;4:4078–4086.
129. Gupta VK, Ali I, Chapter 2 – Water treatment for inorganic pollutants by adsorption technology, in: *Environmental Water – Advances in Treatment*, Remediation and Recycling, 2013, pp. 29–91.
130. Langmuir I. The adsorption of gases on plane surfaces of glass, mica and platinum. *J Am Chem Soc*. 1918;40:1361–1403.
131. Freundlich HMF. Über die Adsorption in Lösungen. *J Phys Chem*. 1906;57:385–470.
132. Worch E. *Adsorption Technology in Water Treatment Fundamentals, Processes, and Modeling*. Walter de Gruyter GmbH & Co. KG, Berlin, Germany, 2012.
133. Yu X, Tong S, Ge M, Wu L, Zuo J, Cao C, Song W. Adsorption of heavy metal ions from aqueous solution by carboxylated cellulose nanocrystals. *J Environ Sci*. 2013;25(5):933–943.
134. Yang R, Aubrecht KB, Ma H, Wang R, Grubbs RB, Hsiao BS, Chu B. Thiol-modified cellulose nanofibrous composite membranes for chromium (VI) and lead (II) adsorption. *Polymer*. 2014;55(5):1167–1176.
135. Zhu H, Jia S, Wan T, Jia Y, Yang H, Li Y, Yan L, Zhong C. Biosynthesis of spherical Fe_3O_4/bacterial cellulose nanocomposites as adsorbents for heavy metal ions. *Carbohydr Polym*. 2011;86(4):1558–1564.
136. Demirbas A. Heavy metal adsorption onto agro-based waste materials: A review. *J Hazard Mat*. 2008;157:220–229.
137. Sakairi N, Suzuki S, Ueno K, Han S-M, Nishi N, Tokura S. Biosynthesis of heteropolysaccharides by *Acetobacter xylinum* – Synthesis and characterization of metal-ion adsorptive properties of partially carboxymethylated cellulose. *Carbohydr Polym*. 1998;37:409–414.
138. Ma H, Hsiao BS, Chu B. Ultrafine cellulose nanofibers as efficient adsorbents for removal of UO_2^{2+} in water. *ACS Macro Lett*. 2012;1:213–216.
139. Carpenter AW, de Lannoy C-F, Wiesner MR. Cellulose nanomaterials in water treatment technologies. *Environ Sci Technol*. 2015;49:5277–5287.
140. Karim Z, Hakalahti M, Tammelin T, Mathew AP. In situ TEMPO surface functionalization of nanocellulose membranes for enhanced adsorption of metal ions from aqueous medium. *RSC Adv*. 2017;7:5232.

141. Ruan C, Strømme M, Lindh J. A green and simple method for preparation of an efficient palladium adsorbent based on cysteine functionalized 2,3-dialdehyde cellulose. *Cellulose*. 2016;23:2627–2638.
142. Sirviö JA, Hasa T, Leiviskä T, Liimatainen H, Hormi O. Bisphosphonate nanocellulose in the removal of vanadium(V) from water. *Cellulose*. 2016;23:689–697.
143. Kokol V, Božič M, Vogrinčič R, Mathew AP. Characterisation and properties of homo- and heterogenously phosphorylated nanocellulose. *Carbohydr Polym*. 2015;125:301–313.
144. Sehaqui H, Mautner A, de Larraya UP, Pfenninger N, Tingaut P, Zimmermann T. Cationic cellulose nanofibers from waste pulp residues and their nitrate, fluoride, sulphate and phosphate adsorption properties. *Carbohydr Polym*. 2016;135:334–340.
145. Hubbe MA, Beck KR, O'Neal WG, Sharma YC., Cellulosic substrates for removal of pollutants from aqueous systems: A review. 2. DYES. *BioResources*. 2012;7:2592–2687.
146. He X, Male KB, Nesterenko PN, Brabazon D, Paull B, Luong JHT. Adsorption and desorption of methylene blue on porous carbon monoliths and nanocrystalline cellulose. *ACS Appl. Mater. Interfaces*. 2013;5:8796–8804.
147. Karim Z, Mathew AP, Grahn M, Mouzon J, Oksman K. Nanoporous membranes with cellulose nanocrystals as functional entity in chitosan: Removal of dyes from water. *Carbohydr Polym*. 2014;112:668–676.
148. Batmaz R, Mohammed N, Zaman M, Minhas G, Berry RM, Tam KC. Cellulose nanocrystals as promising adsorbents for the removal of cationic dyes. *Cellulose*. 2014;21:1655–1665.
149. Ma H, Burger C, Hsiao BS, Chu B. Nanofibrous microfiltration membrane based on cellulose nanowhiskers. *Biomacromolecules*. 2012;13:180–186.
150. Ibrahim (Al-Khateeb) I, Al-Obaidi YM, Hussin SM. Removal of methylene blue using cellulose nanocrystal synthesized from cotton by ultrasonic technique. *Am Chem Sci J*. 2015;9:1–7.
151. Jin LQ, Li WG, Xu QH, Sun QC. Amino-functionalized nanocrystalline cellulose as an adsorbent for anionic dyes. *Cellulose*. 2015;22:2443–2456.
152. Syed S, Alhazzaa MI, Asif M. Treatment of oily water using hydrophobic nano-silica. *Chem Eng J*. 2011;167:99–103.
153. Liu H, Geng B, Chen Y, Wang H. A review on the aerogel-type oil sorbents derived from nanocellulose. *ACS Sustainable Chem Eng*. 2017;5:49–66.
154. Korhonen JT, Kettunen M, Ras RH, Ikkala O. Hydrophobic nanocellulose aerogels as floating, sustainable, reusable, and recyclable oil absorbents. *ACS Appl Mater Inter* 2011;3:1813–1816.
155. Zhou S, Liu P, Wang M, Zhao H, Yang J, Xu F. Sustainable, reusable, and superhydrophobic aerogels from microfibrillated cellulose for highly effective oil/water separation. *ACS Sustainable Chem Eng*. 2016;4:6409–6416.
156. Mulyadi A, Zhang Z, Deng Y. Fluorine-free oil absorbents made from cellulose nanofibril aerogels. *ACS Appl Mater Inter* 2016;8:2732–2740.
157. Zhang Z, Sèbe G, Rentsch D, Zimmermann T, Tingaut P. Ultralightweight and flexible silylated nanocellulose sponges for the selective removal of oil from water. *Chem Mater* 2014;26:2659–2668.
158. Dwivedi AD, Sanandiya ND, Singh JP, Husnain SM, Chae KH, Hwang DS, Chang Y-S. Tuning and characterizing nanocellulose interface for enhanced removal of dual-sorbate (AsV and CrVI) from water matrices. *ACS Sustainable Chem Eng*. 2017;5:518–528.
159. Yang H, Sheikhi A, van de Ven TGM. Reusable green aerogels from cross-linked hairy nanocrystalline cellulose and modified chitosan for dye removal. *Langmuir*. 2016;32:11771–11779.

160. Jin L, Sun Q, Xu Q, Xu Y. Adsorptive removal of anionic dyes from aqueous solutions using microgel based on nanocellulose and polyvinylamine. *Biores Tech.* 2015;197:348–355.
161. Anirudhan TS, Shainy F. Effective removal of mercury(II) ions from chlor-alkali industrial wastewater using 2-mercaptobenzamide modified itaconic acid-grafted-magnetite nanocellulose composite. *J Coll Interf Sci.* 2015;456:22–31.
162. Anirudhan TS, Deepa JR, Shainy F. Thorium(IV) Recovery from water and sea water using surface modified nanocellulose/nanobentonite composite: Process design. *J Polym Environ.* 2016; doi:10.1007/s10924-016-0892-2.
163. Jain P, Varshney S, Srivastava S. Site-specific functionalization for chemical speciation of Cr(III) and Cr(VI) using polyaniline impregnated nanocellulose composite: Equilibrium, kinetic, and thermodynamic modeling. *Appl Water Sci.* 2015;7(4):1827–1839.
164. Olivera S, Muralidhara HB, Venkatesh K, Guna VK, Gopalakrishna K, Kumar KY. Potential applications of cellulose and chitosan nanoparticles/composites in wastewater treatment: A review. *Carbohydr Polym.* 2016;153:600–618.
165. Karim Z, Mathew AP, Kokol V, Wie J, Grahn M. High-flux affinity membranes based on cellulose nanocomposites for removal of heavy metal ions from industrial effluents. *RSC Adv.* 2016;6:20644–20653.
166. Karim Z, Claudpierre S, Grahn M, Oksman K, Mathew AP. Nanocellulose based functional membranes for water cleaning: Tailoring of mechanical properties, porosity and metal ion capture. *J Membr Sci.* 2016;514:418–428.
167. Wei H, Rodriguez K, Renneckar S, Vikesland PJ, Environmental science and engineering applications of anocellulose-based nanocomposites. *Environ Sci Nano.* 2014;1:302–316.
168. Field R, Fundamentals of fouling, in: *Membrane Technology: Membranes for Water Treatment,* Vol. 4, Peinemann K. V., Nunes S. P. (Eds.). Wiley-VCH, Weinheim, Germany, 2010.
169. Luo J, Chan W-B, Wang L, Zhong C-J. Probing interfacial interactions of bacteria on metal nanoparticles and substrates with different surface properties. *Intern J Antimicrob Agents.* 2010;36(6):549–556.
170. Jönsson C, Jönsson A-S. Influence of the membrane material on the adsorptive fouling of ultrafiltration membranes. *J Membr Sci.* 1995;108(1–2):79–87.
171. Li H-J, Cao Y-M, Qin J-J, Jie X-M, Wang T-H, Liu J-H, Yuan Q. Development and characterization of anti-fouling cellulose hollow fiber UF membranes for oil–water separation. *J Membr Sci* 2006;279(1–2):328–335.
172. Meng X, Tang W, Wang L, Wang X, Huang D, Chen H, Zhang N. Mechanism analysis of membrane fouling behaviour by humic acid using atomic force microscopy: Effect of solution pH and hydrophilicity of PVDF ultrafiltration membrane interface. *J Membr Sci.* 2015;487:180–188.
173. Jardak K, Drogui P, Daghrir R. Surfactants in aquatic and terrestrial environment: occurrence, behaviour, and treatment processes. *Environ Sci Pollut Res.* 2016;23:3195–3216.
174. Hunter RJ. *Introduction to Modern Colloid Science.* Oxford University Press, New York, 1993; Berg JC. *An Introduction to Colloids and Interfaces – The Bridge to Nanoscience.* World Scientific Publishing, Singapore, 2012.
175. Ramsden W. Separation of solids in the surface-layers of solutions and 'Suspensions' (Observations on surface-membranes, bubbles, emulsions, and mechanical coagulation) – Preliminary account. *Proc Roy Soc.* 1903;72:156–164.
176. Pickering SU. CXCVI.—Emulsions. *J Chem Soc Trans.* 1907;91:2001–2021; Pickering SU. Über Emulsionen. *Z Chem Ind Koll.* 1910;7:11–16.

177. Lee KY, Bismarck A, Stoyanov SD, Paunov VN. Colloidal and nanocellulose-stabilized emulsions, in: *Handbook of Green Materials Vol 3: Self- and Direct-Assembling of Bionanomaterials, Materials and Energy*, Oksman, K., Mathew, A. P., Bismarck, A., Rojas, O., Sain, M., Qvintus, P. (Eds.), 2014, 5:185–196.
178. Oza KP, Frank SG. Microcrystalline cellulose stabilized emulsions. *J Dispersion Sci Technol*. 1986;7:543–561.
179. Lee KY, Blaker JJ, Murakami R, Heng JYY, Bismarck A. Phase behaviour of medium and high internal phase water-in-oil emulsions stabilized solely by hydrophobized bacterial cellulose nanofibrils. *Langmuir*. 2014;30:452–460.
180. Andresen M, Stenius P. Water-in-oil emulsions stabilized by hydrophobized microfibrillated cellulose. *J Dispers Sci Technol*. 2007;28:837–844.
181. Finkle P, Draper DH, Hildebrand JH. The theory of emulsification. *J Am Chem Soc*. 1923;45:2780–2788.
182. Schulman JH, Leja J. Control of contact angles at the oil-water-solid interfaces. *Trans Faraday Soc*. 1954;50:598–605.
183. Tambe DE, Sharma MM. Factors controlling the stability of colloid-stabilized emulsions. I. An experimental investigation. *J Colloid Interface Sci*. 1993;157:244–253.
184. Xhanari K, Syverud K, Chinga-Carrasco G, Paso K, Stenius P. Structure of nanofibrillated cellulose layers at the O/W interface. *J Colloid Interface Sci*. 2111;356:58–62.
185. Salas C, Nypelö T, Rodriguez-Abreu C, Carrillo C, Rojas OJ. Nanocellulose properties and applications in colloids and interfaces. *Curr Opin Colloid Interface Sci*. 2014;19(5):383–396.
186. Winuprasit T, Suphantharika M. Microfibrillated cellulose from mangosteen (*Garcinia mangostana* L.) rind: Preparation, characterization, and evaluation as an emulsion stabilizer. *Food Hydrocoll*. 2013;32:383–394.
187. Lee KY, Blaker JJ, Heng JYY, Murakami R, Bismarck A. pH-triggered phase inversion and separation of hydrophobised bacterial cellulose stabilised Pickering emulsions. *React Funct Polym*. 2014;85:208–213.
188. Dickinson E. Biopolymer-based particles as stabilizing agents for emulsions and foams. *Food Hydrocoll*. 2017;68:219–231.
189. Ojala J, Sirviö JA, Liimatainen H. Nanoparticle emulsifiers based on bifunctionalized cellulose nanocrystals as marine diesel oil–water emulsion stabilizers. *Chem Eng J*. 2016;288:312–320.
190. Nikfarjam N, Qazvini NT, Deng Y. Surfactant free pickering emulsion polymerisation of styrene in w/o/w system using cellulose nanofibrils. *Europ Polym J*. 2015;64:179–188.

7 Nanocellulose-Based Membranes for Water Purification

Fundamental Concepts and Scale-Up Potential

Aji P. Mathew, Peng Liu, Zoheb Karim and Jessica Lai

CONTENTS

7.1 Introduction 129
7.2 Nanocellulose for Water Treatment: State of the Art 131
7.3 Membranes for Water Treatment 134
7.4 Membrane Processing 136
  7.4.1 Monolayered Membranes/Adsorbents 136
  7.4.2 Multilayered Membranes 137
  7.4.3 Impregnated Electrospun Mats 138
7.5 Membrane Scale-Up 139
7.6 Membrane Module Design Concepts 141
7.7 Conclusion and Future Perspectives 144
Acknowledgements 144
References 144

7.1 INTRODUCTION

Considering the increasing demand of water in various sectors and in geographic regions, there is a clear need to develop new and sustainable technologies. While developing new water cleaning technologies, performance efficiency, energy efficiency, commercial viability, sustainability, resource efficiency and durability are of relevance [1]. Nanotechnology has been identified as a technology that could play an important role in resolving many of the problems involving water purification and quality [2,3]. Nanomaterials and nanostructures have nanoscale dimensions that range from 1 to 100 nm and often exhibit novel and significantly changed physical, chemical and biological properties.

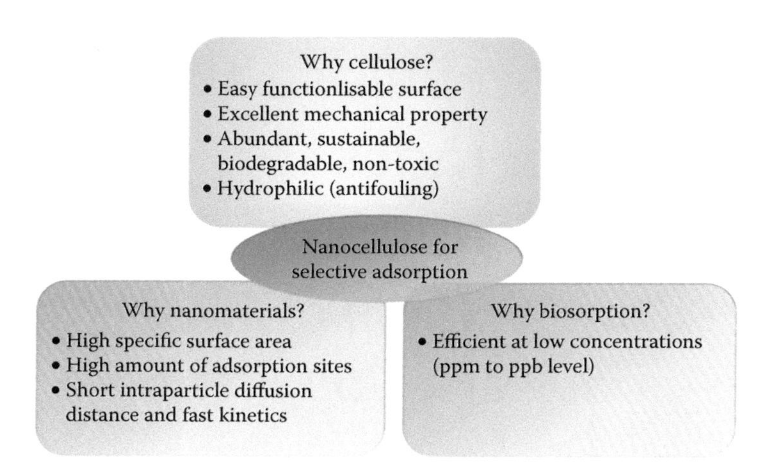

FIGURE 7.1 Summary of the rationale behind using nanocellulose for water cleaning.

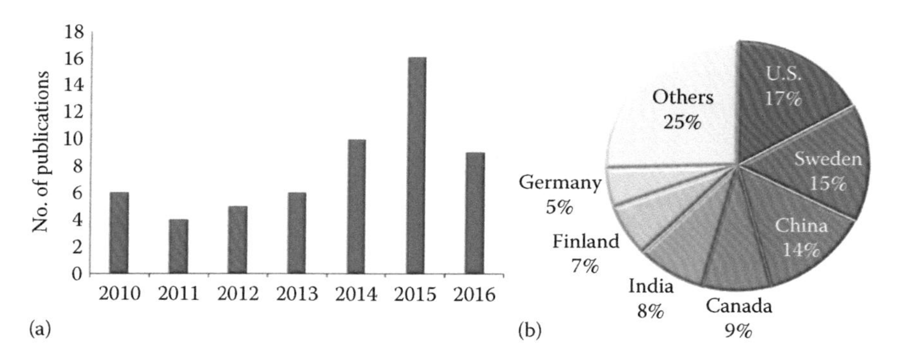

FIGURE 7.2 Overview of the activities on nanocellulose-based water cleaning. (a) Nanocellulose membrane publications, yearwise and (b) Nanocellulose membrane activities, countrywise. (From SCOPUS, October 2016.)

Nanocellulose, which offers a combination of biosorption, nanodimension and unique cellulosic nature, has a potential for a new and green route to solve the current water pollution problems. The rationale behind water cleaning approach using nanocellulose for selective adsorption of contaminants is summarised in Figure 7.1.

The research activity in this topic has increased dramatically in recent years. Figure 7.2 shows the publications available in the area of nanocellulose-based membrane research during the past seven years and the geographic distribution.

Removal of contaminants such as dyes, metal ions and microbes via size exclusion and charge-mediated adsorption are found in the literature [4–9]. The adsorption of organic contaminants, such as oils and cyclohexenes, has also been demonstrated with modified nanocellulose matrix that is grafted with hydrophobic or oleophilic functionalities [10,11]. Some challenges in the application of nanocellulose and its membranes in water purification may be related to its agglomeration, regeneration potential, immobilisation, cost effectiveness and long-term performance [12–14]. In this current review, water purification via adsorption process using nanocellulose and its membranes is focused, and separation or purification via size exclusion is not covered.

7.2 NANOCELLULOSE FOR WATER TREATMENT: STATE OF THE ART

Very few studies reported the contaminant adsorption properties of nanocellulose before 2011, and most of the studies on this topic were published in the past few years. Saito et al. [15] reported that the removal of toxic metal ions can be achieved by using 2,2,6,6-tetramethylpiperidine-1-oxyl (TEMPO)-oxidised cellulose nanofibres (TOCNFs) as adsorbents and showed the potential of these materials in water purification.

Liu et al. [16] reported that both unmodified cellulose nanocrystals (CNCs) and cellulose nanofibres (CNFs) have the capacity to adsorb Ag(I) in aqueous environment. However, CNCs (34 mg/g) have higher Ag(I) adsorption capacity than nanofibres (14 mg/g), due to the $-HSO_3^-$ functional groups on the surface of CNC that are created by sulphuric acid hydrolysis. The adsorption behaviour was found to be pH dependent, and the best adsorption performance was found observed near neutral pH, because H^+ competes with Ag^+ for the adsorption onto nanocrystals and nanofibres at acidic conditions. It was also found that unmodified CNF has the capacity to adsorb a wide range of heavy metal ions, including Pb(II), Cu(II), Cr(II) and Zn(II). However, the adsorption capacities of the metal ions onto CNF differ a lot in different reports [17,18].

In order to increase the adsorption capacities of pristine nanocellulose, a variety of functional groups were introduced onto the surface of different types of nanocellulose via chemical or enzymatic modifications. For example, the major metal binding groups for biosorption, such as carboxyl group, sulphonate group, phosphonate group, were introduced onto nanocellulose via surface modification for the purpose of water remediation [16,19–21].

So far, introduction of carboxylate group onto nanocellulose is the most studied method for increasing their adsorption behaviour. Significant amount of C6 carboxylate groups can be selectively formed on the surface of CNFs by TEMPO-mediated oxidation without changes in the original crystallinity. Especially, Pb(II), La(III) and Ag(I) ions were preferably captured onto TOCNF as counterions of the carboxylate groups with the metal ion/carboxylate molar ratio of about 1:1 [15]. Sehaqui et al. [18] did a systematic study on the adsorption of Cu(II) onto TOCNF as a function of pH conditions and carboxylate group content on the fibril surface (or CNF oxidation degree). The authors revealed that Cu(II) adsorption onto nanofibres increased linearly with the carboxylate content at a given pH condition. In addition, after adsorption, copper nanoparticles (*d*: 200–300 nm) were widely observed on the surface of oxidised nanofibres. The adsorbed Cu(II) ions could be desorbed from nanofibres through an acidic washing so that nanofibre could be easily reused [18,19].

Yu et al. [22] reported that the introduction of carboxylate groups onto its surface dramatically enhances the binding capacity of Pb(II) (461 mg/g) and Cd(II) (364 mg/g). The kinetics of Pb(II) and Cd(II) adsorption process was very fast (within 5 min) and can be well explained by the pseudo second-order kinetic model, and the adsorption isotherm can be well explained by Langmuir isotherm, which likely indicates a chemical adsorption between adsorbate and adsorbent. When the size of bioadsorbents is reduced to nanoscale, short intraparticle diffuse distance is expected to provide a great advantage for metal ion removal kinetics. The fast kinetics also indicates the fast sorption reaction between the functional groups and metal ions [14,22].

The metal adsorption selectivity of succinic anhydride-modified mercerised nanocellulose was investigated by Hokkanen et al. [23]. The maximum metal adsorption ranged from 0.72 to 1.95 mmol/g following the order: Cd(II) > Cu(II) > Zn(II) > Co(II) > Ni(II). Moreover, modified nanocellulose was successfully regenerated via ultrasonic treatment with regeneration efficiencies ranging from 96% to 100%. Zhang et al. [24] grafted carboxylate group on the surface of bamboo CNFs by coupling acrylic acid molecules and sequentially utilised bamboo CNFs-graft-poly(acrylic acid) as biosorbents for removal of Cu(II) from aqueous solutions. The adsorption capacity of modified bamboo CNFs was improved three times higher than that of pristine bamboo CNFs.

Liu et al. [26] reported that enzymatically phosphorylated CNCs and CNFs derived from industrial residue showed great potential for immobilisation of Ag(I), Cu(II) and Fe(II), and both nanocrystals and nanofibres exhibited the same adsorption selectivity (Ag(I) > Fe(III) > Cu(I)). It was claimed that both phosphorylated CNCs and CNFs show the capacity to reduce Cu(II) and Fe(III) concentrations in mirror industry effluent to drinking water level. Figure 7.3 shows the dramatic enhancement of Cu(II) adsorption capacity (≈5,000 times) after 'nanofication' and surface functionalisation of cellulose.

Xanthated nano banana cellulose was reported by Pillai et al. [27] as an efficient biosorbent for the removal of Cd(II). The nano banana cellulose was prepared by the steam explosion of banana fibre followed by acid hydrolysis and xanthation. The biosorption of Cd(II) by nano banana cellulose is done through ion exchange or

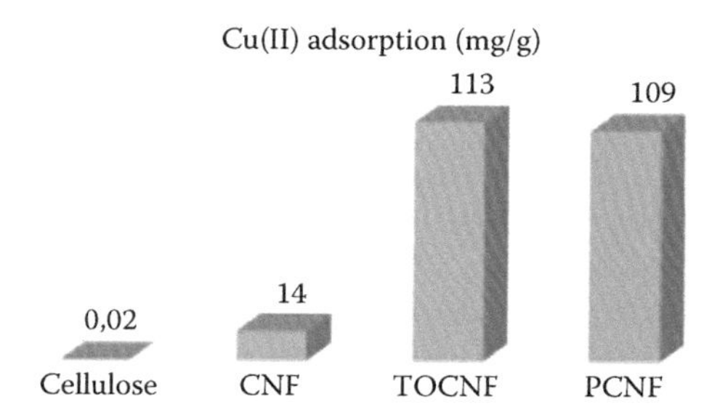

FIGURE 7.3 Comparison of Cu(II) adsorption capacities among cellulose, nanocellulose and surface-modified nanocellulose. (From Sehaqui, H. et al., *Cellulose*, 21, 4, 2831–2844, 2014; Ulmanu, M. et al., *Water Air Soil Poll.*, 142, 1–4, 357–373, 2003; Liu, P. et al., *J. Hazard. Mater.*, 294, 177–185, 2015.)

complexation or by the combination of both processes. Hokkanen et al. [28] investigated the adsorption properties of aminopropyltriethoxysilane (APS)-modified microfibrillated cellulose (MFC) in aqueous solutions containing Ni(II), Cu(II) and Cd(II) ions. According to the Sips isotherm model that provided the best fit to the experimental adsorption data, the maximum adsorption capacities are 3.09, 2.59 and 3.47 mmol/L for Ni(II), Cu(II) and Cd(II), respectively. Unlike other studies that focus on the adsorption behaviour of nanocellulose suspensions, He et al. [29] demonstrated that quaternary ammonium-functionalised CNF aerogel could well retain the large specific surface area, in combination with the unique porous structure, leading to the rapid and effective removal of Cr(VI) from water. Chemical cross-linking made the nanocellulose aerogel adsorbent mechanically robust. Unlike nanocellulose suspensions, aerogel could be easily separated from water after adsorption without complicated centrifugation or filtration process. After four cycles of adsorption–regeneration tests, Cr(VI) adsorption capacity was decreased by 7%.

Apart from positively charged heavy metal ions, modified nanocellulose could be used as adsorbents to capture negatively charged contaminants in water, such as nitrate, phosphate, fluoride, sulphate arsenic(V) and humic acid. Sehaqui et al. [21] prepared cationic CNFs with ammonium content in the range 0.34–1.2 mmol/g. The chemical modification reversed the surface charge of the original CNF from negative to positive values and rendered the nanofibres' displaying ability to adsorb negatively charged ions including nitrate, phosphate, fluoride, sulphate via electrostatic interactions. As expected, the adsorption capacities of all the anions increase with the content of surface charge on the surface of CNF. In the presence of mono- and multivalent ions, the cationic nanofibres displayed a higher selectivity towards multivalent ions (PO_4^{3-} and SO_4^{2-}) than monovalent ions (F^- and NO_3^-). Cationic CNF also displayed a good ability to adsorb humic acid, a widely spread natural organic matter, via electrostatic interactions. Maximum humic acid adsorption onto cationic CNF is 310 mg/g (i.e. 1.2 mmol/g), which stands among the highest capacities reported in the literature. A higher adsorption capacity and faster adsorption kinetics were observed at low pH values, which were attributed to the coiled conformation of humic acid in this pH range allowing a thicker humic acid layer to be adsorbed onto CNF. It is shown that the cationic CNF with adsorbed humic acid can be used for the preparation of porous foams via freeze drying. These foams have a good affinity to pollutants such as positive dyes [30].

Hokkanen et al. [28] reported that the maximum removal capacities of carbonated hydroxyapatite-modified nanocellulose as adsorbent for PO_4^{3-} and NO_3^- were 0.843 and 0.209 mmol/g, respectively. The same group also modified nanocellulose by APS for the removal of hydrogen sulphide (H_2S) from the aqueous solutions. The uptake of H_2S was found to be 103.95 mg/g, and the optimum pH was found to be 6 [28]. In another study [31], magnetic iron nanoparticle-modified nanocellulose was used for arsenate (As(V)) removal from aqueous solutions. The monolayer adsorption capacity of the adsorbent, as obtained from the Langmuir isotherm, is 2.460 mmol/g, which showed great As(V) uptake properties compared to original iron nanoparticles. However, it is worth to mention that although different types of nanocellulose and their modified derivatives can bind a wide range of contaminants in aqueous environment, only those with sufficiently high binding capacity and selectivity for the pollutants are suitable for use in a full-scale biosorption process. In addition, great efforts have to be made to improve immobilisation technique under the condition of continuous wastewater treatment process for industrial application of biosorption [32,33].

7.3 MEMBRANES FOR WATER TREATMENT

Membrane technologies are getting more and more attention nowadays especially in water and wastewater treatment processes due to their reliable contaminant removal without producing any harmful by-products [34]. A good membrane should be designed to have a high and stable flux even at low transmembrane pressures and without any pre-treatment before use; in addition, the selectivity should be high. Membranes may be classified according to characteristic pore size (Figure 7.4). Microfiltration (MF) and ultrafiltration (UF) membrane systems have been the most attractive alternatives because they offer quick and selective separation on removing suspended particles, pathogenic agents and dissolved macromolecules, such as proteins and polysaccharides, from water. Nanofiltration (NF) membranes are relatively new and are sometimes called 'loose' reverse osmosis (RO) membranes. They are generally porous membranes, but as the pores are on the order of nanometres, they show performance in between that of RO and UF membranes [34].

In membrane technology, the term 'nanostructured membrane' (NSM) is used for any membrane with engineered nanosized structures (e.g. pores) according

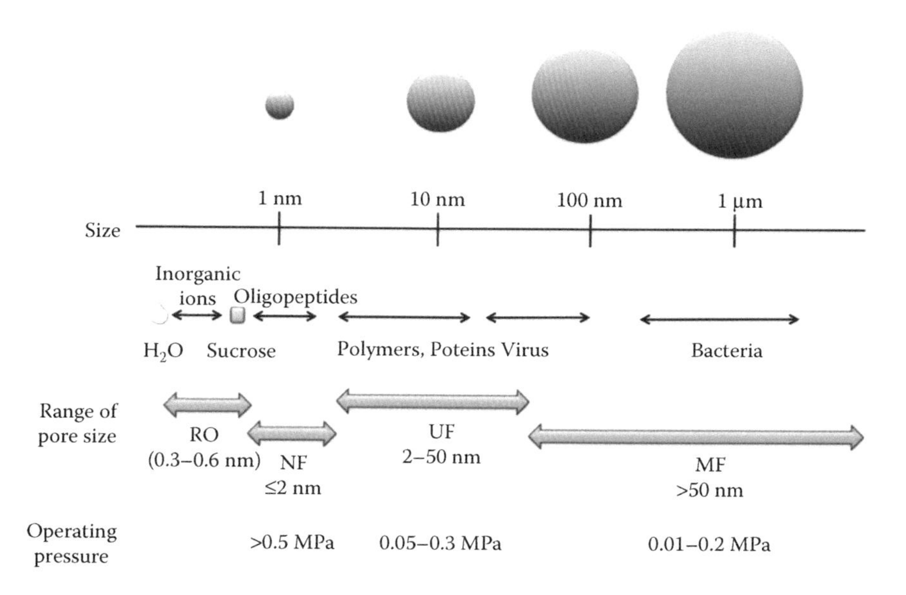

FIGURE 7.4 Membrane characterisation based on pore sizes, target species and used applied pressure for separation. (Reprinted from *Polymer*, 47, Ulbricht M. Advanced functional polymer membranes, 2217–2262, Copyright 2006, with permission from Elsevier.)

to the current ISO definition of nanomaterials, whereas nano-enhanced membranes (NEMs) are those with integrated nanoparticles. NEMs have shown potential in removing particles, bacteria/viruses, (natural) organic matter, hardness, metals (such as arsenic, uranium, lead, chromate), sulphate and a number of organic and inorganic substances (such as pesticides and some pharmaceuticals) sometimes even in one single treatment step. Rather than working purely on sieving mechanism, functional groups present on the surface of membranes are responsible for binding and usually have properties similar to adsorbent materials to bind and concentrate selected pollutants from aqueous solutions [14]. By using nanoparticles as functional entities, porous membranes could be used for very precise and continuous perm-selective separations based on differences in functional groups. Appropriate surface modification will enable these NEMs to enhance the sorption selectivity of the membranes. Because of increasing environmental consciousness and demands of reusable and degradable products, it is desirable to replace traditional membranes that are usually made up of non-degradable and toxic materials with bio-based ones.

The fabrication of membranes/composites by incorporation of nanocellulose as 'functional material' is a new application area in the field of membrane technology. Researchers from Stony Brook University reported in 2010 about the fabrication of multilayered and thin-layered polysaccharide nanofibrous membranes where nano-sized celluloses (fibres and crystals) have been used as functional additives. MF membranes composed of polymeric fibrous scaffold were doped with functional ultrafine CNF. The advantages of using cellulose and/or chitin include enhanced mechanical and thermal stability as well as specific functional properties such as adsorption selectivity. The mechanism of purification is based on the capability of the nanocellulose and/or nanochitin (with or without functionalisation) to selectively adsorb, store and desorb contaminants from industrial water and drinking water while passing through a nanoporous, ultraporous or semi-permeable membrane. The mechanism responsible for binding of pollutants using functional materials, although understood only to a limited extent, may be one or combination of ion exchange, complexation, co-ordination, adsorption, electrostatic interaction, chelation and microprecipitation. Functional groups containing elements such as oxygen, nitrogen, sulphur or phosphorous tend to have affinity and function as ligands. The most important pollutant-binding groups for nanocellulose-based functional membranes are listed in Table 7.1.

Ma et al. [36] hypothesised that in CNF-based membranes/filters, the liquid flux was enhanced due to the formation of directed water channels at the interface between hydrophilic CNF and hydrophobic polymer matrix in barrier layer. The so-called 'directed water channels' in barrier layer have much higher capacity of water passage or permeability than the conventional tortuous channels that are formed inside the matrix of highly cross-linked polymer molecules. Water flux through the highly permeable membrane is 3–10 times higher than those of commercial UF membranes but with a comparable reject ratio. Our studies have also shown high flux through nanocellulose-incorporated membranes and are attributed to (1) the inherent hydrophilicity of nanocellulose and (2) an increase in hydrophilicity after adsorption of charged species on the membranes [19,39]. The membrane

TABLE 7.1
Major Functional Groups Presented in/on Affinity Membranes Responsible for Pollutants Binding

Name	Structural Formula	References
Hydroxyl	–OH	[35,36]
Carboxyl	–COOH	[36]
Sulphonate	$-SO_3^-$	[37]
Phosphoryl	$-PO_3^{2-}$	[38]
Amine	$-NH_2$	[8]

performance is however significantly dependent on the membrane processing routes and on the subsequent pore structure that is obtained through the process. Some examples of membrane processing and its effect on pore structure and membrane performance are summarised in Section 7.4.

7.4 MEMBRANE PROCESSING

7.4.1 Monolayered Membranes/Adsorbents

One concept when using nanocellulose-based membranes as adsorption membranes is to develop structures with high porosity and macropores where the water flux can be high even at low pressures, whereas the surface functionality from the nanocrystals or nanofibres leads to rejection of contaminants. Figure 7.5 shows the porous structure that allows passage of water molecules (blue dots) and rejects/captures contaminants (red dots). Karim et al. [37] developed freeze-dried flat sheets from CNCs with an aim to have a porous structure with easy accessibility to the nanocrystal surfaces. The membranes had thickness of approximately 250 μm, low specific surface area (3 m^2/g) and were in UF range (average pore size of 13–17 nm). Despite the porous structure, the water flux of fabricated membranes was low (64 $L/m^2/h/MPa$) and attributable to the high thickness of fabricated membranes. Low mechanical stability was another drawback and was probably due to the absence of H-bonding after the freeze-drying process. The membranes however showed good adsorption capability towards positively charged dyes as shown in Figure 7.5.

Another approach is to develop a 'network' or 'paper' with functional nanocellulose where the whole structure acts as adsorbent or tight UF membranes with pore sizes in the range of 5–10 nm or NF membranes. These membranes can effectively remove metal ions (Cu^{2+}, Ag^+, Fe^{3+}), nitrates, fluorides, phosphates, sulphates, humic acid and even organic compounds [21,40,41], depending on the functionality on nanocellulose and pore sizes of the membranes.

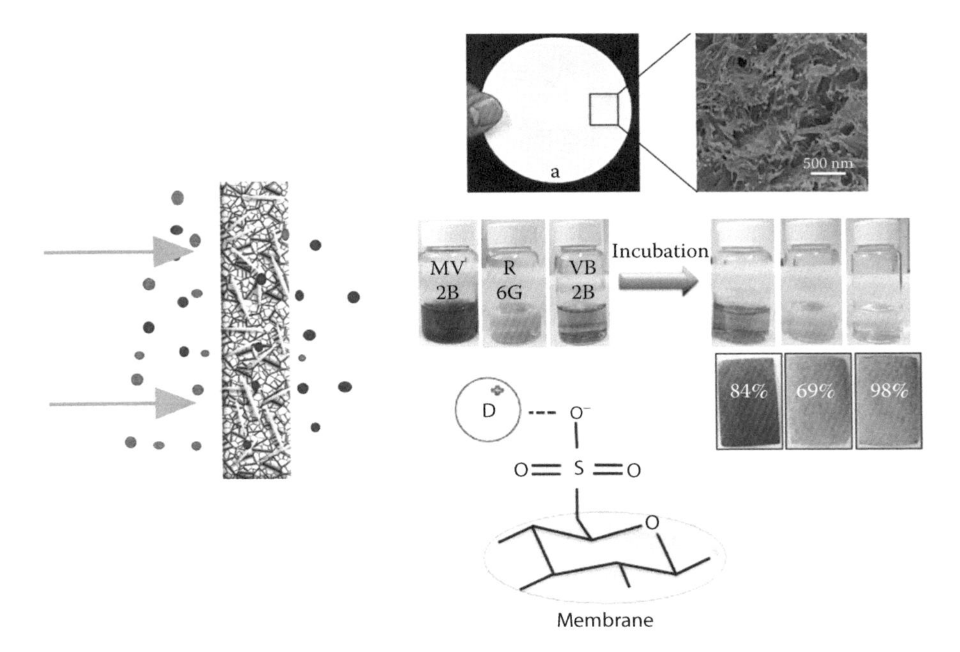

FIGURE 7.5 **(See colour insert.)** The removal of dyes (MV 2B: methyl violet 2B; R 6B: Rhodamine 6B; VB 2B: Victoria blue 2B) was studied in static mode for 24 h. Highest removal of dyes is found for VB 2B (98%). A possible electrostatic bonding (dotted line) between positively charged dyes (D+) and negatively charged membranes due to the presence of SO_3^- groups after sulphuric acid hydrolysis is found. (From Karim, Z. et al., *Carbohydr. Polym.*, 112, 668–676, 2014.)

7.4.2 Multilayered Membranes

Layered membranes are developed as an attempt to increase the membrane flux while providing easy accessibility to functional groups on the nanocellulose. In such membranes, it is highly relevant to tailor a porous support layer to overcome the flux decay that occurs due to the thick support layer. Similarly, a tight but thin functional layer is beneficial to achieve the permeation selectivity or adsorption at the top functional layer.

Karim et al. [38] have shown the potential of these types of membranes in water purification, with special focus on metal ions. The use of processes such as vacuum filtration, dip coating, roll coating and their combinations is found to be efficient in making layered membranes from bio-based nanofibres. The use of microfibres in the support layer, acetone treatment of membranes and so on is employed to tailor the pore structure, whereas use of nanocellulose with sulphate, carboxyl or phosphoryl functionalities provided the adsorption efficiency. Figure 7.6 shows the schematic representation of the multilayered membrane and the scanning electron microscope (SEM) images, showing the surface functional layer with CNCs.

These approaches were found to have potential to tailor the pore sizes of the membranes being in nanometre range (5–50 nm) when CNFs were used or in micrometer range (5–50 μm) when microfibres or microfibres with nanofibres were used to form

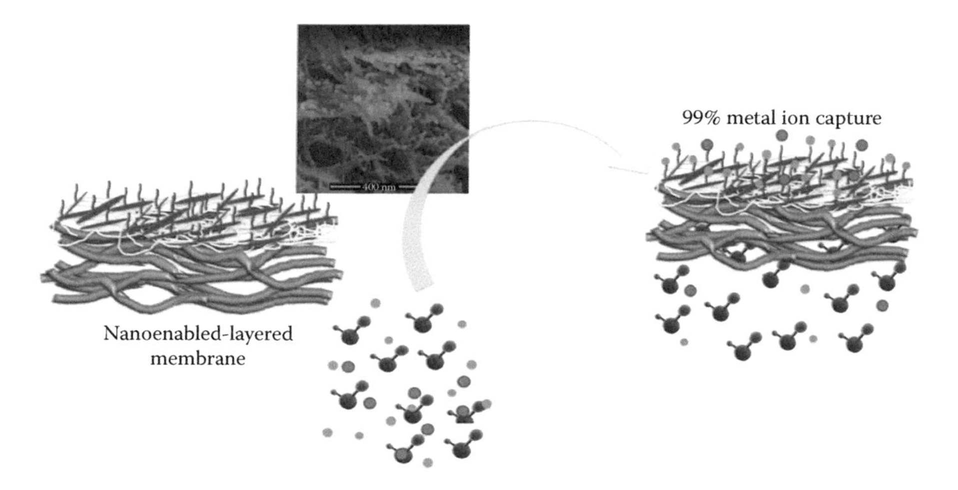

FIGURE 7.6 Schematic representation of the multilayered membrane with coarse support layer and functional layer on the surface. The SEM image shows the functional layer with CNCs. The functional layer captures the pollutants with high efficiency during cross flow filtration. (Karim, Z. et al., *RSC Ad.*, 6, 25, 20644–20653, 2016. Reproduced by permission of The Royal Society of Chemistry.)

the support layer. It is worth mentioning that the flux through the membranes was as low as 19 L/m^2/h/MPa in the case of CNF-based support layer. In the case of these with pores in nanometre range, enhanced ion removal efficiency was observed, which is attributed to a shift to size exclusion following an initial adsorption process and aggregation of metal ions. This phenomenon was supported by density function theory (DFT) modelling of adsorption on functionalized cellulose [42].

7.4.3 Impregnated Electrospun Mats

The electrospun fibrous membranes have also shown potential as an excellent support layer for a secondary active surface – by coating individual fibres or by creating a more complete secondary layer of a bilayered composite membrane. Wang et al. [6] fabricated membranes using ultrafine cellulose nanowhiskers (nanocrystals) infused into an electrospun polyacrylonitrile (PAN) nanofibrous scaffold supported by a mechanically strong polyethylene terephthalate (PET) non-woven substrate. The membranes had pore sizes in the range of 0.22 μm and a water flux of 59 L/m^2/h/kPa. The impregnated CNCs possessed very high negative surface charge density due to surface carboxyl groups that are grafted during TEMPO oxidation, which provided high adsorption capacity to remove positively charged species, such as crystal violet (CV) dye. They also showed that the membranes are capable of retaining bigger pollutants such as bacteria (*E. coli*: 0.5 μm diameter × 2.0 μm length; *Brevundimonas diminuta*: 0.3 μm diameter × 0.9 μm length) by size exclusion, whereas smaller pollutants such as bacteriophage MS2 (pI = 3.9) was retained due to affinity with the negatively charged membrane. Goetz et al. [39] have developed CA-based electrospun mats that were impregnated with

TABLE 7.2
A Comparative Study of Processing Routes and Performance of Fabricated Membranes

Routes	TS (MPa)	Thickness (µm)	Pore Size (nm)	Water Flux ($L/m^2/h/MPa$)	Working Range
Freeze-drying	1.1 ± 0.3	250	10–13	64.0	UF
Vacuum filtration/ Dip coating	53 ± 0.2	200	2.5–78	19.0	Tight UF
Vacuum filtration	19 ± 0.8	440	5,100–6,200[a]	>500	MF
Impregnated electrospun mats	3.3 ± 0.5	170	10–11 nm	2,790	UF

[a] Determined by bubble point method.

CNCs (sulphate functional groups) and chitin nanocrystals (amino groups), giving rise to a highly porous membrane. Our study has shown that stand-alone and bio-based electrospun membranes of cellulose acetate with sufficient mechanical properties and flux as well as antifouling performance can be obtained via impregnation using chitin or CNCs (Table 7.2).

7.5 MEMBRANE SCALE-UP

To develop the commercially competitive and high-performance membranes made from nanocellulose, a strong industry involvement supporting scale-up activities is needed. There are several different routes for the up-scaling of nanocellulose-based membranes. The choice of production process will influence the final properties of the membranes. Very few reports are available for the production of nanocellulose-based film/membranes at pilot-scale setup.

Researchers of Innventia AB, Sweden are currently working together with several industrial companies to identify and implement technologies for pilot production of nanocellulose-based membranes. The development of the test bed that allows a flexible manufacturing and the evolution of different types of nanocellulose-based membranes would be the crucial steps towards up-scaling and commercialisation.

A paper-making approach has been tried by Mathew and co-workers from Lulea University of Technology, Sweden for manufacturing nanocellulose-based membranes using experimental paper-making machine (XPM) in MoRe Research pilot-scale facility, Ornskoldsvik, Sweden under EU FP7 Nanoselect project [43]. CNCs and CNFs with carboxyl surface functionality, namely CNC_{BE} (CNC isolated using integrated step of bioethanol production plant) and TEMPO CNF (TEMPO-oxidised CNF) were used to fabricate the membranes. Two approaches were successfully done with the XPM: (1) support layer fabricated using sludge microfibres collected from Domsjö Fabriker AB, Ornskoldsvik, Sweden coated with functional nanocellulose on both sides of support layer and (2) hybrid membranes, where CNC_{BE} with sludge microfibres fabricate membranes according to desired

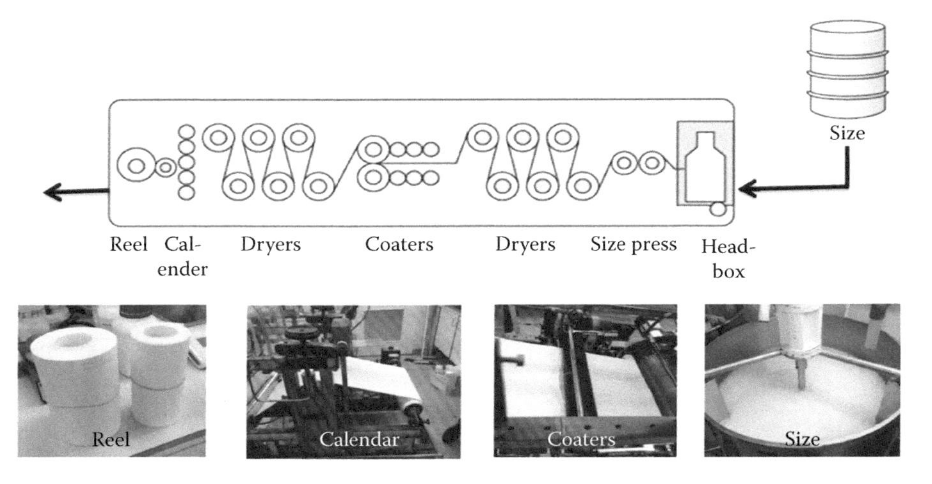

FIGURE 7.7 Diagrammatic representation of XPM having various setups. Images shown here are related to coated membranes that are fabricated by researcher of Lulea University of Technology, Sweden.

grammage (80–130 gsm). The diagrammatic representation of XPM is given in Figure 7.7, and some images have also been shown here to understand the processing steps for coated membranes.

The XPM-fabricated membranes were more homogeneous with controlled pore-size distribution (measured by bubble point technique) compared to lab-scale production setup. Excellent adsorption capacity of metal ions was also recorded, indicating their potential for water purification application. Recently, VTT Technical Research Centre of Finland in collaboration with Aalto University has developed a pilot-scale method to manufacture nanofibrillated cellulose film. This method enables industrial-scale roll-to-roll production of the film, which is suitable for specific packaging applications, or can be used in several added value applications of printed electronics and diagnostics. The nanocellulose films are translucent, showing no shrinkage or defects. The high smoothness of the surface provides excellent printing quality, and the densely packed structure results in a material with outstanding oxygen barrier properties. Based on their properties, the potential applications for these films are numerous, being high-performance packaging, flexible displays and printable electronics or low cost diagnostics. Figure 7.8a displays the pilot-scale setup; it produced nanofibrillated cellulose film in VTT, Finland (Figure 7.8b).

The high aspect ratio of the fibrils makes nanocellulose a gel already at low concentrations and may require addition of inorganic nanoparticles with nanocellulose for production of composite membranes/film [45–47]. Furthermore, 100% loading of nanocellulose on conveyor belt (wire mesh) is a big hurdle if native nanocellulosic membranes/films are produced. The biggest challenge in scaling-up of nanocellulose-based membranes is the removal of water and drying. This challenge could be overcome using VTT method, where nanofibrillated cellulose films were manufactured by evenly coated fibril cellulose on plastic films (to avoid filtration)

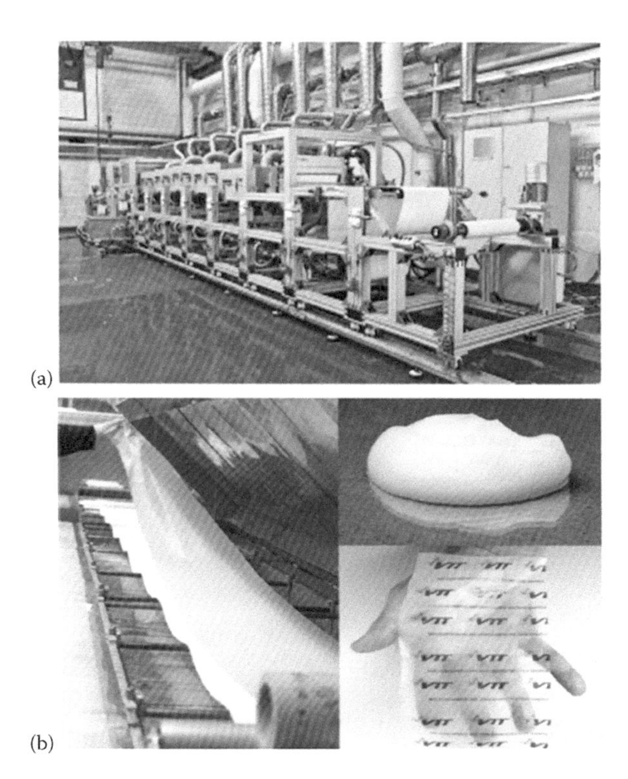

FIGURE 7.8 (a) Pilot-scale setup of production unit in VTT and (b) fabricated nanofibrillated cellulose film. (Courtesy of VTT Technical Research Centre of Finland, Espoo, Finland.)

so that the spreading and adhesion on the surface of the plastic could be controlled. The films were dried in a controlled manner by using a range of existing techniques. The more fibrillated cellulose material was used, the more transparent films could be manufactured. The even surface of the underlying plastic surface was almost completely replicated on nanocellulose film surface as shown in Figure 7.8b.

7.6 MEMBRANE MODULE DESIGN CONCEPTS

Although new membranes with improved permeability and selectivity are being developed, it is essential that these membranes can be assembled for separation processes in industrial applications. To provide large surface area for membrane processes, membranes are packed into individual units known as membrane modules. The design of an efficient and economical module with high surface area is crucial for successful commercialisation of membrane processes [48].

The four most common types of modules are plate-and-frame, spiral wound, tubular and hollow fibre modules. The choice of the module type is largely based

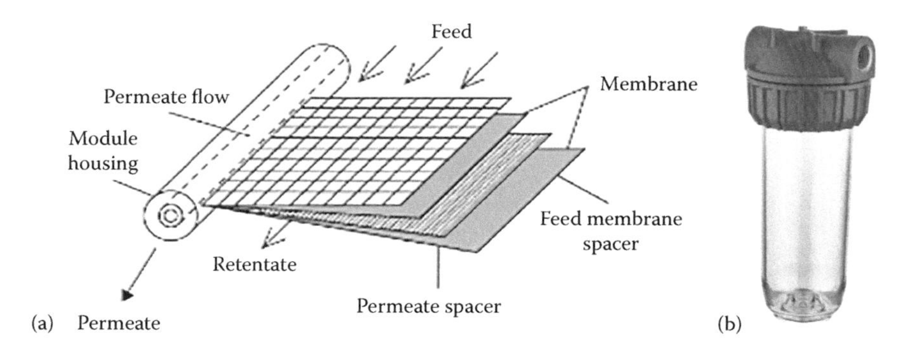

FIGURE 7.9 Schematic diagram of: (a) spiral wound module (From Bottero, J.Y. et al., *Integr. Environ. Assess. Manag.*, 2, 4, 391–395, 2006.) and (b) the cartridge housing.

on the configuration of the membranes and the targeted applications. For flat sheet membranes, the spiral wound modules are usually configured. These are popular due to their low cost, ease of replacement and modularity, which allow scaling-up by simply adding more units in parallel. The construction of a spiral wound module consists of membrane with spacers glued to build an envelope that creates passages and barriers to distribute the feed and to collect the permeate as shown in Figure 7.9 [49,50]. Given the specialised functionalities of the newly formulated nanocellulose flat sheet membranes/adsorbents (also known as nanopaper), the removal of toxic chemicals, heavy metal ions, pesticides and fertilisers from contaminated industrial, ground, surface and drinking water can be customised for individual industry. The development of point-of-use or point-of-entry modules targeting specific contaminants for each industry will be a major advancement in water management. In order to facilitate the introduction of the new product into the water market, the module design is based on existing commercial size and can be placed into standard cartridge housings as shown in Figure 7.9, which are already present in the market. A typical size of a cartridge module is 10″ in height and 63 mm in diameter.

The design of the nanocellulose membrane module is based on a spiral wound structure, given its advantage of high packing density. The module is further customised into two different modes, depending on the formulations of the membranes. For the hybrid nanocellulose membranes fabricated with nanocellulose and sludge fibres, the modules are made with through mode (Figure 7.10) to maximise the capacity for dynamic adsorption as these membranes have considerably high flux. The feed passes through the membrane from the outermost layer, then flows along the spiral wound path that has less resistance. Permeate is collected from the perforated tube.

For the nanopapers coated with functionalised nanocellulose, the modules are made with pass over mode (Figure 7.11) to maximise the contact area for ion adsorption. With minimal restriction to the flow, the chance of pressure building up within the module is minimised. Feed water enters from the side of the module, flows along the entire spiral wound path for adsorption and exits from the perforated tube.

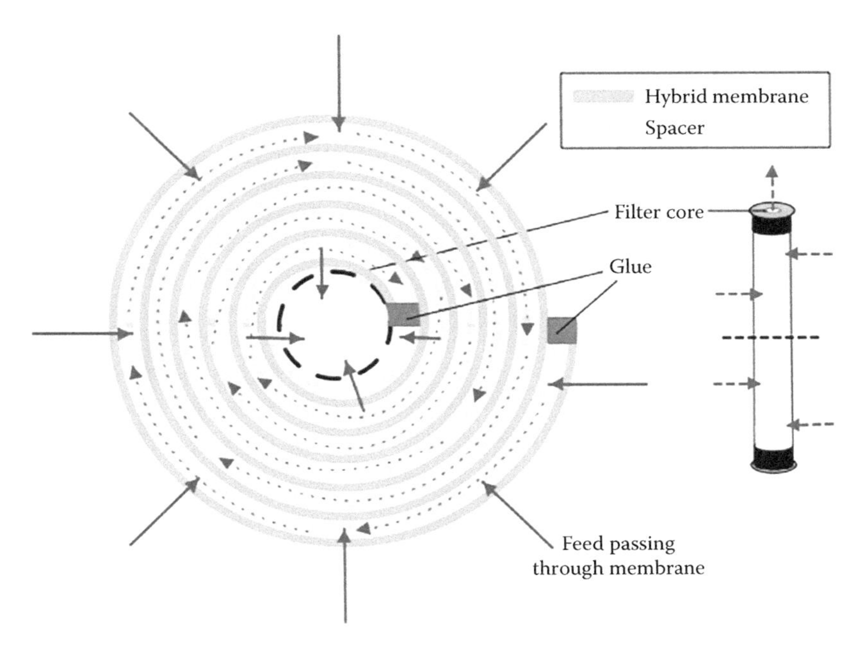

FIGURE 7.10 Schematic diagram of the nanocellulose membrane module with pass through mode.

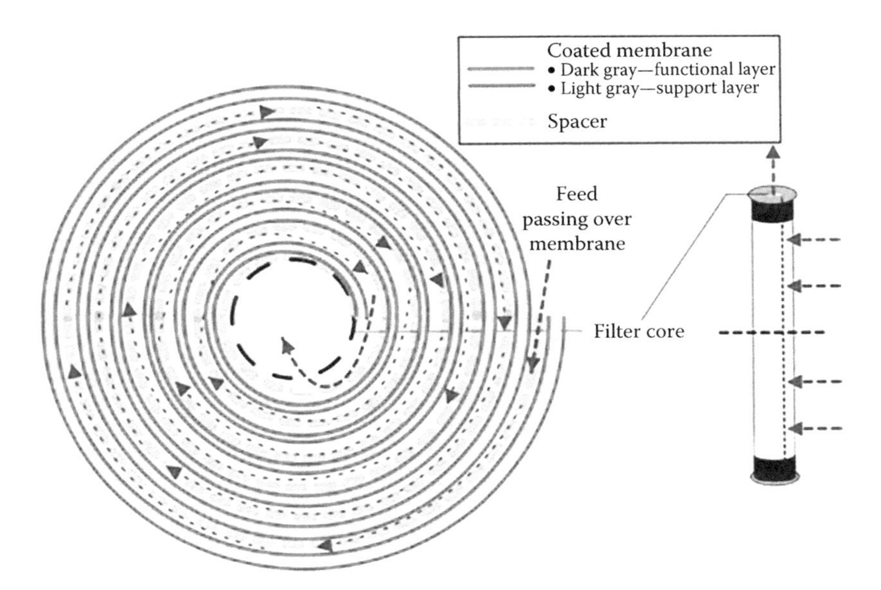

FIGURE 7.11 Schematic diagram of the nanocellulose membrane module with pass over mode.

7.7 CONCLUSION AND FUTURE PERSPECTIVES

The review shows that CNFs, CNCs, their surface-modified versions and the combinations thereof have potential use as adsorbents for water purification. The mechanism of water purification is driven by surface interaction of cationic or anionic nanocellulose entities with charged species in aqueous medium. The surface functionality of nanocellulose combined with its ability to form mechanically stable flat sheets has been exploited to fabricate membranes for water cleaning applications. The scaling-up of the process to pilot scale and fabrication of spiral wound modules have been evaluated for these types of membranes successfully. The challenges for commercial-scale utilisation of these technologies are related to control of pore structure and homogeneity during scaling-up, regeneration of membranes, long-term performance and recovery of valuable resources, and safety aspects related to use of nanosized particles are to be evaluated.

ACKNOWLEDGEMENTS

All authors gratefully acknowledge the financial support of the European Commission, under the NanoSelect Project, EU FP7-NMP4-SL-2012-280519. A. Mathew, P. Liu and Z. Karim also acknowledge the financial support from Wallenberg Wood Science Centre, Sweden.

REFERENCES

1. Weber K, Stahl W. Improvement of filtration kinetics by pressure electrofiltration. *Separation and Purification Technology*. 2002;26(1):69–80.
2. Savage N, Diallo MS. Nanomaterials and water purification: Opportunities and challenges. *Journal of Nanoparticle Research*. 2005;7(4–5):331–342.
3. Bottero JY, Rose J, Wiesner MR. Nanotechnologies: Tools for sustainability in a new wave of water treatment processes. *Integrated Environmental Assessment and Management*. 2006;2(4):391–395.
4. Carpenter AW, De Lannoy CF, Wiesner MR. Cellulose nanomaterials in water treatment technologies. *Environmental Science and Technology*. 2015;49(9):5277–5287.
5. Ma H, Hsiao BS, Chu B. Ultrafine cellulose nanofibers as efficient adsorbents for removal of UO_2^{2+} in water. *ACS Macro Letters*. 2012;1(1):213–216.
6. Wang R, Guan S, Sato A, Wang X, Wang Z, Yang R et al. Nanofibrous microfiltration membranes capable of removing bacteria, viruses and heavy metal ions. *Journal of Membrane Science*. 2013;446:376–382.
7. Salama A, Shukry N, El-Sakhawy M. Carboxymethyl cellulose-g-poly(2-(dimethylamino) ethyl methacrylate) hydrogel as adsorbent for dye removal. *International Journal of Biological Macromolecules*. 2015;73(1):72–75.
8. Jin L, Li W, Xu Q, Sun Q. Amino-functionalized nanocrystalline cellulose as an adsorbent for anionic dyes. *Cellulose*. 2015;22(4):2443–2456.
9. Zhu C, Dobryden I, Rydén J, Öberg S, Holmgren A, Mathew AP. Adsorption behavior of cellulose and its derivatives toward Ag(I) in aqueous medium: An AFM, spectroscopic, and DFT Study. *Langmuir*. 2015;31(45):12390–12400.
10. Zhang Z, Sèbe G, Rentsch D, Zimmermann T, Tingaut P. Ultralightweight and flexible silylated nanocellulose sponges for the selective removal of oil from water. *Chemistry of Materials*. 2014;26(8):2659–2668.

11. Wang X, Yeh TM, Wang Z, Yang R, Wang R, Ma H et al. Nanofiltration membranes prepared by interfacial polymerization on thin-film nanofibrous composite scaffold. *Polymer (United Kingdom)*. 2014;55(6):1358–1366.
12. Wang J, Chen C. Biosorbents for heavy metals removal and their future. *Biotechnology Advances*. 2009;27(2):195–226.
13. Qu X, Alvarez PJJ, Li Q. Applications of nanotechnology in water and wastewater treatment. *Water Research*. 2013;47(12):3931–3946.
14. Volesky B. Biosorption and me. *Water Research*. 2007;41(18):4017–4029.
15. Saito T, Isogai A. Ion-exchange behavior of carboxylate groups in fibrous cellulose oxidized by the TEMPO-mediated system. *Carbohydrate Polymers*. 2005;61(2):183–190.
16. Liu P, Sehaqui H, Tingaut P, Wichser A, Oksman K, Mathew AP. Cellulose and chitin nanomaterials for capturing silver ions (Ag^{+}) from water via surface adsorption. *Cellulose*. 2014;21(1):449–461.
17. Kardam A, Raj KR, Arora JK, Srivastava S. Artificial neural network modeling for biosorption of Pb (II) ions on nanocellulose fibers. *BioNanoScience*. 2012;2(3):153–160.
18. Sehaqui H, de Larraya UP, Liu P, Pfenninger N, Mathew AP, Zimmermann T et al. Enhancing adsorption of heavy metal ions onto biobased nanofibers from waste pulp residues for application in wastewater treatment. *Cellulose*. 2014;21(4):2831–2844.
19. Liu P, Oksman K, Mathew AP. Surface adsorption and self-assembly of Cu(II) ions on TEMPO-oxidized cellulose nanofibers in aqueous media. *Journal of Colloid and Interface Science*. 2016;464:175–182.
20. Isogai A, Saito T, Fukuzumi H. TEMPO-oxidized cellulose nanofibers. *Nanoscale*. 2011;3(1):71–85.
21. Sehaqui H, Mautner A, Perez DLU, Pfenninger N, Tingaut P, Zimmermann T. Cationic cellulose nanofibers from waste pulp residues and their nitrate, fluoride, sulphate and phosphate adsorption properties. *Carbohydrate Polymers*. 2016;135:334–340.
22. Yu X, Tong S, Ge M, Zuo J, Cao C, Song W. One-step synthesis of magnetic composites of cellulose@iron oxide nanoparticles for arsenic removal. *Journal of Materials Chemistry A*. 2013;1(3):959–965.
23. Hokkanen S, Repo E, Sillanpää M. Removal of heavy metals from aqueous solutions by succinic anhydride modified mercerized nanocellulose. *Chemical Engineering Journal*. 2013;223:40–47.
24. Zhang X, Zhao J, Cheng L, Lu C, Wang Y, He X et al. Acrylic acid grafted and acrylic acid/sodium humate grafted bamboo cellulose nanofibers for Cu^{2+} adsorption. *RSC Advances*. 2014;4(98):55195–55201.
25. Ulmanu M, Marañón E, Fernández Y, Castrillón L, Anger I, Dumitriu D. Removal of copper and cadmium ions from diluted aqueous solutions by low cost and waste material adsorbents. *Water, Air, and Soil Pollution*. 2003;142(1–4):357–373.
26. Liu P, Borrell PF, Božič M, Kokol V, Oksman K, Mathew AP. Nanocelluloses and their phosphorylated derivatives for selective adsorption of Ag^{+}, Cu^{2+} and Fe^{3+} from industrial effluents. *Journal of Hazardous Materials*. 2015;294:177–185.
27. Pillai SS, Deepa B, Abraham E, Girija N, Geetha P, Jacob L et al. Biosorption of Cd(II) from aqueous solution using xanthated nano banana cellulose: Equilibrium and kinetic studies. *Ecotoxicology and Environmental Safety*. 2013;98:352–360.
28. Hokkanen S, Repo E, Suopajärvi T, Liimatainen H, Niinimaa J, Sillanpää M. Adsorption of Ni(II), Cu(II) and Cd(II) from aqueous solutions by amino modified nanostructured microfibrillated cellulose. *Cellulose*. 2014;21(3):1471–1487.
29. He X, Cheng L, Wang Y, Zhao J, Zhang W, Lu C. Aerogels from quaternary ammonium-functionalized cellulose nanofibers for rapid removal of Cr(VI) from water. *Carbohydrate Polymers*. 2014;111:683–687.

30. Sehaqui H, Perez DLU, Tingaut P, Zimmermann T. Humic acid adsorption onto cationic cellulose nanofibres for bioinspired removal of copper(II) and a positively charged dye. *Soft Matter*. 2015;11(26):5294–5300.
31. Hokkanen S, Repo E, Lou S, Sillanpää M. Removal of arsenic(V) by magnetic nanoparticle activated microfibrillated cellulose. *Chemical Engineering Journal*. 2015;260:886–894.
32. Vijayaraghavan K, Yun YS. Bacterial biosorbents and biosorption. *Biotechnology Advances*. 2008;26(3):266–291.
33. Park D, Yun YS, Park JM. The past, present, and future trends of biosorption. *Biotechnology and Bioprocess Engineering*. 2010;15(1):86–102.
34. Ulbricht M. Advanced functional polymer membranes. *Polymer*. 2006;47(7):2217–2262.
35. Ma H, Burger C, Hsiao BS, Chu B. Ultra-fine cellulose nanofibers: New nano-scale materials for water purification. *Journal of Materials Chemistry*. 2011;21(21):7507–7510.
36. Ma H, Burger C, Hsiao BS, Chu B. Highly permeable polymer membranes containing directed channels for water purification. *ACS Macro Letters*. 2012;1(6):723–726.
37. Karim Z, Mathew AP, Grahn M, Mouzon J, Oksman K. Nanoporous membranes with cellulose nanocrystals as functional entity in chitosan: Removal of dyes from water. *Carbohydrate Polymers*. 2014;112:668–676.
38. Karim Z, Mathew AP, Kokol V, Wei J, Grahn M. High-flux affinity membranes based on cellulose nanocomposites for removal of heavy metal ions from industrial effluents. *RSC Advances*. 2016;6(25):20644–20653.
39. Goetz LA, Jalvo B, Rosal R, Mathew AP. Superhydrophilic anti-fouling electrospun cellulose acetate membranes coated with chitin nanocrystals for water filtration. *Journal of Membrane Science*. 2016;510:238–248.
40. Mautner A, Lee KY, Tammelin T, Mathew AP, Nedoma AJ, Li K et al. Cellulose nanopapers as tight aqueous ultra-filtration membranes. *Reactive and Functional Polymers*. 2015;86:209–214.
41. Mautner A, Maples HA, Kobkeatthawin T, Kokol V, Karim Z, Li K et al. Phosphorylated nanocellulose papers for copper adsorption from aqueous solutions. *International Journal of Environmental Science and Technology*. 2016;13(8):1861–1872.
42. Karim Z, Claudpierre S, Grahn M, Oksman K, Mathew AP. Nanocellulose based functional membranes for water cleaning: Tailoring of mechanical properties, porosity and metal ion capture. *Journal of Membrane Science*. 2016;514:418–428.
43. Mathew AP. NanoSelect. 2017. www.nanoselect.eu
44. VTT. Technial Research Centre of Finland. www.vttresearch.com
45. Carosio F, Kochumalayil J, Cuttica F, Camino G, Berglund L. Oriented clay nanopaper from biobased components – mechanisms for superior fire protection properties. *ACS Applied Materials and Interfaces*. 2015;7(10):5847–5856.
46. Cowie J, Bilek EMT, Wegner TH, Shatkin JA. Market projections of cellulose nanomaterial-enabled products – Part 2: Volume estimates. *Tappi Journal*. 2014;13(6):57–69.
47. Shatkin JA, Wegner TH, Bilek EM, Cowie J. Market projections of cellulose nanomaterial-enabled products – Part 1: Applications. *Tappi Journal*. 2014;13(5):9–16.
48. Baker RW. *Membrane Technology and Applications*. West Sussex, UK: John Wiley & Sons; 2012.
49. Wang LK, Hung Y-T, Shammas NK. Membrane filtration. In: Wang LK, Hung Y-T, Shammas NK, (Eds.). *Advanced Physicochemical Treatment Processes*. Totowa, NJ: Humana Press; 2012. pp. 203–259.
50. Mohanty K, Purkait MK. *Membrane Technologies and Applications*. Boca Raton, FL: CRC Press; 2011.

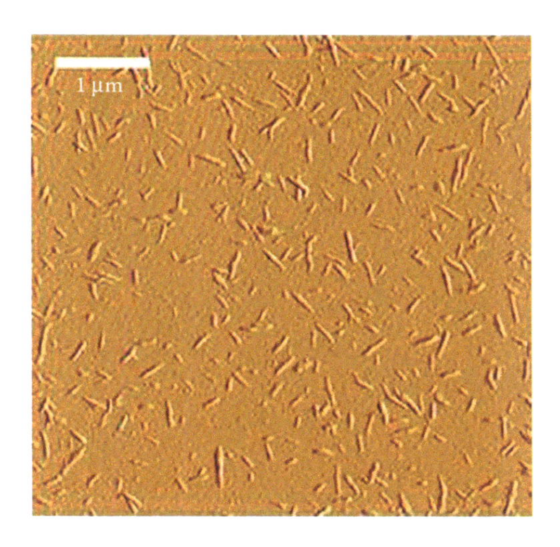

FIGURE 3.1 $5 \times 5\ \mu m^2$ atomic force microscopy image of cellulose nanocrystals prepared from Whatman No. 1 filter paper (cotton linter source).

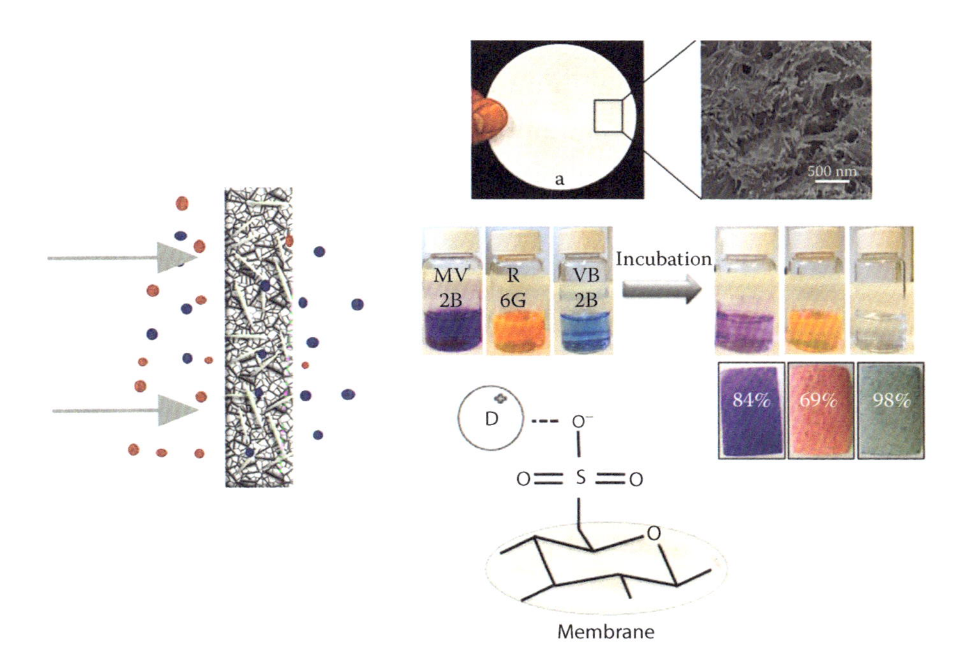

FIGURE 7.5 The removal of dyes (MV 2B: methyl violet 2B; R 6B: Rhodamine 6B; VB 2B: Victoria blue 2B) was studied in static mode for 24 h. Highest removal of dyes is found for VB 2B (98%). A possible electrostatic bonding (dotted line) between positively charged dyes (D+) and negatively charged membranes due to the presence of SO_3^- groups after sulphuric acid hydrolysis is found. (From Karim, Z. et al., *Carbohydr. Polym.*, 112, 668–676, 2014.)

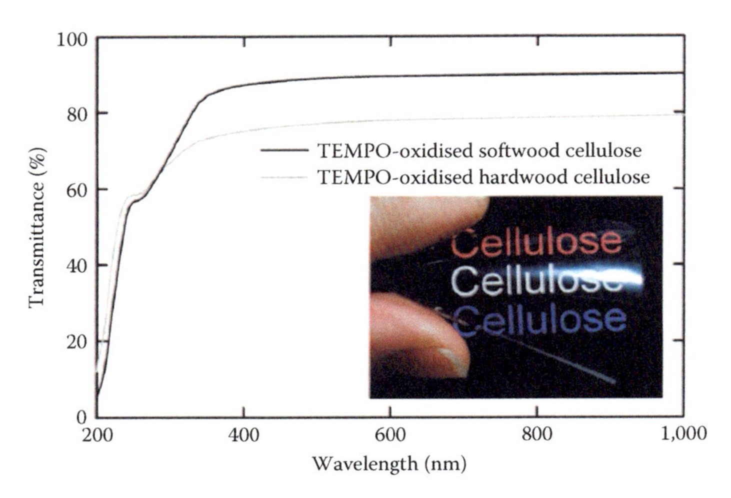

FIGURE 8.1 Ultraviolet–visible transmittance of nanofibre films prepared from TEMPO-oxidised softwood and hardwood celluloses. The photograph shows light transmittance behaviour of the nanofibre film prepared from the TEMPO-oxidised softwood cellulose. (Reprinted with permission from Fukuzumi, H. et al., *Biomacromolecules*, 10, 162–165, 2009. Copyright 2009 American Chemical Society.)

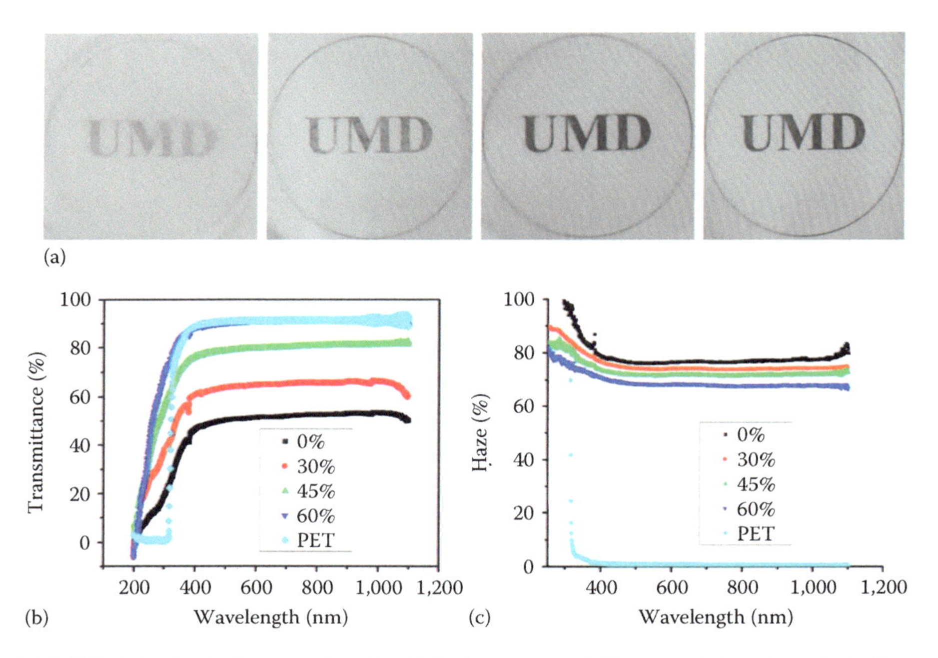

FIGURE 8.3 Optical properties of cellulosic paper at different weight ratios of NFC to whole paper. (a) The visual appearance of the cellulosic paper. The weight ratio of NFC to paper from left to right is 0%, 30%, 45% and 60%. (b) Total light transmittance of the cellulosic paper at various weight ratios of NFC. (c) The haze value versus wavelength for transparent paper with different NFC ratios and PET. (Fang, Z. et al., *J. Mater. Chem. C*, 1, 6191–6197, 2013. Reproduced by permission of The Royal Society of Chemistry.)

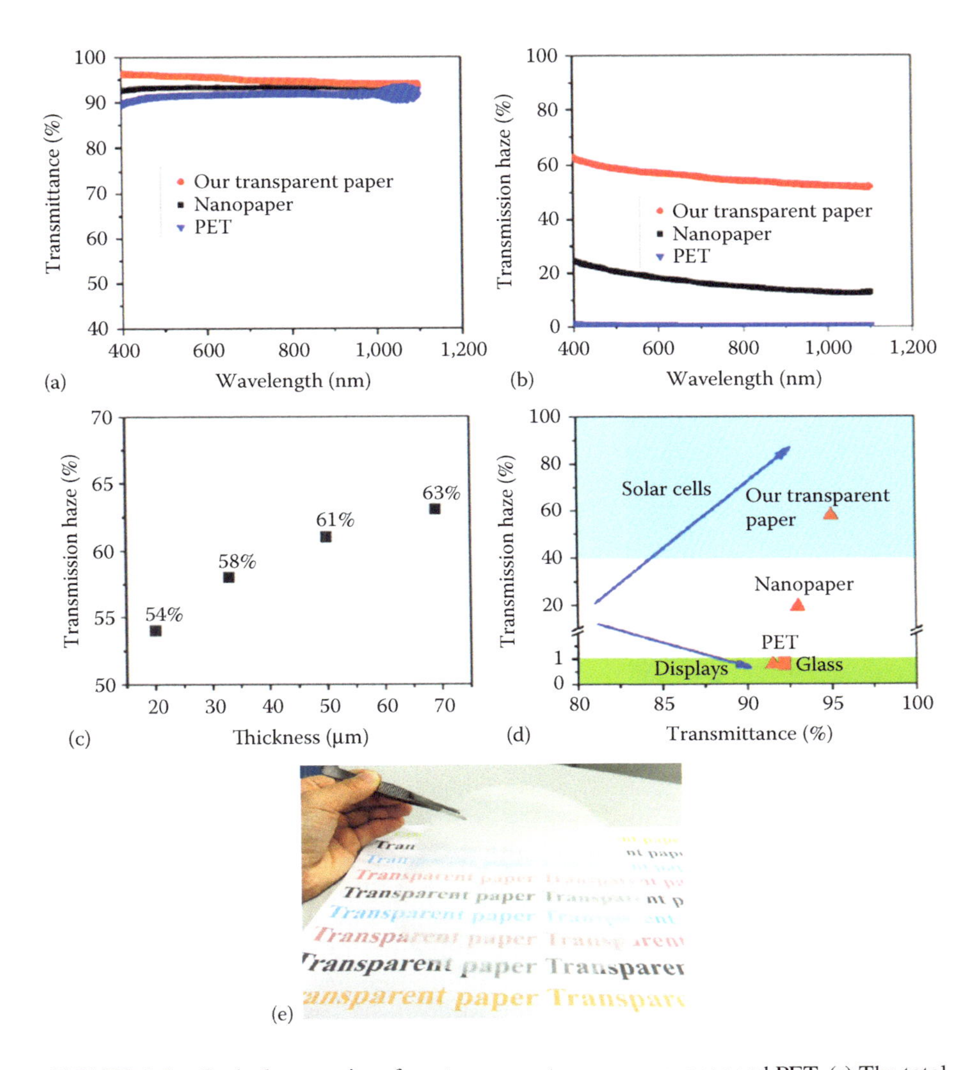

FIGURE 8.5 Optical properties of our transparent paper, nanopaper and PET. (a) The total optical transmittance versus wavelength measured with an integrating sphere setup. (b) The transmission haze versus wavelength. (c) The transmission haze of transparent paper with varying thicknesses at 550 nm. (d) Optical transmission haze versus transmittance for different substrates at 550 nm. Glass and PET are in the green area, which are suitable for displays due to their low haze and high transparency; transparent paper developed in this work located in the cyan area is the most suitable for solar cells and (e) A digital image of transparent paper produced from TEMPO-oxidised wood fibres with a diameter of 20 cm. Transparent paper is made of mesoscale fibres. The primary fibres have an average diameter of ~26 μm. (Reprinted with permission from Fang, Z. et al., *Nano Lett.*, 14, 765–773, 2014. Copyright 2014 American Chemical Society.)

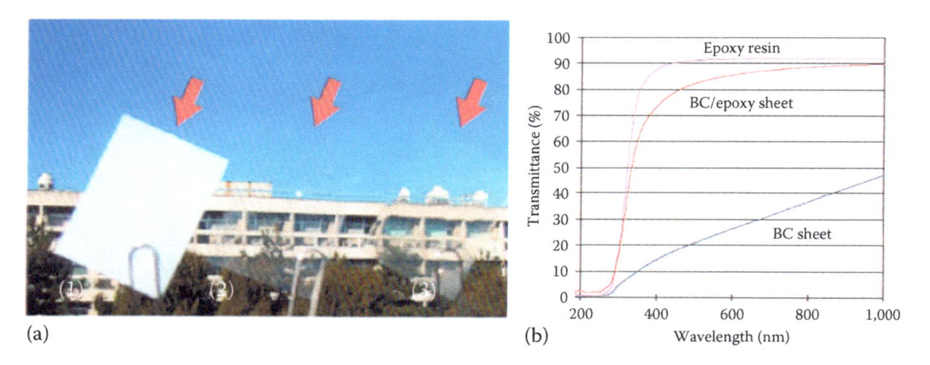

FIGURE 8.8 (a) Appearance of a 65 μm thick bacterial cellulose sheet: (1) with no added resin, (2) with acrylic resin, 62 wt.% fibre content and (3) with epoxy resin, 65 wt.% fibre content. (b) Light transmittance of a 65 μm thick bacterial cellulose/epoxy-resin sheet (65 wt.% fibre content), BC sheet and epoxy-resin sheet. (From Yano, H. et al.: Optically Transparent Composites Reinforced with Networks of Bacterial Nanofibres. *Advanced Materials*. 2005. 17. 153–155. Copyright Wiley-VCH Verlag GmbH & Co. KGaA. Reproduced with permission.)

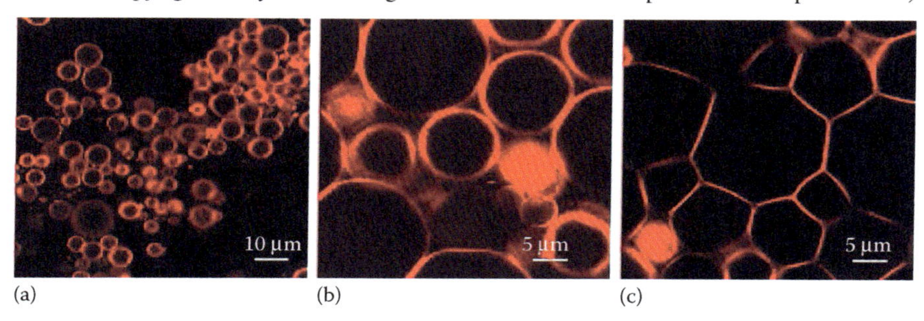

FIGURE 9.10 Confocal laser scanning microscopic images of emulsions stabilised by cotton CNCs containing increasing amounts of hexadecane stained with Bodipy from (a) the original 10/90 oil–water Pickering emulsion; (b) 65% of internal phase; (c) 85.6% of internal phase. (Reprinted with permission from Capron, I. and Cathala, B., *Biomacromolecules*, 14, 291–296, 2013. Copyright 2013 American Chemical Society.)

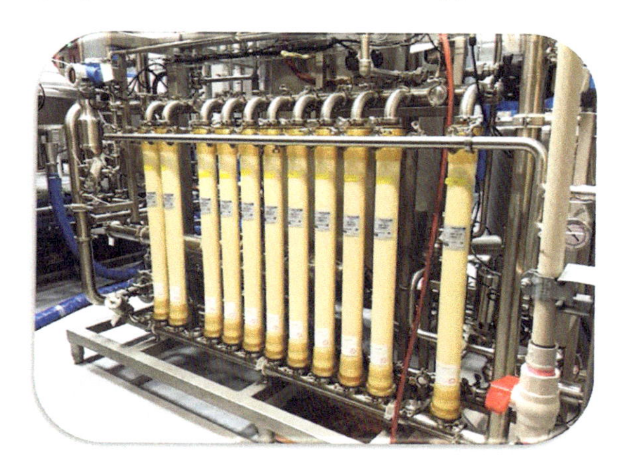

FIGURE 12.8 GEA-Niro ultrafiltration unit.

8 Applications of Nanocellulose as Optically Transparent Papers and Composites

Franck Quero

CONTENTS

8.1 Introduction 147
8.2 Optically Transparent Nanocellulose Papers 148
  8.2.1 Generalities on Nanocellulose 148
  8.2.2 Fabrication Methods of Nanocellulose Papers 149
  8.2.3 Nanofibrillated Cellulose 151
  8.2.4 Bacterial Cellulose 158
8.3 Optically Transparent Nanocellulose Composites 158
  8.3.1 Generalities on Nanocellulose Composites 158
  8.3.2 Fabrication Methods of Nanocellulose Polymer Composites 158
  8.3.3 Nanofibrillated Cellulose Composites 160
  8.3.4 Bacterial Cellulose Composites 163
8.4 Conclusion and Future Directions 168
Acknowledgements 168
References 168

8.1 INTRODUCTION

Transparent materials are suitable for a variety of industrial sectors including optoelectronics, packaging and biomedical applications. Due to environmental concerns and governmental pressure, there is a growing need to replace oil-sourced materials by renewable materials to design consumer goods. Cellulose has attracted intense fundamental and applied research efforts to try to find new applications for this natural polymer. Cellulose has the advantage of being an environmentally friendly, renewable, abundant and a relatively cheap material among other benefits.

In the past 30 years, a new class of nanomaterial, referred to as nanocellulose, has emerged, which includes cellulose nanocrystals or nanowhiskers, nanofibrillated cellulose (NFC) and bacterial cellulose (BC). The physical properties of nanocellulose were investigated, such as mechanical, rheological, thermal and optical properties showing that this nanomaterial has high potential to be used for multiple applications such as for optoelectronic devices [1,2], biomedicine [3], packaging [4] and as a food additive [5,6] to cite a few only. The properties of nanocellulose have been evaluated in the form of individual nanofibres, dispersed in a polymer matrix but also as papers, coatings and laminated composites. One of the main challenges remains in fully realising the potential of nanocellulose to be used as paper or when introduced in a polymeric matrix.

The evaluation of the optical properties of papers and composites made from nanocellulose revealed that it is possible to obtain translucent to highly optically transparent materials from nanocellulose, comparable to the optical transparency of some thermoplastic and thermoset polymers as well as glass. In addition, nanocellulose was found to have similar thermal expansion coefficient compared to quartz glass with a value of ~0.1 ppm/K [7], which again demonstrated the potential of nanocellulose to replace glass substrates for specific applications such as optoelectronic applications. In addition, the mechanical properties of nanocellulose papers [1] and composites [8] were found to be close to that of conventional polymers, with improved levels of flexibility and foldability, showing the way for the design of high-performance flexible electronics.

Cellulose nanofibres (CNFs), and more particularly BC nanofibres, have demonstrated great potential to be used for biomedical applications due to its low toxicity, biocompatibility, hemocompatibility and cytocompatibility [3]. Wet nanocellulose papers also display good optical transparency. In some specific biomedical applications such as wound healing [9] and tympanic membrane regeneration [10], optical transparency is an important benefit for quality control of tissue regeneration, whereas optical transparency is crucial for the design of artificial cornea [11,12].

8.2 OPTICALLY TRANSPARENT NANOCELLULOSE PAPERS

8.2.1 Generalities on Nanocellulose

Cellulose is the most abundant natural polymer on earth [13]. It is mostly found in the cell walls of plants as nanofibres having a diameter of 15–20 nm. Other sources of cellulose have been reported in the literature, namely from algae [14], bacteria [15,16] and tunicates [17]. These nanofibres are made of an arrangement of cellulose microfibrils having a diameter of ~4 nm. CNF, a form of nanocellulose, can be isolated from plants. Briefly, a cellulose pulp is obtained by successive mechanical and chemical treatments, which is followed by an energy-intensive mechanical disintegration process to disassemble cellulose microfibres into CNF. At the end of the process, a material referred to as microfibrillated cellulose (MFC) or NFC is obtained [18,19]. This terminology is related to the diameter range of the generated fibres, which are referred to as NFC if most of the fibres have a diameter below 100 nm. For the fabrication of optically transparent nanocellulose papers, the terminology NFC might be more appropriate because very high level of transparency can be reached for CNF having a diameter typically below 15 nm.

The physical properties of nanocellulose have been largely evaluated, demonstrating its high performance and potential for a wide range of applications. Young's modulus of CNF and nanowhiskers obtained from plant, tunicate and bacteria, in the form of single nanofibres, networks and nanocomposites have been estimated by atomic force microscopy (AFM) [20,21] and Raman spectroscopy [22–24]. The tensile strength of single wood cellulose fibres was found to be in the range of 2–3 GPa [25,26]. CNFs also display a low coefficient of thermal expansion (CTE) of ~0.1 ppm/K [7] presumably due to their high crystallinity, which is similar to that of quartz glass. They also show high optical transparency due to their nanosized diameter, typically 10 times lower than the wavelength of visible light. These physical properties are particularly relevant for the use of nanocellulose in the form of transparent papers and nanocomposites for flexible optoelectronic applications. They are also a relevant nanomaterial to be used for biomedical applications owing to their low toxicity, biocompatibility, hemocompatibility and cytocompatibility [3]. These characteristics make nanocellulose an ideal material to be used for tissue engineering applications. The use of ultrapure wood CNF has, however, been very recently recommended for wound healing applications to lower the cytotoxicity of CNF with respect to human skin cells [27].

Another source of nanocellulose, referred to as BC, can be used to generate dry and wet nanocellulose papers. This form of nanocellulose is produced in specific conditions by bacteria [15,16]. It has the advantage of being obtained in a very pure and fine form and without the need for high energy-intensive fibrillation processes. It has attractive features over NFC such as higher crystallinity, Young's modulus, water-holding capacity, biocompatibility, hemocompatibility and cytocompatibility to cite a few only [3]. BC is, however, relatively expensive compared to NFC, so its use is, for now, not recommended for cheap and high volume applications. BC is currently thought to be more relevant to be used for high-value biomedical applications in order to justify its higher cost compared to NFC obtained from plant sources.

8.2.2 Fabrication Methods of Nanocellulose Papers

Several factors have been reported to be crucial to fabricate highly optically transparent nanocellulose papers, also referred to as cellulose nanopapers. These factors are directly related to the fabrication routes of nanocellulose papers. Characteristics such as fibre diameter, optical index of refraction, packing density, surface roughness and pore size of nanocellulose papers have been reported to be the most crucial factors to control optical transparency [28]. For example, the nanofibre size and the pore size should be small enough to avoid light scattering as much as possible [28].

There is one main method to fabricate optically transparent nanocellulose papers. It has been applied for the fabrication of nanocellulose papers using mostly NFC (more specific comments on BC will be provided) and includes the following main steps:

- *Production of CNFs*: This step consists in converting a never-dried aqueous cellulose pulp containing micron-size diameter cellulose fibres into an aqueous nanocellulose suspension using a high-pressure microfluidizer [19], an ultrafine friction grinder [29] or a high-pressure water-jet system [30]. Several passes through these processing tools are usually needed to fine-tune the nanosized diameter of cellulose fibres. Before processing, the cellulose pulp can be subjected to chemical or enzymatic modification to reduce the number of passes, facilitating and reducing the amount of energy involved in the mechanical disintegration of cellulose microfibres into CNF. The most common and efficient chemical modification of cellulose, reported to date, to facilitate the nanofibrillation process is by 2,2,6,6-tetramethylpiperidine-1-oxyl (TEMPO)-mediated oxidation [31]. More detailed information on these mechanical disintegration methods can be found in the literature [32]. At the end of the nanofibrillation process, a typical aqueous nanocellulose suspension is obtained, with a typical concentration of 1–2 wt.%. This concentration can be subsequently adjusted to a concentration of 0.1–0.2 wt.% for the following vacuum filtration step for the fabrication of nanocellulose paper. The suspension is then degassed by sonication to remove air bubbles that may have formed during the disintegration process. It is important to mention that this step is performed using never-dried plant cellulose pulp and not using BC. This is because at the end of the BC synthesis a wet membrane that already contains CNF is obtained, and so there is no need to use any energy-intensive process to obtain nanofibres. A simple disintegration step using a kitchen blender can be used to disintegrate BC membranes into a nanocellulose suspension.

- *Vacuum filtration*: This step consists in filtering out water under vacuum, usually using a Büchner funnel, a nanocellulose suspension having a typical concentration of 0.1–0.2 wt.% at room temperature. Nitrocellulose ester or polytetrafluoroethylene filters, having micron-sized pore diameters, are usually used to carry out this step. At the end of the process, a wet nanocellulose filter cake that contains little free water is obtained.
- *Extrusion papermaking process*: This step can be used instead of using a vacuum filtration process. It consists in pressuring the nanocellulose suspension using a gas to squeeze out the excess of water through a micron-sized filter [33]. Similar to the vacuum filtration process, a nanocellulose filter cake is obtained at the end of the process, which will be subsequently oven-dried and/or hot-pressed.
- *Drying and/or hot-pressing*: This step is sometimes reported as a separate or as a single step. It consists in removing the excess of water that is still present after the previous vacuum filtration step. A nanocellulose filter cake is basically submitted to a temperature cycle. At the same time, a pressure cycle can be applied to the sample to squeeze out remaining water and to increase the packing density of the nanocellulose filter cake.
- *Solvent exchange*: This step has been reported recently as an alternative to the slow and costly supercritical drying or freeze-drying processes to obtain nanocellulose papers [34]. After the vacuum filtration step, the material is solvent exchanged from water to 2-propanol and subsequently to octane. The material is then air-dried in ambient conditions, and a transparent and robust nanocellulose paper can be obtained.
- *Surface treatments*: Some additional surface post-treatments have been reported in the literature to enhance optical transparency of nanocellulose papers [35]. These surface treatments involved polishing of the surface of

nanocellulose papers, laminating using polycarbonate and acrylic resin deposition to smoothen the surface and reducing surface roughness of the nanocellulose papers. These strategies were found to turn translucent or semi-transparent nanocellulose papers into highly transparent nanocellulose papers.

8.2.3 Nanofibrillated Cellulose

NFC has been the most evaluated source of cellulose so far to fabricate nanocellulose papers. This is because plant cellulose is highly available in nature, and it is a cheaper source of cellulose compared to bacterial and tunicate cellulose. Consequently, most of the literature reports application of NFC as optically transparent nanocellulose papers.

In 2007, TEMPO-mediated oxidation has been reported to facilitate the nanofibrillation process of cellulose [36], which allows to obtain CNF having a diameter of 3–4 nm. A subsequent work reported the fabrication of the first optically transparent nanocellulose paper using TEMPO-mediated oxidised CNF having a diameter of 3–4 nm [31]. The use of two cellulose sources, namely softwood and hardwood was compared. Nanocellulose papers fabricated using TEMPO-mediated oxidised CNF obtained from softwood resulted in a transmittance of ~90% at a wavelength of 600 nm as shown in Figure 8.1. This work, however, demonstrated that TEMPO-mediated oxidation of CNF considerably reduced the thermal stability of nanocellulose papers, which is an issue if one needs to process these nanofibres at high temperature in a thermoplastic melt or for any other high temperature process where these nanocellulose papers could be submitted to, including roll-to-roll fabrication

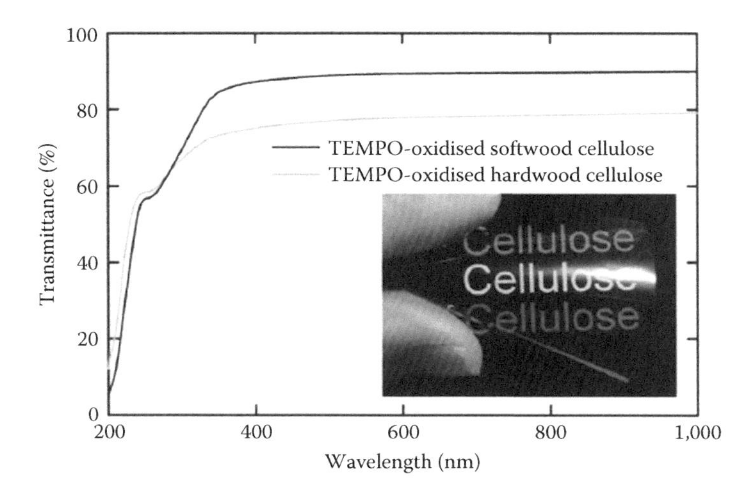

FIGURE 8.1 **(See colour insert.)** Ultraviolet–visible transmittance of nanofibre films prepared from TEMPO-oxidised softwood and hardwood celluloses. The photograph shows light transmittance behaviour of the nanofibre film prepared from the TEMPO-oxidised softwood cellulose. (Reprinted with permission from Fukuzumi, H. et al., *Biomacromolecules*, 10, 162–165, 2009. Copyright 2009 American Chemical Society.)

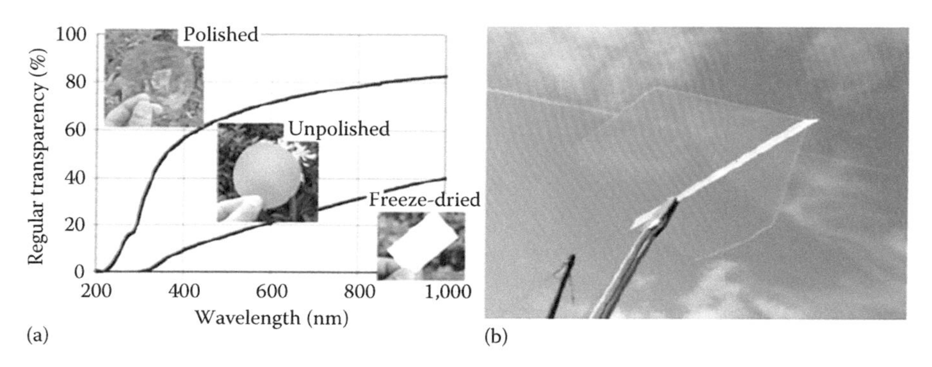

FIGURE 8.2 (a) Light transmittance of the cellulose nanofibre sheets. The thicknesses of the oven-dried nanofibre sheet were 60 μm before and 55 μm after polishing. (b) The sheet is foldable as conventional paper. (From Nogi, M. et al.: Optically Transparent Nanofibre Paper. *Advanced Materials*. 2009. 21. 1595–1598. Copyright Wiley-VCH Verlag GmbH & Co. KGaA. Reprinted with permission.)

process or heat-pressing. A recent study reported a strategy to improve the thermal stability of TEMPO-oxidised CNF by a heat-induced conversion of ionic bonds into amide bonds [37]. Results suggested an impressive improvement of thermal stability by up to 90°C. A suggestion for future work would be to fabricate nanocellulose papers from those nanofibres and evaluate their CTE and optical properties to estimate their potential to be used for the fabrication of flexible electronics.

The first optically transparent nanocellulose paper using natural CNF, and not a cellulose derivative, was reported by Nogi et al. [28]. In this work, densely packed nanocellulose papers were fabricated by using 15 nm diameter plant CNF, which was previously obtained from a grinding process [29]. A traditional papermaking process, followed by surface polishing, allowed the fabrication of optically transparent nanocellulose papers with a regular transmittance of ~70% at a wavelength of 600 nm as reported in Figure 8.2. This work highlighted the fact that CNF having a diameter of 15 nm that is combined with a low surface roughness have to be used to generate optically transparent nanocellulose papers. Several strategies to lower the surface roughness treatments of nanocellulose papers were proposed. Low surface roughness was obtained by surface polishing, by deposition of a polymer layer or by lamination [35]. These conditions were found to reduce surface light scattering, and so materials with high optical transparency could be obtained. This work was made possible owing to a previous work that focused on obtaining CNF having a uniform diameter of 15 nm using a single-step grinding process [29]. The nanofibres were, however, not yet used to fabricate optically transparent nanocellulose papers. Instead, optically transparent nanocellulose composites were fabricated using the uniform 15 nm diameter nanofibres by embedding them into an acrylic resin. More details on this evaluation will be given later.

The comparison of these two first reports on optically transparent nanocellulose papers using TEMPO-oxidised CNF [31] and natural CNF [28] clearly highlighted the importance of using CNF as small as possible to avoid light scattering to obtain nanocellulose papers with ultrahigh optical transparency, comparable to traditional optically transparent thermoplastics and thermosets. The deposition process used to reduce the surface roughness of nanocellulose papers to improve their optical transparency was further investigated [28]. The potential of fabricating these materials by a simple roll-to-roll process was also discussed. Several commercial acrylic resins were deposited on the surface of nanocellulose papers to study the influence of the refractive index of the acrylic resins on the optical transparency of coated nanocellulose papers. The results revealed that despite the wide range of refractive index, the materials displayed a high light transmittance as well as a low CTE, which was still comparable to that of quartz glass.

Although the first optically transparent nanocellulose papers were obtained by using CNF that was previously obtained using a grinding method [28], another processing method was then proposed to produce CNF. A high-pressure water-jet system [30] was proposed to generate CNF to be used to fabricate optically transparent nanocellulose papers [38]. CNF with a diameter of 15 nm can be obtained from that approach [39], which is comparable to what has been obtained previously using the grinding method [35]. Consequently, optically transparent nanocellulose papers owing to their high optical transparency, low CTE and good mechanical properties such as tensile strength, flexibility and foldability have the potential to replace transparent glass and plastic substrates currently used in the optoelectronic industry and allow the fabrication of flexible and foldable optoelectronic goods.

Most of the subsequent studies reported have been focusing on providing conductive properties to optically transparent papers for flexible optoelectronic applications [40] with the aim of trying to replace brittle indium tin oxide (ITO) in optoelectronic systems including solar cells. One of the first conductive optically transparent nanopaper was reported by Hu et al. [41]. In this work, nanocellulose papers were rendered conductive by depositing tin-doped indium oxide, carbon nanotubes or silver nanowires on their surface. The benefit of using nanocellulose papers having light scattering or haze effect for flexible solar cell applications was also demonstrated.

As previously mentioned, glass is commonly used in optoelectronic displays. It has high levels of optical transparency and a low CTE of ~1–10 ppm/K. It has the drawback, however, of having a high density and of being brittle. It is also not suitable for roll-to-roll manufacturing process. As a result, glass is not a suitable candidate to be used for the development of flexible optoelectronics. Polymers such as polyethylene terephthalate (PET) have the advantage of being flexible, display high optical transparency (~90%) and have low density but their CTE is much higher than glass, ~50 ppm/K, which is a drawback for the fabrication of optoelectronic devices where the use of relatively high temperature is involved. Nanocellulose papers fabricated using NFC were found to display high optical transparency, 80%–90%, as well as low CTE, typically <8.5 ppm/K and high flexibility and foldability. This makes nanocellulose papers ideal candidates to be used for the design of flexible optoelectronic applications.

High optical transparency is sometimes not the only important optical property parameter. For some specific applications, both high haze and high optical transmittance are necessary. A wood fibre–NFC nanopaper hybrid has been reported to

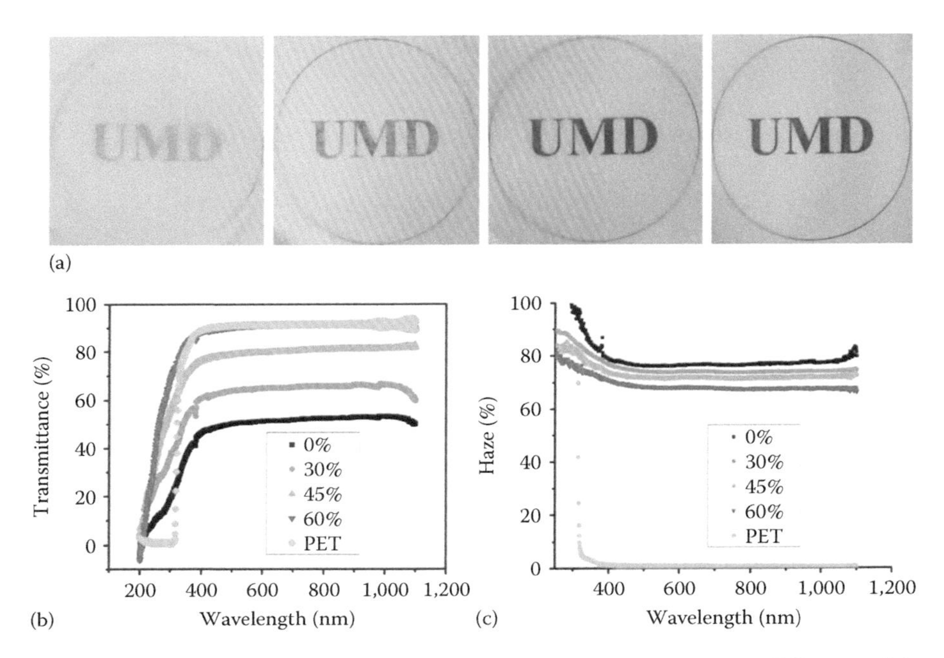

FIGURE 8.3 **(See colour insert.)** Optical properties of cellulosic paper at different weight ratios of NFC to whole paper. (a) The visual appearance of the cellulosic paper. The weight ratio of NFC to paper from left to right is 0%, 30%, 45% and 60%. (b) Total light transmittance of the cellulosic paper at various weight ratios of NFC. (c) The haze value versus wavelength for transparent paper with different NFC ratios and PET. (Fang, Z. et al., *J. Mater. Chem. C*, 1, 6191–6197, 2013. Reproduced by permission of The Royal Society of Chemistry.)

combine high optical transmittance, low roughness and tunable haze as shown in Figure 8.3 [42]. In this work, the material was also found to be writable and useful for touchscreen applications owing to its anti-glare effect. The nanocellulose papers could also be combined with carbon nanotubes to create conductive nanopaper hybrids. Another work on similar hybrids demonstrated their usefulness for the design of solar cells [43]. The combined high optical transparency and high haze was found to suit that application. High haze was found to enhance light trapping, which resulted in a significant improvement of power conversion efficiency. More details on this study will be given later in this section.

Most of the nanocellulose papers reported in the literature are fabricated by vacuum filtration, followed by heat-pressing. A recent study demonstrated that it is possible to simply ambient-air-dry NFC wet filter cake to obtain highly transparent nanocellulose papers [34]. This was made possible by carrying out a solvent exchange from water to 2-propanol and finally to octane. The mechanical properties of the nanocellulose papers were improved by compacting with water after drying from octane. The potential of these materials to be used as conductive transparent templates for flexible devices was evaluated; it showed good potential.

Another work studied the influence of the fabrication methods, including drying, on the physical properties of nanocellulose papers made with NFC [44]. This work was performed to try to obtain optically transparent nanocellulose papers in a less tedious way. The study demonstrated that drying conditions significantly affect the physical properties of nanocellulose papers, among them the optical transparency. The highest light transmittance was obtained for papers modified with TEMPO-oxidised NFC. A straightforward method involving vacuum filtration and air- or oven-drying was sufficient to obtain nanocellulose papers having high mechanical strength and high optical transparency. The use of holocellulose nanofibres (HCNF) was also evaluated to fabricate nanocellulose papers [45]. In this work, a mild delignification process was performed using paracetic acid to obtain CNF on applying several passes through a mechanical disintegration process. The nanocellulose papers displayed a high optical transparency of ~70% at 600 nm. This light transmittance was much higher compared to nanocellulose papers prepared by an enzymatic treatment as shown in Figure 8.4. In addition, the nanocellulose papers prepared using paracetic acid displayed high tensile strength. The optical performance of these nanopapers was lower compared to those made with TEMPO-oxidised CNF. Their tensile modulus and strength were, however, higher than the value reported by Fukuzumi et al. [31]. Values of tensile modulus and strength of 16 GPa and 330 MPa and 7 GPa and 233 MPa were reported, respectively by Galland et al. [45] and Fukuzumi et al. [31]. Unfortunately, the thermal properties of HCNF were not evaluated for further comparison with TEMPO-oxidised CNF.

As previously mentioned, one drawback of nanocellulose papers made with TEMPO-oxidised CNF is their low thermal stability [31]. Thermal stability is particularly important when aiming to use nanocellulose papers for the production of flexible electronics by roll-to-roll process. A recent study focused on

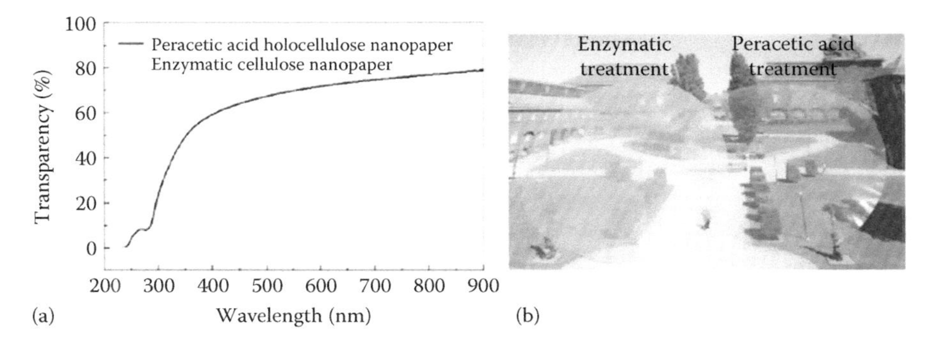

FIGURE 8.4 (a) Molecular absorption spectroscopy data for cellulose nanopaper in the ultraviolet (UV) and visible (VIS) range of wavelengths. (b) Spruce nanopaper prepared from peracetic acid holocellulose and enzymatic CNFs. The photograph demonstrates optical transparency of nanopaper with a thickness of ~45 μm. (Reprinted with permission from Galland, S. et al., *Biomacromolecules*, 16, 2427–2435, 2015. Copyright 2015 American Chemical Society.)

improving the thermal resistance of nanocellulose papers made with TEMPO-oxidised CNF [46]. This study demonstrated that by decreasing the carboxylate content of the CNF, the thermal resistance of these chemically modified nanocellulose papers was significantly improved while maintaining high optical transparency (~80% at 600 nm). In addition, the low CTE of nanocellulose paper was preserved, and a relatively low energy process was involved in the production of the CNF.

A series of very recent works focusing on the application of NFC nanopapers for flexible electronic device applications [40], including solar cells [43], have been recently reported. In these works, the optical properties of the nanocellulose paper were found to be extremely important to enhance the yield of solar cells. Transmittance is of course very important to provide optical transparency to the nanopaper. The challenge is, however, to fabricate a nanopaper with high transmittance and high haze. Figure 8.5a reports the transmittance of the transparent nanopaper designed in this work. The transmittance was ~95% at a wavelength of 600 nm. As displayed in Figure 8.5b, the transmission haze was found to be high, ~60% at 600 nm. Haze was found to be dependent on the thickness of the transparent paper, which increases when thickness increases. This transparent nanocellulose paper was found to be an excellent candidate to be used as a flexible and transparent solar cell substrate as shown in Figure 8.5d. Figure 8.5e illustrates both the high optical transmittance and high haze of the transparent nanocellulose paper.

Nanocellulose papers were also found to be suitable for the fabrication of ionic diodes [47], foldable antenna [48], transistors [49–51], luminescent NFC nanopapers [33,52,53], iridescent NFC nanopapers [54], non-volatile memory [55], optical sensing platforms [56], printed circuit boards [57], Li-ion battery [58], self-powered human interactive transparent nanopaper [59] and writable NFC nanopapers [42]. In addition, some very recent works reported the fabrication of flexible conductive nanocellulose papers based on NFC [41,60–62]. These works involved various strategies such as the use of silver nanowires and carbon nanotubes, which were added directly during the nanopaper fabrication process and in sufficient amounts to provide conductivity through the formation of a three-dimensional network within the nanocellulose paper structure.

Another example of nanocellulose paper based on NFC with improved wet strength has been recently reported [63]. In this work, chitosan was used as a physical cross-linker. The solubility of chitosan depends on pH. At low pH, chitosan was found to be soluble in an aqueous medium, whereas at higher pH, it was not soluble. By adjusting the pH of an aqueous chitosan–NFC suspension using hydrochloric acid or sodium hydroxide, it was possible to immobilise the chitosan phase possibly through the formation of multivalent physical interaction of chitosan with NFC, and so chitosan acted as a physical cross-linker for NFC. The materials showed high mechanical strength in the wet state and also high optical transparency, depending on the volume fraction of NFC and chitosan introduced in the nanocellulose paper. This material presents interesting characteristics to be used for biomedical applications where high tensile strength and transparency are needed in the wet state.

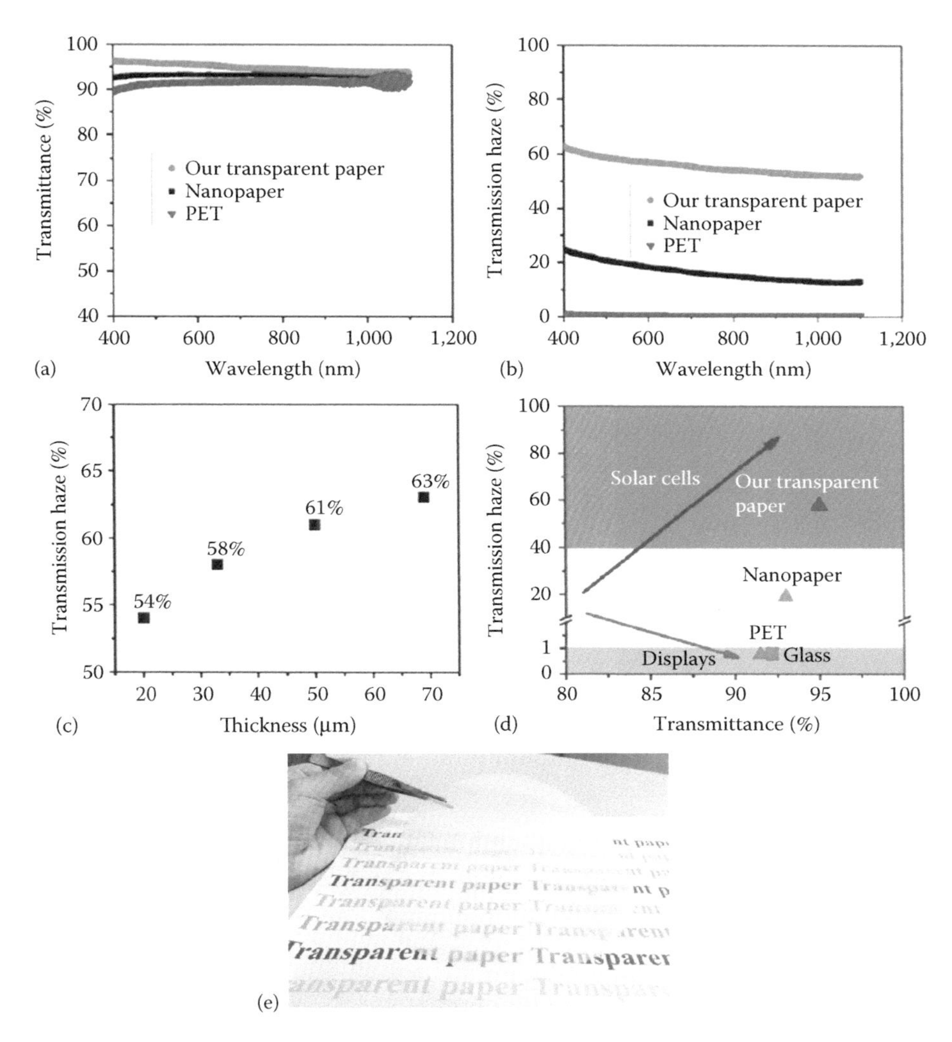

FIGURE 8.5 **(See colour insert.)** Optical properties of our transparent paper, nanopaper and PET. (a) The total optical transmittance versus wavelength measured with an integrating sphere setup. (b) The transmission haze versus wavelength. (c) The transmission haze of transparent paper with varying thicknesses at 550 nm. (d) Optical transmission haze versus transmittance for different substrates at 550 nm. Glass and PET are in the gray area, which are suitable for displays due to their low haze and high transparency; transparent paper developed in this work located in the dark gray area is the most suitable for solar cells and (e) A digital image of transparent paper produced from TEMPO-oxidised wood fibres with a diameter of 20 cm. Transparent paper is made of mesoscale fibres. The primary fibres have an average diameter of ~26 μm. (Reprinted with permission from Fang, Z. et al., *Nano Lett.*, 14, 765–773, 2014. Copyright 2014 American Chemical Society.)

8.2.4 Bacterial Cellulose

Only a very few studies focused on using BC to produce nanocellulose papers. The main reason is because BC is still less available and more expensive compared to plant cellulose, making BC less attractive for high-volume and low-cost applications. BC, however, has shown great potential for tissue engineering applications including wound healing [64], artificial cornea [11] and tympanic membrane regeneration where good optical transparency is needed in order to visually observe regeneration [10]. In these applications, nanocellulose papers are usually used in the wet state. These applications are of high added value where the relatively high cost of BC could be justified.

8.3 OPTICALLY TRANSPARENT NANOCELLULOSE COMPOSITES

8.3.1 Generalities on Nanocellulose Composites

In this section, nanocellulose composites are referred to as materials where one phase is the nanocellulose material (NFC or BC) and the other phase is either a thermoplastic or a thermoset polymer. By combining these materials, high-performance polymeric composite materials with unique properties may be obtained. Important efforts in this area of research have been already reported in the literature [65]. As already discussed in Section 8.2.1, nanocellulose has shown to have great potential to be used as a reinforcement nanomaterial in order to mechanically reinforce polymer thermoplastics and thermosets. In addition, its low CTE could help reducing the relatively high CTE of most thermoplastics and thermoset polymeric resins. Another challenge that was addressed was the preservation of the high optical transparency of some thermoplastics and thermosets upon the addition of high volume fractions of nanocellulose. Owing to the nanosized diameter of CNF and the relatively close refractive index of cellulose and carefully selected thermoplastics and thermosets, one can obtain high-performance composite materials with high optical transparency. Section 8.3.2 describes the fabrication methods reported to fabricate optically transparent nanocellulose composites and the detailed research carried out using either NFC or BC to design this relatively new class of polymeric composite materials.

8.3.2 Fabrication Methods of Nanocellulose Polymer Composites

Several factors have been reported to be crucial for the successful production of highly optically transparent nanocellulose composites. The recommendation to design highly optically transparent nanocellulose polymer composites is to use CNF with a diameter of less than one-tenth of the wavelength of visible light and a highly optically transparent polymer matrix with an optical index of refraction as close as possible to the CNF [66,67]. The level of impregnation of the nanocellulose paper by the polymer matrix is also extremely important. Voids or porosities, if present, should also be less than one-tenth of the wavelength of visible light to avoid light scattering as much as possible [66,67]. This is usually achieved using a low viscosity resin or polymer solution to impregnate the nanocellulose paper efficiently. Surface roughness should also be controlled and should be as small as possible to avoid surface light scattering.

There are five main methods that have been reported to successfully fabricate transparent nanocellulose polymer composites. The choice on whether using one of these methods depends on when considering polymers that are water soluble, thermoplastics or thermosets, high or low viscosity in the molten, liquid or solution state. Some of these methods can be used for both NFC and BC, whereas others are specific to BC only. The fabrication methods reported in the literature to produce optically transparent nanocellulose composites are the following:

- *Impregnation*: This method consists of impregnating a dry or wet nanocellulose paper with a liquid polymer or a polymer solution in standard atmospheric conditions or under vacuum [28,68–71]. When using this method, it is crucial to use a low-viscosity polymer or a low-concentration polymer solution in order to fully impregnate the nanocellulose papers and so limit the presence of voids. In addition, the compatibility and refractive index matching between the polymer and nanocellulose are crucial to obtain highly optically transparent nanocellulose composites [68].
- *Compression moulding*: This method has been reported in order to smoothen the surface of optically translucent nanocellulose papers, reducing the surface roughness and so light scattering [35]. This treatment allowed the fabrication of highly optically transparent nanocellulose paper.
- *Melt compounding*: This method consists in mixing CNF in a thermoplastic polymer melt at high temperature by extrusion. It usually results in poor dispersion of CNF in the thermoplastic polymer, especially if a hydrophobic polymer thermoplastic is used. Another issue is related to thermal stability. Some interesting and promising strategies, which do not involve chemical modification, have been reported to try to overcome these issues [72,73].
- *Solvent casting*: This method consists of adding controlled amounts of a nanocellulose filter cake into a polymer solution [74–76]. After subsequent mixing, which can also include ultrasonication, a specific volume of the mixture is poured into a mould or petri dish and is dried at a controlled temperature. After drying, the polymer film containing CNF is delicately pilled off from the mould for characterisation purposes.
- *In situ biosynthesis*: This method is specific to BC. It consists in adding controlled amounts of a water-soluble and low toxicity polymer to the culture medium of cellulose-producing bacteria [70,77]. During the biosynthesis, the BC membrane will integrate the selected water-soluble polymer into its structure. This method has, however, the disadvantage of being limited by the viscosity increase of the culture medium due to the addition of the water-soluble polymer, which reduces the yield of BC biosynthesis.
- *Surface selective dissolution*: This method consists in dissolving partially nanocellulose papers using a good cellulose solvent such as ionic liquids. This method results in the fabrication of all-cellulose composites because both the reinforcement and matrix phases are cellulose, and so strong interfaces are formed. This method has been reported for both NFC [7,78] and BC [79].

8.3.3 Nanofibrillated Cellulose Composites

NFC has attractive physical properties including mechanical, thermal and optical properties. The elastic modulus of cellulose I, which constitutes NFC and BC, was found to be ~138 GPa [80] and the tensile strength of wood fibres was ~2–3 GPa [25,26], which demonstrates the potential of NFC to be used as reinforcement for optically transparent thermoplastic and thermoset polymers. In addition, NFC has low CTE, which makes it an ideal potential candidate to reduce the CTE of thermoplastics and thermosets. In addition, NFC, when not chemically modified by TEMPO-mediated oxidation, has a relatively high thermal stability, which makes it an ideal additive to improve the thermal stability of most polymeric thermoplastic and thermoset resins.

The first optically transparent composites reinforced with plant CNF were reported by Iwamoto et al. [81]. In this work, the composite materials were fabricated by vacuum impregnation of NFC sheets using a low-viscosity neat acrylic resin. The optical transparency of the composite materials, as shown in Figure 8.6a, was evaluated by ultraviolet–visible spectrophotometry. At a wavelength of 600 nm, the regular transmittance of NFC/acrylic-resin composites, where the CNF having a diameter of 50–100 nm were prepared by a grinding method, was found to be ~70% as shown in Figure 8.6b. This was found to be lower compared to composites manufactured using BC nanofibres [82] where the regular transmittance at a wavelength of 600 nm was found to be higher than 80%. The main reason for that difference was most likely to be due to the smaller size of BC nanofibres compared to the plant CNF produced in this work.

Further efforts, in order for plant-CNF/acrylic and epoxy-resin composites to display light transmittance performance as high as BC/acrylic and epoxy-resin

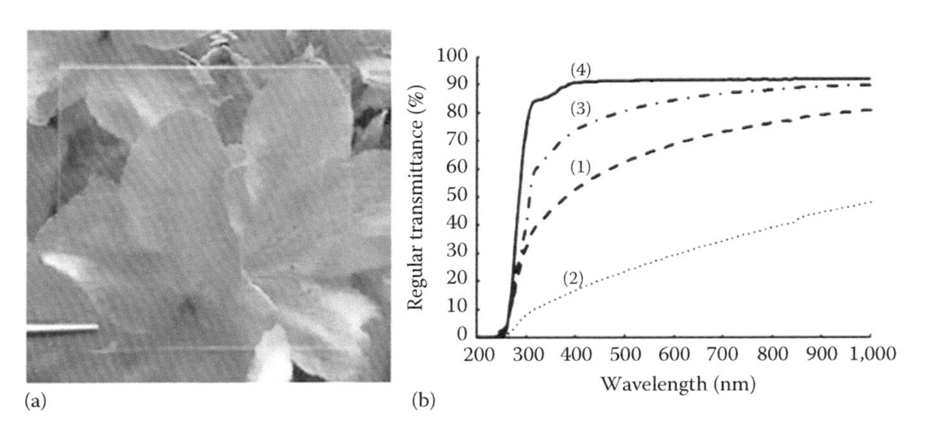

FIGURE 8.6 (a) Appearance of 45 μm thick grinder-treated pulp/acrylic resin sheet. (b) Regular light transmittance of (1) 45 μm thick grinder-treated pulp/acrylic resin sheet, (2) 53-μm thick high-pressure homogeniser-treated pulp/acrylic resin sheet, (3) 60 μm thick BC/acrylic resin sheet and (4) 40 μm thick acrylic resin sheet. (Adapted from Iwamoto, S. et al., *App. Phys. A*, 81, 1109–1112, 2005. With permission.)

composites, were carried out. The effect of various fibrillation conditions, using a grinder, on the optical transparency of plant-CNF/acrylic-resin composites was investigated [83]. The results suggested an improvement of light transmittance on increasing the number of passes of the nanocellulose suspension through the grinder. The light transmittance of the plant-CNF/acrylic-resin composites was found to increase strongly when the plant-CNFs were processed using a number of passes, from one to five passes. The value of the light transmission when the nanocellulose suspension was processed using five passes was found to be ~80% at a wavelength of 600 nm for the resulting composite materials containing ~11 wt.% of CNF. At higher number of passes, the light transmission was found to increase linearly, up to a value of ~85%, which is as good as the light transmittance of BC/acrylic-resin composites containing 62 wt.% BC nanofibres [82]. It is, however, unlikely that these very same plant-CNFs are as competitive as BC nanofibres with respect to light transmission if they are incorporated in an acrylic resin at similar weight fraction. The increase in the number of passes was, however, found to significantly reduce the crystallinity and degree of polymerisation of the nanofibres, which resulted in an increase in CTE and a decrease in the mechanical reinforcement effect. In addition, an augmentation in the number of passes increases the amount of energy involved in the fibrillation process, which may affect the cost of production of the CNF.

Another work, focusing on obtaining uniform plant-CNF having a diameter of 15 nm, evaluated and compared the optical transparency of plant-CNF/acrylic-resin composites [29]. At 600 nm, the light transmittance was ~80%, which was higher than for BC/acrylic composites at a similar weight fraction of CNF of 60 wt.% and at a similar thickness of ~50 μm. These BC nanofibres had a diameter of ~50 nm. This evaluation demonstrated, for the first time, that with respect to light transmittance, acrylic-resin-matrix composites reinforced with plant-CNF having a uniform diameter of 15 nm can compete with acrylic-resin matrix composites reinforced with BC nanofibres with respect to optical transparency.

Wood powder was also evaluated as a source of cellulose to obtain CNF for the production of mats of nanocellulose and nanocomposites by subsequent impregnation with acrylic resins [71]. The nanocomposite materials, having a CNF content of ~40 wt.% and a thickness range of 90–100 μm, were found to have a regular transmittance of ~80% at 600 nm. This demonstrated the suitability of these materials, with respect to optical properties to be used for the fabrication of organic light-emitting diodes. In addition, the materials were found to be suitable with respect to tensile mechanical properties and CTE. In addition, the material resins having a low Young's modulus were found to be more suitable for obtaining low CTE, high flexibility and ductile nanocomposite materials to be used for OLEDS in flexible and transparent displays compared to resins having higher Young's moduli.

The effect of delignification in the production of plant-based CNF for optically transparent composites has also been investigated [71]. In this work, several treatments for the removal of lignin from cellulose extracted from immature and mature bamboo were investigated. The results suggested a higher optical transparency of CNF/acrylic-resin composites when CNFs obtained from immature bamboo were used. These composite materials showed light transmittance almost as high as BC/acrylic-resin composites. This result is encouraging for the cheap and high volume

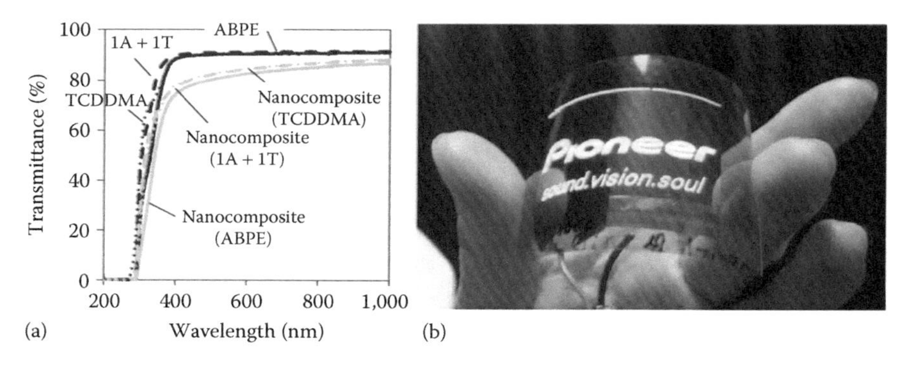

FIGURE 8.7 (a) Regular transmittance spectra of neat resins and nanocomposites. (b) Luminescence of an organic light-emitting diode deposited on a flexible, low CTE and optically transparent wood–cellulose nanocomposite. (Reprinted from Okahisa, Y et al., Optically transparent wood-cellulose nanocomposite as a base substrate for flexible organic light-emitting diode displays. *Composites Science and Technology*, 69(11–12), 1958–1961, 2009. With permission.)

production of CNF/acrylic-resin composite for optoelectronic applications. In this work, the idea of replacing glass substrates by high-performance nanocellulose composites and not polymers was illustrated. The need to add NFC to polymers to improve flexibility and reduce CTE of polymers while maintaining a high degree of optical transparency was demonstrated (Figure 8.7).

Other matrices including cellulose triacetate were introduced recently for the production of highly transparent, low-birefringent and highly tough nanocellulose composites [84]. The introduction of only 2.5 wt.% TEMPO-oxidised CNF modified by amine-terminated polyethylene glycol (PEG) chains into cellulose triacetate resulted in a significant improvement of Young's modulus and toughness owing to the presence of TEMPO-oxidised CNF but also due to surface-grafted PEG chains into cellulose triacetate. This compatibilisation may have improved filler–matrix interaction. Matrix effects, in other words the transcrystallisation of the cellulose triacetate matrix around TEMPO-oxidised CNF, may have also contributed to that impressive reinforcement effect. The CTE of cellulose triacetate was, however, maintained to a value of ~50 ppm/K. The study suggested the usefulness of using PEG-modified TEMPO-oxidised CNF for the reinforcement of optically transparent polymeric materials.

NFC also has attractive properties to be used for biomedical applications [3] such as tissue engineering where optical transparency is sometimes needed. Nanocomposite films made of chitosan and NFC were reported to be transparent [74]. Those materials were fabricated by adding NFC to a 1.5% w/v solution of chitosan that was prepared by dissolving chitosan powder into an aqueous acetic acid solution (1% v/v). The dispersions were then allowed to dry at 30°C for 16 h. The nanocomposite films were found to have high optical transparency varying from 80% to 20%, depending on the type of chitosan and concentration of NFC. NFC and chitosan being biocompatible materials, these nanocomposite materials could have the potential to be used for biomedical applications.

The combination of NFC and cellulose acetate butyrate (CAB) was also evaluated [85]. Nanocomposites, having a NFC weight fraction of 60%, were obtained by impregnation of NFC network by a dilute CAB solution. NFC was found to improve the mechanical properties of CAB while maintaining a high level of optical transparency. The transmittance of the nanocomposite materials was ~44% at a wavelength of 600 nm. These optically transparent nanocomposites could also find applications in the biomedical sector where optical transparency, biocompatibility and good mechanical properties are needed.

8.3.4 Bacterial Cellulose Composites

BC was mostly used as a model material to fabricate the first optically transparent nanocomposites, probably because no energy-intensive process is needed to obtain very fine CNF. BC was found to efficiently improve the mechanical properties of optically transparent thermoplastics and thermoset polymers as well as lowering their CTE while maintaining a high level of transparency.

The first report on optically transparent composites reinforced with bacterial CNF was published by Yano et al. [82]. In this work, BC networks were impregnated using low-viscosity acrylic and epoxy resins to fabricate BC/acrylic-resin and BC/epoxy-resin composites as shown in Figure 8.8a. The light transmittance of BC/epoxy-resin composites was found to be higher than 80% at a wavelength of 600 nm as displayed in Figure 8.8b. This value was very high considering that this composite material contained 65 wt.% of BC nanofibres, and its light transmittance was close to that of the light transmittance of an epoxy-resin sheet.

In a subsequent work, the sensitivity to refractive index of the polymer matrix of optically transparent BC bionanofibre composites was studied [86]. In this work, BC composites were fabricated with a range of acrylic resins, having refractive

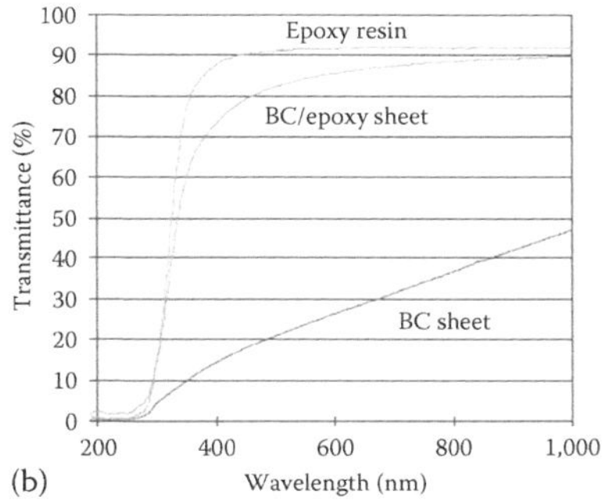

FIGURE 8.8 **(See colour insert.)** (a) Appearance of a 65 μm thick bacterial cellulose sheet: (1) with no added resin, (2) with acrylic resin, 62 wt.% fibre content and (3) with epoxy resin, 65 wt.% fibre content. (b) Light transmittance of a 65 μm thick bacterial cellulose/epoxy-resin sheet (65 wt.% fibre content), BC sheet and epoxy-resin sheet. (From Yano, H. et al.: Optically Transparent Composites Reinforced with Networks of Bacterial Nanofibres. *Advanced Materials*. 2005. 17. 153–155. Copyright Wiley-VCH Verlag GmbH & Co. KGaA. Reproduced with permission.)

index from 1.492 to 1.636. The effect of the refractive index on the transmittance of the composite materials was investigated. The results revealed low sensitivity of the transmittance of the composite materials at 20°C when acrylic resins having a refractive index ranging from 1.492 to 1.636 were used. In addition, the effect of temperature (from 20°C to 80°C) on the light transmittance of the composite materials was investigated. Interestingly, the results revealed that the temperature had no effect on the ability of the composite materials to transmit light.

In another work, the effect of the fibre content on the optical transparency of bacterial nanofibre-reinforced composites was investigated [87]. In this study, the fibre content range of the composite materials was ranging from 7.4 to 66.1 wt.%. This wide range of fibre contents was achieved by using a composite fabrication method that consists of a combination of heat drying and organic solvent exchange. The results revealed a linear decrease in the light transmission upon increase of BC fibre content of the composite materials. The reduction of light transmittance for the composite materials containing 66.1 wt.% was as small as ~13.7%. This demonstrated the exceptional potential of BC nanofibres to produce optically transparent composite materials having a wide range of fibre content. This is particularly important when trying to improve the mechanical properties of these composite materials while maintaining light transmittance as high as possible.

Cellulose and consequently BC can be chemically modified owing to its hydroxyl group-rich backbone structure. This is usually performed in order to render cellulose and cellulose-based composites less hydrophilic and often to improve its compatibility with hydrophobic polymeric matrices. A typical chemical modification of the surface of CNF is acetylation, which is performed through an esterification reaction. A first study on the effect of acetylation on the optical transparency of BC/acrylic-resin composites was reported by Nogi et al. [88]. BC nanofibres were slightly acetylated with acetic anhydride. This resulted in a significant reduction in hygroscopicity of the composite materials while maintaining high optical transparency. In addition, the acetylation of the BC nanofibres was found to prevent the degradation of the optical properties upon exposure to a temperature of 200°C for 3 h.

The effect of the acetylation of the surface of BC nanofibres on the optical transparency of BC/acrylic-resin composites was further optimised [68]. Acetylation of BC nanofibres resulted in a significant enhancement of the optical transparency of BC/acrylic-resin composites. Regular transmittance of the composite materials was found to significantly increase, up to a maximum degree of substitution of the hydroxyl groups of BC to the acetyl groups of 0.74. This improvement of regular light transmission was attributed to the change of refractive index of BC nanofibres due to acetylation, which was getting closer to the refractive index of the acrylic resin. At higher degree of substitution, the regular light transmittance was found to decrease, probably due to the refractive index of acetylated BC nanofibres that was getting too low compared to the refractive index of the acrylic resin. In addition to the improvement of optical transparency, the hygroscopicity of the composite materials was significantly reduced with water absorption of less than 0.5% at 20°C.

Another study investigated the potential of using BC/acrylic-resin composites in the electronic devices industry [8]. In this work, BC/acrylic-resin composites containing 5 wt.% of BC nanofibres were evaluated for the design of organic

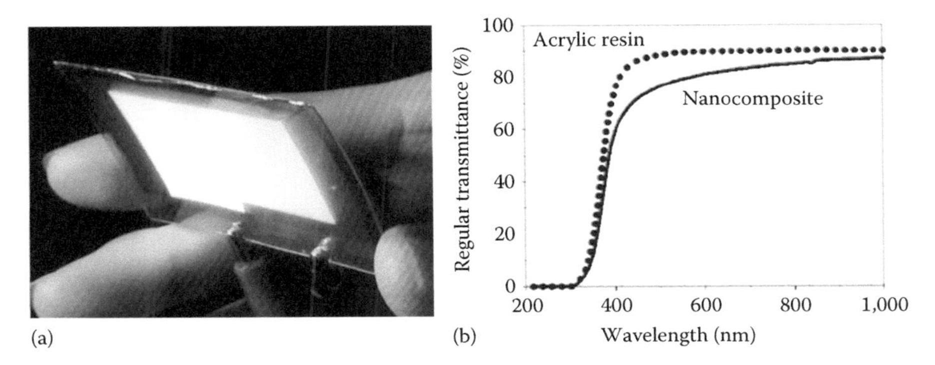

FIGURE 8.9 (a) Luminescence of an organic light-emitting diode deposited onto a bacterial cellulose/acrylic-resin composite. The luminescence area is 40 × 25 mm². (b) Regular transmittance spectra of neat acrylic resin matrix and a bacterial cellulose/acrylic-resin composite containing 5 wt.% of bacterial CNF. (From Nogi, M. and Yano, H.: Transparent Nanocomposites Based on Cellulose Produced by Bacteria Offer Potential Innovation in the Electronics Device Industry. *Advanced Materials*. 2008. 20. 1849–1852. Copyright Wiley-VCH Verlag GmbH & Co. KGaA. Reproduced with permission.)

light-emitting diodes as shown in Figure 8.9a. The suitability of these composite materials to replace glass in application where high optical transparency is needed is potentially possible owing to the high optical transparency of BC/acrylic-resin composites as shown in Figure 8.9b. In addition, these composite materials have the advantage of presenting ultralow CTE and high foldability, which make them suitable for the potential design of flexible electronics by roll-to-roll manufacturing process.

Two other works, focusing on the development of nanocellulose-based flexible and optically transparent substrates for organic light-emitting diode (OLED) display, was also reported [89,90]. Ummartyotin et al. [89] fabricated nanocomposite materials using BC (10–50 wt.%) and a polyurethane matrix material. The nanocomposite materials were found to have suitable optical, thermal and flexible properties to be used as substrate for OLED devices. This work also evaluated the successful fabrication of OLED onto the nanocomposite film substrates, showing promising performance such as the possibility to emit light while being bent.

Organic–inorganic nanocomposite membrane, fabricated using never-dried and dried BC, boehmite and epoxy-modified siloxane were found to have high optical transparency due to their homogeneous structure [91]. As shown in Figure 8.10, the transmittance of the nanocomposite material was ~80% at 600 nm. In addition, the nanocomposites were found to have good mechanical properties including flexibility. These nanocomposite materials could have potential to be used not only as substrate for optoelectronic applications but also as membranes for biomedical applications.

Besides showing great potential for optoelectronic applications, optically transparent BC-based nanocomposites have potential to be used for low-volume and high-value biomedical applications where the high cost of BC could be justified. Transparent BC–(PLA) polylactic acid nanocomposites were successfully fabricated

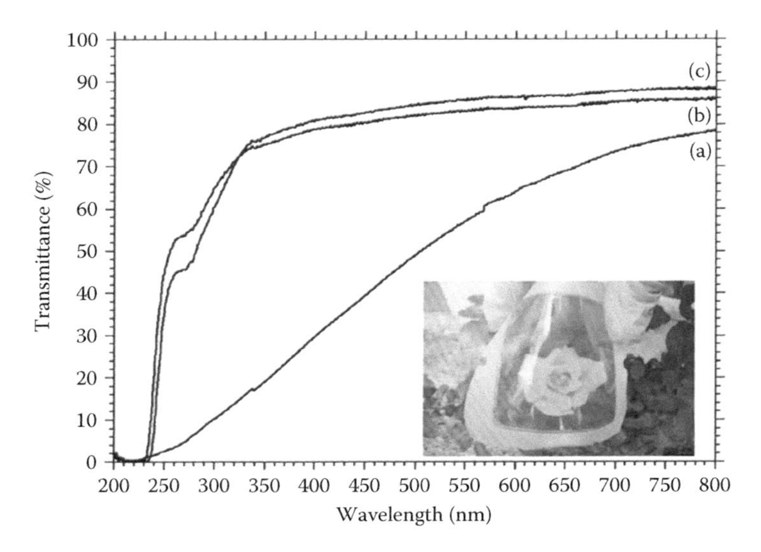

FIGURE 8.10 Optical transmission spectra for the (a) bacterial cellulose membrane, (b) Never-dried bacterial cellulose (NDBC) composite membrane and (c) DBC membrane. The inset is an image of a transparent NDBC composite membrane showing flexibility. The initial dried BC membrane is visible at the border. (Reprinted from *Compos. Part A: Appl. Sci. Manufact.*, 43, Barud, H.S. et al., Transparent bacterial cellulose–boehmite–epoxi-siloxane nanocomposites, 973–977, Copyright 2012, with permission from Elsevier.)

by solvent impregnation of BC sheets in a solution of PLA/chloroform at 1 wt.% and by subsequent drying at room temperature [69]. The incorporation of PLA in the porous structure of BC sheets resulted in an improvement of transparency with a transmittance of ~50% at a wavelength of 600 nm. In addition, the mechanical properties of PLA were successfully improved due to the presence of well-dispersed BC. This material may have potential to be used for biomedical applications due to the biocompatibility of both BC and PLA.

The concept of all-cellulose nanocomposites by selective surface dissolution [7,78] has also been reported for BC [79]. In this work, BC sheets were immersed in lithium chloride/*N*,*N*-dimethylacetamide for various times. Nanocomposite materials having a wide range of structures and physical properties were obtained. The authors also mention that the nanocomposite materials showed improved transparency due to surface dissolution compared to BC sheets although they did not quantify the percentage of light transmittance of their materials.

BC–(PVA) polyvinyl alcohol nanocomposites prepared by in situ and impregnation process were reported to be transparent, as shown in Figure 8.11, with only a small volume fraction of PVA added to the BC sheets [70]. The nanocomposite materials also displayed excellent mechanical properties including high ductility and work of fracture. This demonstrated the suitability of these materials to be used for biomedical applications where good optical transparency and mechanical properties are needed.

The incorporation of poly(3-hydroxybutyrate) (PHB) was also found to improve the transparency of BC sheets by impregnation of the BC sheets in a 2 wt.% PHB/

FIGURE 8.11 Optical photograph of a (a) pure bacterial cellulose, (b) in situ bacterial cellulose–polyvinyl alcohol nanocomposite, (c) impregnated bacterial cellulose–polyvinyl alcohol nanocomposite, and (d) pure PVA sheet. (Reprinted from *Mater. Lett.*, 64, Gea, S. et al., Bacterial cellulose–polyvinyl alcohol nanocomposites prepared by an in-situ process, 901–904, Copyright 2010, with permission from Elsevier.)

chloroform solution [92]. The resulting nanocomposite materials were found to have not only improved light transmittance compared to BC but also improved mechanical and thermal properties compared to PHB. The authors suggested the potential of these nanocomposite materials for tissue engineering.

Chemically cross-linked BC–PVA nanocomposites produced by in situ biosynthesis with interesting optical properties have also been reported [93]. In this work, glyoxal was introduced into the culture medium of *Gluconacetobacter xylinus* as a cross-linking agent not only to immobilise PVA during the purification steps of the nanocomposites but also to provide stability for possible applications of these materials for biomedical applications. The cross-linked nanocomposite materials displayed improved optical transparency compared to pure BC sheets.

Another work reported the fabrication of physically cross-linked BC–PVA nanocomposites [77]. It is indeed important to cross-link the composite system in order to stabilise the material when it is exposed to body fluids and temperature. The physical cross-linking was induced *via* a cyclic freezing and thawing, promoting mechanical entanglement and hydrogen-bonding interactions. The nanocomposite materials were found to display a porous architecture within the PVA matrix. In addition, the materials were found to have high compressive strength due to a synergistic effect between PVA and BC. The presence of BC significantly improved the thermo-mechanical properties as well as moisture and dimensional stability. The presence of PVA rendered the BC sheets transparent and more extensible. The presence of BC also helped in promoting cell proliferation, demonstrating the potential of these materials to be used for biomedical applications including artificial cornea, veins, arteries and tympanic membranes.

8.4 CONCLUSION AND FUTURE DIRECTIONS

This chapter reports the development of optically transparent nanocellulose papers and composites during the past 12 years. The fabrication methods of nanocellulose papers and composites as well as their optical properties and applications are reported. Papers and composites that are designed using NFC are particularly good candidates to be used for low-cost and high-volume flexible optoelectronics and packaging applications owing to their high availability and relatively low cost. On the other hand, nanopapers and composites based on BC show great potential to be used for low-volume and high-value biomedical applications including skin and tympanic membrane regeneration as well as for the design of artificial cornea. More applications need to be explored for these materials and more particularly with respect to the use of NFC-based papers and composites for biomedical applications.

ACKNOWLEDGEMENTS

The author acknowledges Fondecyt (Chile) for providing funding under the postdoctoral grant N° 3140036. This chapter was written in the framework of this research grant.

REFERENCES

1. Koga H, Nogi M. Flexible paper electronics. In: Ogawa S, (Ed.). *Organic Electronics Materials and Devices*, Tokyo, Japan: Springer; 2015. pp. 101–115.
2. Jung YH, Chang T-H, Zhang H, Yao C, Zheng Q, Yang VW et al. High-performance green flexible electronics based on biodegradable cellulose nanofibril paper. *Nature Communications*. 2015;6:1–11.
3. Lin N, Dufresne A. Nanocellulose in biomedicine: Current status and future prospect. *European Polymer Journal*. 2014;59:302–325.
4. Li F, Mascheroni E, Piergiovanni L. The potential of nanocellulose in the packaging field: A review. *Packaging Technology and Science*. 2015;28(6):475–508.
5. Wüstenberg T. *Nanocellulose, Cellulose and Cellulose Derivatives in the Food Industry*, Weinheim, Germany: Wiley-VCH Verlag GmbH & Co. KGaA; 2014. pp. 491–510.
6. Gómez HC, Serpa A, Velásquez-Cock J, Gañán P, Castro C, Vélez L et al. Vegetable nanocellulose in food science: A review. *Food Hydrocolloids*. 2016;57:178–186.
7. Nishino T, Matsuda I, Hirao K. All-cellulose composite. *Macromolecules*. 2004;37(20):7683–7687.
8. Nogi M, Yano H. Transparent nanocomposites based on cellulose produced by bacteria offer potential innovation in the electronics device industry. *Advanced Materials*. 2008;20(10):1849–1852.
9. Bielecki S, Kalinowska H, Krystynowicz A, Kubiak K, Kołodziejczyk M, de Groeve M. Wound dressings and cosmetic materials from bacterial nanocellulose. In: M. Gama, P. Gatenholm and D. Klemm, (Eds.). *Bacterial NanoCellulose*, Boca Raton, FL: CRC Press; 2012. pp. 157–174.
10. Kim J, Kim SW, Park S, Lim KT, Seonwoo H, Kim Y et al. Bacterial cellulose nanofibrillar patch as a wound healing platform of tympanic membrane perforation. *Advanced Healthcare Materials*. 2013;2(11):1525–1531.
11. Jia H, Jia Y, Wang J, Hu Y, Zhang Y, Jia S. Potentiality of bacterial cellulose as the scaffold of tissue engineering of cornea. *2009 2nd International Conference on Biomedical Engineering and Informatics*, IEEE, Phuket, Thailand, 2009. pp. 1–5.

12. Wang J, Gao C, Zhang Y, Wan Y. Preparation and in vitro characterization of BC/PVA hydrogel composite for its potential use as artificial cornea biomaterial. *Materials Science and Engineering: C*. 2010;30(1):214–218.
13. Klemm D, Philipp B, Heinze T, Heinze U, Wagenknecht W. *Comprehensive Cellulose Chemistry: Fundamentals and Analytical Methods*, Vol. 1, Weinheim, Germany: Wiley-VCH; 1998.
14. Astbury WT, Marwick TC, Bernal JD. X-Ray analysis of the structure of the wall of valonia ventricosa.–I. *Proceedings of the Royal Society of London B: Biological Sciences*. 1932;109(764):443–450.
15. Brown AJ. On an acetic ferment which forms cellulose. *Journal of the Chemical Society*. 1886;49:432–439.
16. Brown AJ. XIX.–The chemical action of pure cultivations of bacterium aceti. *Journal of the Chemical Society, Transactions*. 1886;49:172–187.
17. Rånby BG. Physico-chemical investigations on animal cellulose (Tunicin). *Arkiv Kemi*. 1952;4:241–248.
18. Herrick FW, Casebier RL, Hamilton JK, Sandberg KR. Microfibrillated cellulose: Morphology and accessibility. *Journal of Applied Polymer Science: Applied Polymer Symposium*. 1983;37:797–813.
19. Turbak AF, Snyder FW, Sandberg KR. Microfibrillated cellulose, a new cellulose product: Properties, uses, and commercial potential. *Journal of Applied Polymer Science: Applied Polymer Symposium*. 1983;37:815–827.
20. Guhados G, Wan W, Hutter JL. Measurement of the elastic modulus of single bacterial cellulose fibers using atomic force microscopy. *Langmuir*. 2005;21(14):6642–6646.
21. Iwamoto S, Kai W, Isogai A, Iwata T. Elastic modulus of single cellulose microfibrils from tunicate measured by atomic force microscopy. *Biomacromolecules*. 2009;10(9):2571–2576.
22. Tanpichai S, Quero F, Nogi M, Yano H, Young RJ, Lindström T et al. Effective young's modulus of bacterial and microfibrillated cellulose fibrils in fibrous networks. *Biomacromolecules*. 2012;13(5):1340–1349.
23. Šturcová A, Davies GR, Eichhorn SJ. Elastic modulus and stress-transfer properties of tunicate cellulose whiskers. *Biomacromolecules*. 2005;6(2):1055–1061.
24. Rusli R, Eichhorn SJ. Determination of the stiffness of cellulose nanowhiskers and the fiber-matrix interface in a nanocomposite using Raman spectroscopy. *Applied Physics Letters*. 2008;93(3):033111.
25. Page DH, El-Hosseiny F. The mechanical properties of single wood pulp fibres, Part IV. Fibril angle and the shape of the stress-strain curve. *Journal of Pulp and Paper Science*. 1983;9:99–100.
26. Page DH, El-Hosseiny F, Winkler K. Behaviour of single wood fibres under axial tensile strain. *Nature*. 1971;229(5282):252–253.
27. Nordli HR, Chinga-Carrasco G, Rokstad AM, Pukstad B. Producing ultrapure wood cellulose nanofibrils and evaluating the cytotoxicity using human skin cells. *Carbohydrate Polymers*. 2016;150:65–73.
28. Nogi M, Iwamoto S, Nakagaito AN, Yano H. Optically transparent nanofiber paper. *Advanced Materials*. 2009;21(16):1595–1598.
29. Abe K, Iwamoto S, Yano H. Obtaining cellulose nanofibers with a uniform width of 15 nm from wood. *Biomacromolecules*. 2007;8(10):3276–3278.
30. Watanabe Y, Kitamura S, Kawasaki K, Kato T, Uegaki K, Ogura K et al. Application of a water jet system to the pretreatment of cellulose. *Biopolymers*. 2011;95(12):833–839.
31. Fukuzumi H, Saito T, Iwata T, Kumamoto Y, Isogai A. Transparent and high gas barrier films of cellulose nanofibers prepared by TEMPO-mediated oxidation. *Biomacromolecules*. 2009;10(1):162–165.
32. Nechyporchuk O, Belgacem MN, Bras J. Production of cellulose nanofibrils: A review of recent advances. *Industrial Crops and Products*. 2016;93:2–25.

33. Zhao J, Wei Z, Feng X, Miao M, Sun L, Cao S et al. Luminescent and transparent nanopaper based on rare-earth up-converting nanoparticle grafted nanofibrillated cellulose derived from garlic skin. *ACS Applied Materials & Interfaces*. 2014;6(17):14945–14951.
34. Toivonen MS, Kaskela A, Rojas OJ, Kauppinen EI, Ikkala O. Ambient-dried cellulose nanofibril aerogel membranes with high tensile strength and their use for aerosol collection and templates for transparent, flexible devices. *Advanced Functional Materials*. 2015;25(42):6618–6626.
35. Nogi M, Yano H. Optically transparent nanofiber sheets by deposition of transparent materials: A concept for a roll-to-roll processing. *Applied Physics Letters*. 2009;94(23):233117.
36. Saito T, Kimura S, Nishiyama Y, Isogai A. Cellulose nanofibers prepared by TEMPO-mediated oxidation of native cellulose. *Biomacromolecules*. 2007;8(8):2485–2491.
37. Lavoine N, Bras J, Saito T, Isogai A. Improvement of the thermal stability of TEMPO-oxidized cellulose nanofibrils by heat-induced conversion of ionic bonds to amide bonds. *Macromolecular Rapid Communications*. 2016:37(13):1033–1039.
38. Nogi M, Kim C, Sugahara T, Inui T, Takahashi T, Suganuma K. High thermal stability of optical transparency in cellulose nanofiber paper. *Applied Physics Letters*. 2013;102(18):181911.
39. Nogi M, Karakawa M, Komoda N, Yagyu H, Nge TT. Transparent conductive nanofiber paper for foldable solar cells. *Scientific Reports*. 2015;5:17254.
40. Zhu H, Fang Z, Preston C, Li Y, Hu L. Transparent paper: Fabrications, properties, and device applications. *Energy & Environmental Science*. 2014;7(1):269–287.
41. Hu L, Zheng G, Yao J, Liu N, Weil B, Eskilsson M et al. Transparent and conductive paper from nanocellulose fibers. *Energy & Environmental Science*. 2013;6(2):513–518.
42. Fang Z, Zhu H, Preston C, Han X, Li Y, Lee S et al. Highly transparent and writable wood all-cellulose hybrid nanostructured paper. *Journal of Materials Chemistry C*. 2013;1(39):6191–6197.
43. Fang Z, Zhu H, Yuan Y, Ha D, Zhu S, Preston C et al. Novel nanostructured paper with ultrahigh transparency and ultrahigh haze for solar cells. *Nano Letters*. 2014;14(2):765–773.
44. Qing Y, Sabo R, Wu Y, Zhu JY, Cai Z. Self-assembled optically transparent cellulose nanofibril films: Effect of nanofibril morphology and drying procedure. *Cellulose*. 2015;22(2):1091–1102.
45. Galland S, Berthold F, Prakobna K, Berglund LA. Holocellulose nanofibers of high molar mass and small diameter for high-strength nanopaper. *Biomacromolecules*. 2015;16(8):2427–2435.
46. Yagyu H, Saito T, Isogai A, Koga H, Nogi M. Chemical modification of cellulose nanofibers for the production of highly thermal resistant and optically transparent nanopaper for paper devices. *ACS Applied Materials & Interfaces*. 2015;7(39):22012–22017.
47. Zhang W, Zhang X, Lu C, Wang Y, Deng Y. Flexible and transparent paper-based ionic diode fabricated from oppositely charged microfibrillated cellulose. *The Journal of Physical Chemistry C*. 2012;116(16):9227–9234.
48. Zhu H, Narakathu BB, Fang Z, Aijazi AT, Joyce M, Atashbar M et al. A gravure printed antenna on shape-stable transparent nanopaper. *Nanoscale*. 2014;6(15):9110–9115.
49. Huang J, Zhu H, Chen Y, Preston C, Rohrbach K, Cumings J et al. Highly transparent and flexible nanopaper transistors. *ACS Nano*. 2013;7(3):2106–2113.
50. Grau G, Kitsomboonloha R, Swisher SL, Kang H, Subramanian V. Printed transistors on paper: Towards smart consumer product packaging. *Advanced Functional Materials*. 2014;24(32):5067–5074.
51. Fujisaki Y, Koga H, Nakajima Y, Nakata M, Tsuji H, Yamamoto T et al. Transparent nanopaper-based flexible organic thin-film transistor array. *Advanced Functional Materials*. 2014;24(12):1657–1663.

52. Xue J, Song F, Yin XW, Wang XL, Wang YZ. Let it shine: A transparent and photoluminescent foldable nanocellulose/quantum dot paper. *ACS Applied Materials & Interfaces*. 2015;7(19):10076–10079.
53. Miao M, Zhao J, Feng X, Cao Y, Cao S, Zhao Y et al. Fast fabrication of transparent and multi-luminescent TEMPO-oxidized nanofibrillated cellulose nanopaper functionalized with lanthanide complexes. *Journal of Materials Chemistry C*. 2015;3(11):2511–2517.
54. Xiong R, Han Y, Wang Y, Zhang W, Zhang X, Lu C. Flexible, highly transparent and iridescent all-cellulose hybrid nanopaper with enhanced mechanical strength and writable surface. *Carbohydrate Polymers*. 2014;113:264–271.
55. Nagashima K, Koga H, Celano U, Zhuge F, Kanai M, Rahong S et al. Cellulose nanofiber paper as an ultra flexible nonvolatile memory. *Scientific Reports*. 2014;4:5532.
56. Morales-Narváez E, Golmohammadi H, Naghdi T, Yousefi H, Kostiv U, Horák D et al. Nanopaper as an optical sensing platform. *ACS Nano*. 2015;9(7):7296–7305.
57. Liu J, Yang C, Wu H, Lin Z, Zhang Z, Wang R et al. Future paper based printed circuit boards for green electronics: Fabrication and life cycle assessment. *Energy & Environmental Science*. 2014;7(11):3674–3682.
58. Hu L, Liu N, Eskilsson M, Zheng G, McDonough J, Wågberg L et al. Silicon-conductive nanopaper for Li-ion batteries. *Nano Energy*. 2013;2(1):138–145.
59. Zhong J, Zhu H, Zhong Q, Dai J, Li W, Jang S-H et al. Self-powered human-interactive transparent nanopaper systems. *ACS Nano*. 2015;9(7):7399–7406.
60. Preston C, Fang Z, Murray J, Zhu H, Dai J, Munday JN et al. Silver nanowire transparent conducting paper-based electrode with high optical haze. *Journal of Materials Chemistry C*. 2014;2(7):1248–1254.
61. Koga H, Nogi M, Komoda N, Nge TT, Sugahara T, Suganuma K. Uniformly connected conductive networks on cellulose nanofiber paper for transparent paper electronics. *NPG Asia Materials*. 2014;6:e93.
62. Hsieh M-C, Kim C, Nogi M, Suganuma K. Electrically conductive lines on cellulose nanopaper for flexible electrical devices. *Nanoscale*. 2013;5(19):9289–9295.
63. Toivonen MS, Kurki-Suonio S, Schacher FH, Hietala S, Rojas OJ, Ikkala O. Water-resistant, transparent hybrid nanopaper by physical cross-linking with chitosan. *Biomacromolecules*. 2015;16(3):1062–1071.
64. Kucińska-Lipka J, Gubanska I, Janik H. Bacterial cellulose in the field of wound healing and regenerative medicine of skin: Recent trends and future prospectives. *Polymer Bulletin*. 2015;72(9):2399–2419.
65. Eichhorn SJ, Dufresne A, Aranguren M, Marcovich NE, Capadona JR, Rowan SJ et al. Review: Current international research into cellulose nanofibres and nanocomposites. *Journal of Materials Science*. 2010;45(1):1–33.
66. Beecroft LL, Ober CK. Nanocomposite materials for optical applications. *Chemistry of Materials*. 1997;9(6):1302–1317.
67. Novak BM. Hybrid nanocomposite materials—between inorganic glasses and organic polymers. *Advanced Materials*. 1993;5(6):422–433.
68. Ifuku S, Nogi M, Abe K, Handa K, Nakatsubo F, Yano H. Surface modification of bacterial cellulose nanofibers for property enhancement of optically transparent composites: Dependence on acetyl-group DS. *Biomacromolecules*. 2007;8(6):1973–1978.
69. Kim Y, Jung R, Kim H-S, Jin H-J. Transparent nanocomposites prepared by incorporating microbial nanofibrils into poly(L-lactic acid). *Current Applied Physics*. 2009;9(1, Supplement):S69–S71.
70. Gea S, Bilotti E, Reynolds CT, Soykeabkeaw N, Peijs T. Bacterial cellulose–poly(vinyl alcohol) nanocomposites prepared by an in-situ process. *Materials Letters*. 2010;64(8):901–904.

71. Okahisa Y, Yoshida A, Miyaguchi S, Yano H. Optically transparent wood-cellulose nanocomposite as a base substrate for flexible organic light-emitting diode displays. *Composites Science and Technology*. 2009;69(11–12):1958–1961.
72. Ben Azouz K, Ramires EC, Van den Fonteyne W, El Kissi N, Dufresne A. Simple method for the melt extrusion of a cellulose nanocrystal reinforced hydrophobic polymer. *ACS Macro Letters*. 2012;1(1):236–240.
73. Sapkota J, Jorfi M, Weder C, Foster EJ. Reinforcing poly(ethylene) with cellulose nanocrystals. *Macromolecular Rapid Communications*. 2014;35(20):1747–1753.
74. Fernandes SCM, Freire CSR, Silvestre AJD, Neto CP, Gandini A, Berglund LA et al. Transparent chitosan films reinforced with a high content of nanofibrillated cellulose. *Carbohydrate Polymers*. 2010;81(2):394–401.
75. Quero F, Coveney A, Lewandowska AE, Richardson RM, Díaz-Calderón P, Lee K-Y et al. Stress transfer quantification in gelatin-matrix natural composites with tunable optical properties. *Biomacromolecules*. 2015;16(6):1784–1793.
76. Virtanen S, Vartianen J, Setala H, Tammelin T, Vuoti S. Modified nanofibrillated cellulose-polyvinyl alcohol films with improved mechanical performance. *RSC Advances*. 2014;4(22):11343–11350.
77. Castro C, Zuluaga R, Rojas OJ, Filpponen I, Orelma H, Londono M et al. Highly percolated poly(vinyl alcohol) and bacterial nanocellulose synthesized in situ by physical-crosslinking: Exploiting polymer synergies for biomedical nanocomposites. *RSC Advances*. 2015;5(110):90742–90749.
78. Nishino T, Arimoto N. All-cellulose composite prepared by selective dissolving of fiber surface. *Biomacromolecules*. 2007;8(9):2712–2716.
79. Soykeabkaew N, Sian C, Gea S, Nishino T, Peijs T. All-cellulose nanocomposites by surface selective dissolution of bacterial cellulose. *Cellulose*. 2009;16(3):435–444.
80. Nishino T, Takano K, Nakamae K. Elastic modulus of the crystalline regions of cellulose polymorphs. *Journal of Polymer Science Part B: Polymer Physics*. 1995;33(11):1647–1651.
81. Iwamoto S, Nakagaito AN, Yano H, Nogi M. Optically transparent composites reinforced with plant fiber-based nanofibers. *Applied Physics A*. 2005;81(6):1109–1112.
82. Yano H, Sugiyama J, Nakagaito AN, Nogi M, Matsuura T, Hikita M et al. Optically transparent composites reinforced with networks of bacterial nanofibres. *Advanced Materials*. 2005;17(2):153–155.
83. Iwamoto S, Nakagaito AN, Yano H. Nano-fibrillation of pulp fibres for the processing of transparent nanocomposites. *Applied Physics A*. 2007;89(2):461–466.
84. Soeta H, Fujisawa S, Saito T, Berglund L, Isogai A. Low-birefringent and highly tough nanocellulose-reinforced cellulose triacetate. *ACS Applied Materials & Interfaces*. 2015;7(20):11041–11046.
85. Jonoobi M, Aitomäki Y, Mathew AP, Oksman K. Thermoplastic polymer impregnation of cellulose nanofibre networks: Morphology, mechanical and optical properties. *Composites Part A: Applied Science and Manufacturing*. 2014;58:30–35.
86. Nogi M, Handa K, Nakagaito AN, Yano H. Optically transparent bionanofiber composites with low sensitivity to refractive index of the polymer matrix. *Applied Physics Letters*. 2005;87(24):243110.
87. Nogi M, Ifuku S, Abe K, Handa K, Nakagaito AN, Yano H. Fiber-content dependency of the optical transparency and thermal expansion of bacterial nanofiber reinforced composites. *Applied Physics Letters*. 2006;88(13):133124.
88. Nogi M, Abe K, Handa K, Nakatsubo F, Ifuku S, Yano H. Property enhancement of optically transparent bionanofiber composites by acetylation. *Applied Physics Letters*. 2006;89(23):233123.

89. Ummartyotin S, Juntaro J, Sain M, Manuspiya H. Development of transparent bacterial cellulose nanocomposite film as substrate for flexible organic light emitting diode (OLED) display. *Industrial Crops and Products*. 2012;35(1):92–97.
90. Pinto ERP, Barud HS, Silva RR, Palmieri M, Polito WL, Calil VL et al. Transparent composites prepared from bacterial cellulose and castor oil based polyurethane as substrates for flexible OLEDs. *Journal of Materials Chemistry C*. 2015;3(44):11581–11588.
91. Barud HS, Caiut JMA, Dexpert-Ghys J, Messaddeq Y, Ribeiro SJL. Transparent bacterial cellulose–boehmite–epoxi-siloxane nanocomposites. *Composites Part A: Applied Science and Manufacturing*. 2012;43(6):973–977.
92. Zhijiang C, Guang Y. Optical nanocomposites prepared by incorporating bacterial cellulose nanofibrils into poly(3-hydroxybutyrate). *Materials Letters*. 2011;65(2):182–184.
93. Castro C, Vesterinen A, Zuluaga R, Caro G, Filpponen I, Rojas O et al. In situ production of nanocomposites of poly(vinyl alcohol) and cellulose nanofibrils from Gluconacetobacter bacteria: Effect of chemical cross-linking. *Cellulose*. 2014;21(3):1745–1756.

9 Application of Nanocellulose as Pickering Emulsifier

Isabelle Capron

CONTENTS

9.1 Introduction 175
9.2 Pickering Emulsions 177
9.3 Nanocellulose and Emulsions 180
 9.3.1 Adsorption Mechanism 185
 9.3.2 Impact of the Shape of Cellulose Nanocrystals 187
 9.3.3 High Internal Phase Emulsions 188
9.4 Towards New Formulations 189
9.5 Conclusion 192
References 192

9.1 INTRODUCTION

As a chemical raw material, cellulose has been used in the form of fibres or derivatives for nearly 150 years for a wide spectrum of products and materials in daily life. More specifically, development of the nanotechnology during the last decades has resulted in a renewed interest in nanomaterials. As being part of replacement material for petroleum-based structures, nanocelluloses are currently being considered as construction units for nanoscale materials engineering [1–4]. The materials community has paid a tremendous level of attention that does not appear to be relenting because nanocelluloses offer unsurpassed physical and chemical properties, as well as inherent renewability, sustainability and abundance. The designation 'nanocellulose' commonly refers to various types of cellulosic substrates with at least one dimension in nanoscale. Depending on the source and production conditions, which influence the dimensions, composition and properties, nanocelluloses can be divided into three types:

- Cellulose nanofibrils (CNFs) have a cross-section in the nanometre range. They are obtained from multiple mechanical shearing actions to delaminate individual microfibrils, having both crystalline and amorphous domains, from cellulosic fibres (Figure 9.1).

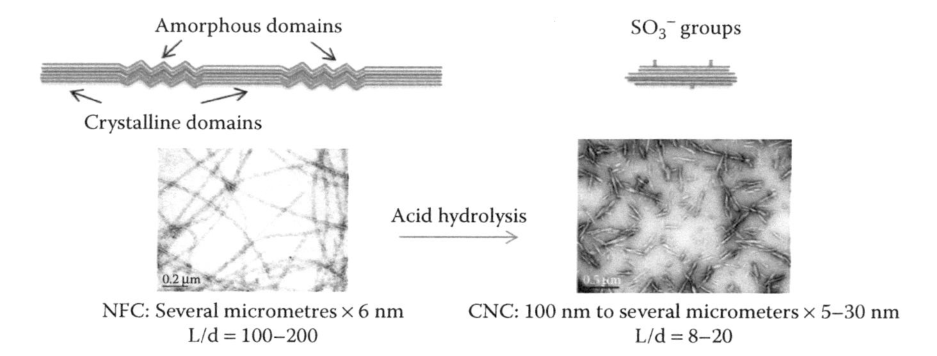

FIGURE 9.1 Schematic representation and transmission electron microscopic images of NFC obtained from wood, and crystalline CNC resulting from removal of amorphous domains by acid hydrolysis. Dimensions as well as the most usual aspect ratio (length/cross-section) are indicated.

- Cellulose nanocrystals (CNCs) are issued from chemically induced destructuring strategy, such as acid hydrolysis, through the removal of amorphous regions and preservation of highly crystalline structure. Released rod-like nanoparticles (CNCs) present a diameter of 5–30 nm and length of 100 nm up to several micrometres, according to the source (Figure 9.1).
- Bacterial cellulose is synthesised by bacteria (such as *Acetobacter xylinum*) in a pure form, which requires no intensive processing to remove unwanted impurities or contaminants such as lignin, pectin and hemicellulose. It is the source of cellulose the best adapted to biomedical applications.

Nanocelluloses have been the subject of intense research since their discovery in the early 1950s [5–7], yielding numerous reports that have tremendously increased in the last two decades [2,3,8,9]. The impressive interest for these nano-objects is of course owing to their biocompatibility, biodegradability and renewability. Besides, crystalline celluloses have a greater axial elastic modulus than Kevlar, and their mechanical properties make them comparable to reinforcement materials. They have high aspect ratio, low density (1.5–1.6 g/cm^3) and a large panel of interaction possibilities through low-energy forces that can be a powerful tool to implement new and attractive materials. Indeed, due to their chemical structure and their intrinsic organisation, unmodified nanocellulose crystalline parts can interact with other components through the following bonds:

- Hydrogen bonds, mainly due to the three hydroxyl (OH) groups on equatorial position of glucose monomers (5–40 kJ/mol).
- Van der Waals interactions, due to the aliphatic groups of glucose monomers. These interactions are weak (2–10 kJ/mol) but are complementary to hydrogen bonds to stabilise the intermolecular association.

- Electrostatic interactions, due to the occurrence of ionic moieties on the crystal surface, usually arising from preparation procedures.
- Uneven exposure of chemical functions on the crystal faces, resulting in hydrophobic or hydrophilic character of the faces.

Cellulose nanoparticles (CNs) have a reactive surface of hydroxyl side groups that is known to facilitate grafting chemical species to achieve different surface properties (surface functionalisation). Such modifications that might be covalent or non-covalent have led to several reviews attesting the large range of possibilities [10–14]. Moreover, due to intrinsic organisation, nanocelluloses and mainly nanocrystals display relevant asset such as excellent thermal properties, or, as asymmetric rod-like particles, they form liquid crystals [15], allowing optical properties. As a consequence, CNs are ideal materials on which to base a new biopolymer composites industry. Design of such new materials requires structuration at a nanoscale level. Particularly, three-dimensional interconnected networks with many open pores in the aerogels allow the quick access and diffusion of ions and molecules, which make aerogels suitable as electrode materials for batteries and supercapacitors [16], elastic conductors [17], catalyst supports [6] and oil sorbent in pollution [18]. In addition to its unique mechanic and surface characteristics, nanocellulose has shown biological properties. As nanocellulose is biocompatible (very slowly degraded within the human body) and shows no or very low toxicity [19], it has attracted a lot of interest for biomedicine applications such as substitute biomaterials for blood vessels or soft tissues replacement, tissues regeneration and nanocellulose-based drug delivery systems [20].

Furthermore, as already noticed earlier, nanocelluloses exhibit distinct crystalline planes exposed at the surface that are not chemically equivalent. Our group showed in 2011 that unmodified CNCs can effectively adsorb at the oil–water interfaces and produce highly stable and deformable oil-in-water emulsions [21], revealing affinity for both aqueous and hydrophobic phases. Such amphiphilic character of unmodified CNCs has been extensively developed in biphasic systems, revealing that CNCs can be used without any additional surfactant and form solid-stabilised emulsions, the so-called Pickering emulsions.

9.2 PICKERING EMULSIONS

Emulsions are out-of-thermodynamic equilibrium systems and are usually kinetically stabilised by surface-active species known as surfactants. Some solid particles of colloidal size proved to be able to strongly adsorb at an oil–water interfaces, forming highly stable emulsions. Referring to the name of one of the pioneering researchers in this field [22,23], such solid-stabilised emulsions are known as Pickering emulsions. After their description in the early twentieth century, these emulsions have been overlooking to regain interest a century later [24,25]. A large range of nanoparticles of various chemical surfaces and shapes can lead to Pickering emulsions. To be adsorbed at the interface, the particle should display a double affinity for both phases, in order to be wetted by the two liquids—the more amphiphilic (i.e. hydrophilic/lipophilic ratio close to one) and the better stabilised. This is revealed by the energy required to desorb a spherical particle from the interface, which can be written as follows [24]:

$$E_{\text{desorption}}(\text{sphere}) = \gamma_{o/w} \times \pi R^2 \times (1 - |\cos\theta|)^2 \tag{9.1}$$

where:

$\gamma_{o/w}$ is the interfacial tension of the pristine oil–water interface
θ is the contact angle between the particle tangent and the interface through the water phase
R is the particle radius

The energy required to desorb the particle from the interface will depend on the hydrophilic/hydrophobic character of the surface as defined by the contact angle that qualifies the particle affinity for either phase. Equivalently to the Bancroft rule for surfactant-stabilised emulsions, the continuous phase for Pickering emulsions is the one in which the particles are preferentially dispersed. This means that mainly hydrophilic particles are able to stabilise direct oil-in-water emulsions, whereas predominantly lipophilic particles are used to stabilise reverse water-in-oil emulsions (Figure 9.2).

Consequently, the strength of the desorption energy in k_BT (where k_B is the Boltzmann constant and T is the absolute temperature) is strongly dependent on contact angle. As an example, for a spherical particle with a given radius R of 25 nm and

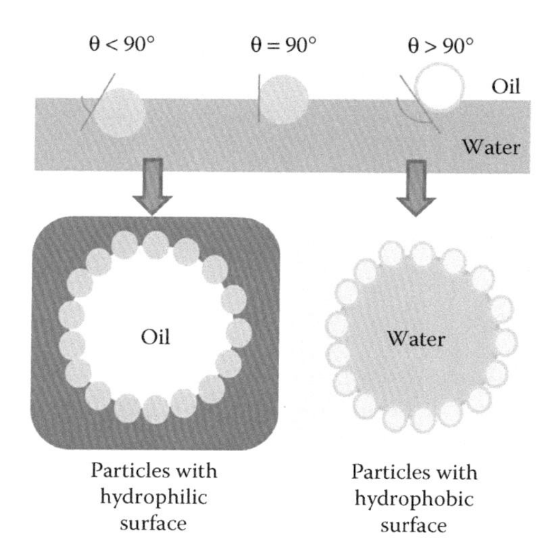

FIGURE 9.2 Probable positioning of spherical particles at oil–water interface according to their surface nature. Hydrophilic particles with a contact angle θ taking values lower than 90° lead to oil-in-water emulsion, whereas hydrophobic particles with a contact angle θ taking values higher than 90° lead to water-in-oil emulsion.

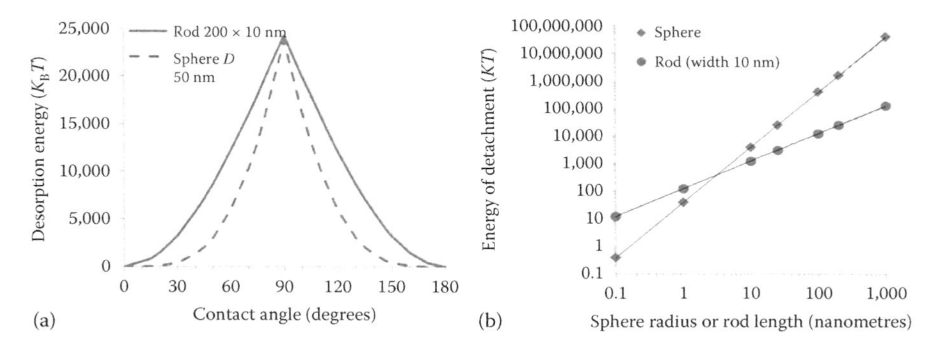

FIGURE 9.3 Comparative energy required for detaching a particle from a planar interface for different aspect ratios as a function of (a) the contact angle for a sphere (50-nm diameter) and rod (200 × 10 nm), and (b) as a sphere or as a rod with e thickness of 10 nm, varying in diameter and length, respectively, taking θ as 90° and $\gamma_{o/w}$ as 0.05 N/m at 298 K.

water–hexadecane interfacial tension $\gamma_{o/w}$ of 0.05 N/m, Equation 9.1 shows that the particle is strongly held in the interface for the most amphiphilic condition (contact angle of 90°) (Figure 9.3a). Desorption energy decreases when contact angle varies on either side of 90° but is still higher than 400 k_BT on a large range between 30° and 150°. It means that small modifications of hydrophilic or hydrophobic particles might be enough to produce particles susceptible to stabilise an oil–water interface.

When rod-like particles adsorb at the interface, the surface in contact with the interface leads to a new expression. Taking into account that the particles adsorb on the larger surface available, the adsorption energy is then given by

$$E_{\text{desorption}}(\text{rod}) = -l \times b \times \gamma_{o/w} \times \left(1 - |\cos\theta|\right) \tag{9.2}$$

where:

l is the length
b is the width of the rods

Taking dimensions of a rod 200 nm in length and 10 nm in width (approximately CNCs dimensions), the same trend is observed (Figure 9.3a) with extremely high energy. As the peak is less sharp, high energy is obtained on an even larger range. The same desorption energy calculated from Equation 9.2 followed for widths of 1–100 nm, lengths 0.1–200 nm and an interfacial tension $\gamma_{o/w}$ of 0.05 N/m.

As the desorption energy required for detaching surfactants is of several k_BT, it is thousands of times larger in the case of particles, in general. Consequently, once adsorbed at the interface, nanoparticles, either sphere and rods, are strongly anchored, and the adsorption can be considered irreversible. Such a strong adsorption confers an outstanding stability to Pickering emulsions as compared with surfactant-stabilised emulsions. A long-term stability is of great interest for storage.

Beyond storage, this strong adsorption leads to several advantages over surfactants. A control of the drop diameter is notably possible by exploiting the so-called limited coalescence phenomenon that occurs in a stabiliser-poor regime [26,27].

Since the particles are irreversibly adsorbed, the coalescence process occurs as long as the oil–water interface is insufficiently covered. The resulting emulsions exhibit a drop diameter that is controlled by the mass of particles and their packing at the interface.

Particles able to stabilise Pickering emulsions generally exhibit spherical geometries; however, a few particles with non-spherical shapes have also demonstrated the interest of varying the type of coverage and association, such as neighboured cubes [28], clay platelets [29,30], bacteria that exhibit interfacial activity when associated to chitosan [31] and particles resulting from aggregates [32]. Nanocellulose proved to display such a property to produce surfactant-free highly stable emulsions. Compared with others, they are non-spherical particles that have been deployed as emulsion stabilisers by taking advantage of their self-assembling ability at the oil–water interface, forming ultra-stable Pickering systems.

9.3 NANOCELLULOSE AND EMULSIONS

Oza and Franck [33] first studied the use of microcrystalline cellulose blended with sodium carboxymethyl cellulose (Avicel® RC-591) to stabilise surfactant-free oil–water emulsions. They described the mechanical barrier at the oil–water interface stabilising the emulsion. Ougiya et al. [34] pointing out the high stability of such emulsions using bacterial cellulose, with stability against changes in pH, temperature and salt concentration. In these two studies, emulsions stabilised with microcrystalline cellulose showed better properties in terms of stability and mechanical resistance compared to emulsions prepared with surfactants or polymers such as xanthan. However, the long length of the cellulose samples compared with the size of the droplets produced network of droplets rather than individual drops. The emulsion characteristics can be better controlled by using better-defined cellulosic particles, with size reduced. In the first decade of the 2000s, a number of patents and articles have been filed that involve the use of hydrophobically modified fibrillated cellulose for long-term stabilisation of Pickering emulsions and foams with microparticles from cellulose [35–37]. The use of unmodified CNCs as Pickering emulsion stabiliser was then first demonstrated in 2011 [21,38]. It was shown that unmodified CNCs dispersed in water, mixed with an apolar phase, such as hexadecane, can adsorb at the oil–water interface, preventing destabilisation from coalescence for a very long period of more than a year (Figure 9.4).

An emulsion is commonly considered stable if it is resistant to physical changes. It can be tested by several methods, including centrifugation, filtration, shaking or stirring, low-intensity ultrasonic vibrations and heating. Emulsion destabilisation generally induces creaming or sedimentation, flocculation and, eventually, coalescence. Concretely, creaming forces a contact (collision and sticking) between the droplets and allows coalescence to occur, whereby majority of the droplets merge, creating fewer larger droplets, thereby reducing the total interface area of the system. As oil generally has a lower density than water, a creaming process is usually observed. However, an emulsion can be considered stable as long as no coalescence occurs, that is, as long as the droplet size and size distribution do not change. When stabilised by CNCs, an unchanged drop diameter (stable for more

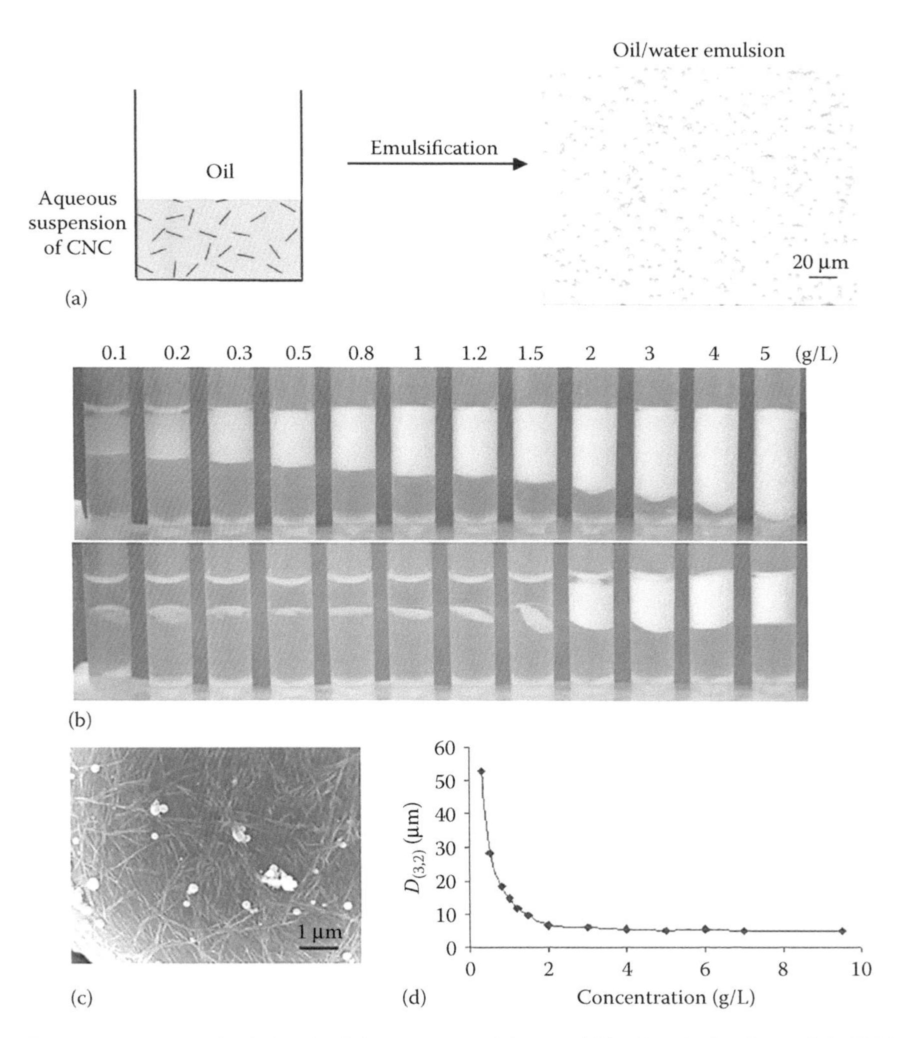

FIGURE 9.4 (a) Principle of oil-in-water emulsion stabilised only by bacterial CNC, (b) emulsion volume for increasing CNC concentration in the water phase from 0.1 to 5 g/L for an oil–aqueous phase ratio of 30:70 before (top) and after (bottom) centrifugation at 4,000 g. (c) Scanning electron micrographs of the surface of a polymerised drop and evolution of (d) the average $D_{(3,2)}$ drop diameter with increasing CNC concentration. (Reprinted with permission from Kalashnikova, I. et al., *Langmuir*, 27, 7471–7479, 2011. Copyright 2011 American Chemical Society.)

than a year) was observed; however, natural creaming occurred. Centrifugation accelerates the aging process, forcing the droplets to concentrate. The ability of CNCs to stabilise the droplets was assessed by varying the concentration of CNCs in the aqueous suspension from 0 to 5 g/L for identical oil volume and oil–water volume ratio (Figure 9.4). Before centrifugation, the emulsion volume increased regularly with the amount of CNCs added. Centrifugation at 4,000 g was performed to follow the resistance of droplets to coalescence. Two parameters were used to characterise and evaluate the stability of the resulting emulsions: (1) the average drop size obtained by granulometry and (2) the volume of the emulsion after centrifugation. The excess water was excluded from the emulsion, leading to close-packing conditions. After centrifugation, emulsions with CNCs concentrations lower than 2 g/L broke up, and the emulsion volume could no longer be measured. For concentrations greater than 2 g/L of CNCs, identical volumes of dense white emulsions were obtained that resisted to the stress induced by centrifugation. The explanation was that compression induced during centrifugation produced drop deformation, increasing the interfacial surface and promoting contact between unprotected surfaces. As a result, coalescence occurred, leading to the oil–water phase separation. At higher CNCs concentrations, the percentage of oil trapped into the cream phase increased sharply to reach a plateau value of 74%. This value is worth noting, as it refers to the theoretical close-packing conditions for monodispersed spheres. It means that it is not theoretically possible to pack these drops more densely, and the same limit was observed at 4,000 g and 10,000 g, leading to the same packing conditions, whatever be the CNCs concentrations. These droplets, which resist to centrifugation, can also be dispersed in a higher volume of water and shaken without disruption, revealing high resistance. The centrifugation deformed the droplets and excluded the water layers present at the interface, without breaking the droplets. The total volume of oil remained trapped within a dense emulsion, with no modification of emulsion volume, whatever be the concentration of CNCs, when centrifugation was stopped (i.e. after relaxation of the droplets). For concentrations greater than 2 g/L, the emulsions displayed excellent mechanical resistance towards coalescence, revealing excellent storage properties.

Simultaneously, the evolution of the average drop diameter varies with increasing the concentration of CNCs in the aqueous phase (Figure 9.4d). The sonication process involves a rather high-energy input, producing submicron-sized droplets. After stabilisation, droplet size increases by coalescence up to a homogeneous drop size population. This average drop size decreases when increasing the CNCs concentration down to a plateau value of 4- to 5-μm diameter for all emulsions containing more than 2 g/L of CNCs in the water phase. For concentrations lower than 2 g/L, the increasing amount of nanocrystals resulted in the ability to stabilise a larger interface area. This coalescence at low concentration in colloidal particles is known as the limited coalescence process [39,40]. It occurs when sonication produces a much larger area of oil–water interface than what can potentially be covered by the CNCs, inducing further coalescence. When sonication is stopped, the partially unprotected droplets coalesce. This coalescence results in a decrease of the oil–water interface and stops as soon as the interface is sufficiently protected. For the samples prepared with the highest CNCs concentrations, greater than 2 g/L, a shear-resistant emulsion of droplet diameter of 4–5 μm is obtained, irrespective of the amount of particles added. The drop size limit has rarely been discussed in the literature. This size limitation might be due to the flexibility of the solid particles in competition with capillary forces.

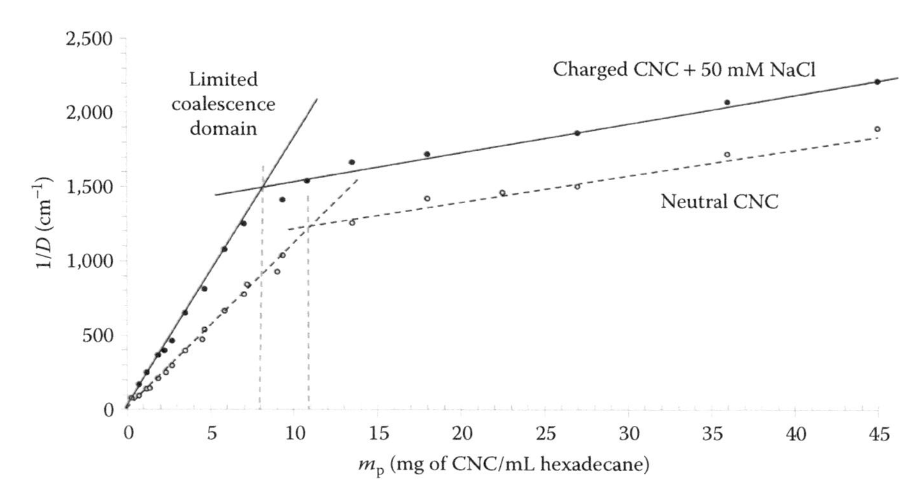

FIGURE 9.5 Limited coalescence process visualised by the inverse $D_{(3,2)}$ diameter plotted against the amount of particles included in the aqueous phase per millilitre of hexadecane. Two CNCs are presented, charged ones with half ester sulphate issued from acid hydrolysis and desulphated CNC, leading to neutral CNC. (Reprinted with permission from Cherhal, F. et al., *Biomacromolecules*, 17, 496–502, 2016. Copyright 2016 American Chemical Society.)

This is better visualised by the inverse diameter versus the concentration:

$$\frac{1}{D} = \frac{m_p}{6hV_dC} \tag{9.3}$$

where:

C is the surface coverage, that is, the percentage of droplet surface area covered by cellulosic particle
h is the thickness of the CNC-covering layer (7 nm for a monolayer)
V_d is the total volume of the dispersed phase

The evolution of the inverse diameter (1/D) for the various amounts of CNCs introduced results in a linear increase, as long as the limited coalescence process occurs, followed by stabilisation at the higher concentrations (Figure 9.5). In the first part of the curve, in the limited coalescence domain, the diameter is then controlled by the amounts of CNCs introduced at a constant surface coverage. This coverage is the minimum required to get a stable emulsion at rest. Once centrifuged, it would not resist to deformation, as described earlier (Figure 9.4b). Above the limit, a different coverage, more resistant to shear, is reached.

Such stable emulsions were shown to be possible when repulsions are screened. Adding salt is the easiest way to decrease repulsions [42]. Moreover, different nanostructures can be obtained at the surface of the drops by changing the attractive-repulsive balance between the CNCs at the interface. Aggregation process in aqueous suspensions has been widely investigated by varying salinity, surface charge density and concentration [43–46]. Following such process at an interface is more difficult. Previous studies showed that the thickness of the layer can be measured using small-angle neutron scattering (SANS), a non-destructive scattering technique with the method of phase contrasts [47]. Easy tuning parameters are ionic strength, surface charge density and concentration. But what occurs when charges are removed? The structuration of the oil–water interface of CNC-stabilised Pickering emulsions was investigated according to CNCs concentration, with surface charges and ionic strength, or without charges. It appeared that the drop size and coverage are influenced by the structure of the CNCs (Figure 9.5). In the limited coalescence domain, one can estimate that aggregated CNCs lead to an estimated lower concentration. The thickness of the interfacial layer was calculated by the SANS. The contrast variation method consists of appropriate mixtures of hydrogenated and deuterated liquids for both the aqueous and oily phases. This technique allows selective observation of the different components of the mixture: aqueous phase, alkane (oil) phase and CNCs. The drop diameters were large compared with the typical spatial scale probed by the SANS ($D_{(3,2)} >> 1/q_{min}$). As a result, only a small surface area of the droplets is probed, and the interface appears as a thin two-dimensional flat film. Focussing on this CNCs layer, the measurements revealed a 7-nm monolayer, irrespective of the amount of CNCs. More aggregated neutral CNCs form a more porous and heterogeneous surface, 18-nm thick. Consequently, only monolayers were observed, with thickness that varies from 7 nm for charged CNCs at low ionic strength to 18 nm when aggregating conditions were used (low charge density and/or high ionic strength), but no thickness variation could be detected with the concentration increase between 8 mg and 45 mg of CNCs per millilitre of hexadecane. This was interpreted as a densification of the interface with increasing concentration that might correspond to a local orientation or alignment of the CNCs at the basis of a tuning of the coverage, according to the aggregation conditions (Figure 9.6). Consequently, a variation of the coverage appeared, depending on the aggregation conditions, and this is a due to the anisotropic shape of CNCs. Similarly to what is observed for liquid crystals in bulk and contrary to the case of spherical particles, CNCs may align, highly varying the surface density.

To visualise the nanoparticles at the surface of the drops, emulsion polymerisation might be carried out by using a hydrophobic unpolymerised monomer as dispersed oil phase for CNC-stabilised oil–water emulsions. The stable emulsion when further polymerised leads to solid representative droplets. Therefore, styrene-in-water

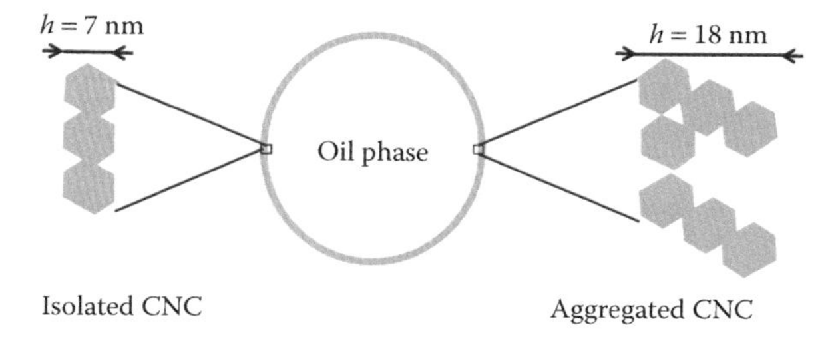

FIGURE 9.6 A schematic representation illustrating the thickness of the CNC layer adsorbed at the oil–water interface in the case of charged isolated CNC (sulphated) and aggregated CNC (desulphated). (Reprinted with permission from Cherhal, F. et al., *Biomacromolecules*, 17, 496–502, 2016. Copyright 2016 American Chemical Society.)

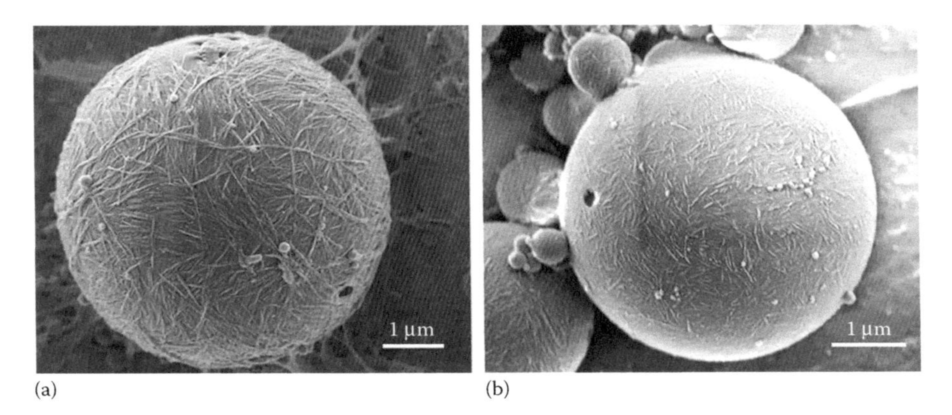

FIGURE 9.7 SEM images of polymerised styrene/water emulsions stabilised by (a) bacterial cellulose nanocrystals and (b) cotton cellulose nanocrystals. (Reprinted with permission from Kalashnikova, I. et al., *Biomacromolecules*, 13, 267–275, 2012. Copyright 2012 American Chemical Society.)

emulsions have been fabricated. Visualisation using scanning electron microscopy of individual droplets revealed that nanocrystals bended along the beads [38]. However, usually recognised as highly rigid, the CNCs rods are subject to torsion at the interface (Figure 9.7).

9.3.1 Adsorption Mechanism

Cellulose is generally considered as fully hydrophilic. However, such an interfacial property undoubtedly reveals a non-negligible affinity for apolar phases. Several evidences have demonstrated a hydrophobic character of CNCs. Indirectly, through adsorption of cellulose-binding module (CBM) in the case of cellulase from *Trichoderma Reesei* on *Valonia* nanocrystals. Those CBMs display some aromatic residues that are critical for binding on cellulose surface. Moreover, dyes were used as molecular probes to characterise cellulose substrates [48], or calcofluor could be used to intercalate the lattice [49]. Such assumption was later confirmed by molecular modelling [50,51]. This amphiphilic character has been attributed to a surface heterogeneity induced by the crystalline organisation of cellulose chains at the nanoscale [42]. As demonstrated in earlier works [52,53], CNCs shows uneven crystalline planes, whatever be the crystal type, assigned to well-known crystallographic planes, indexed according to Sugiyama et al. [54]. Surfaces can be divided into three categories: one is hydrophilic and relatively rough, whereas the other two present less pronounced hydrophilic or hydrophobic characters. These experiments tend to demonstrate that some parts of the nanocrystals expose less polar groups at the surface. A more hydrophobic edge plane composed of only CH groups that might not bear charges was identified as (200) crystalline plane for Iβ allomorph and (110) Iα allomorph. The SANS experiments carried out with a contrast variation method probed the arrangement of CNCs at the interface (Figure 9.8). Q^{-4} decay was observed for both sulphated and desulphated samples when CNCs were contrast-matched to water. This decay is known as a typical Porod behaviour, revealing a net interface without roughness.

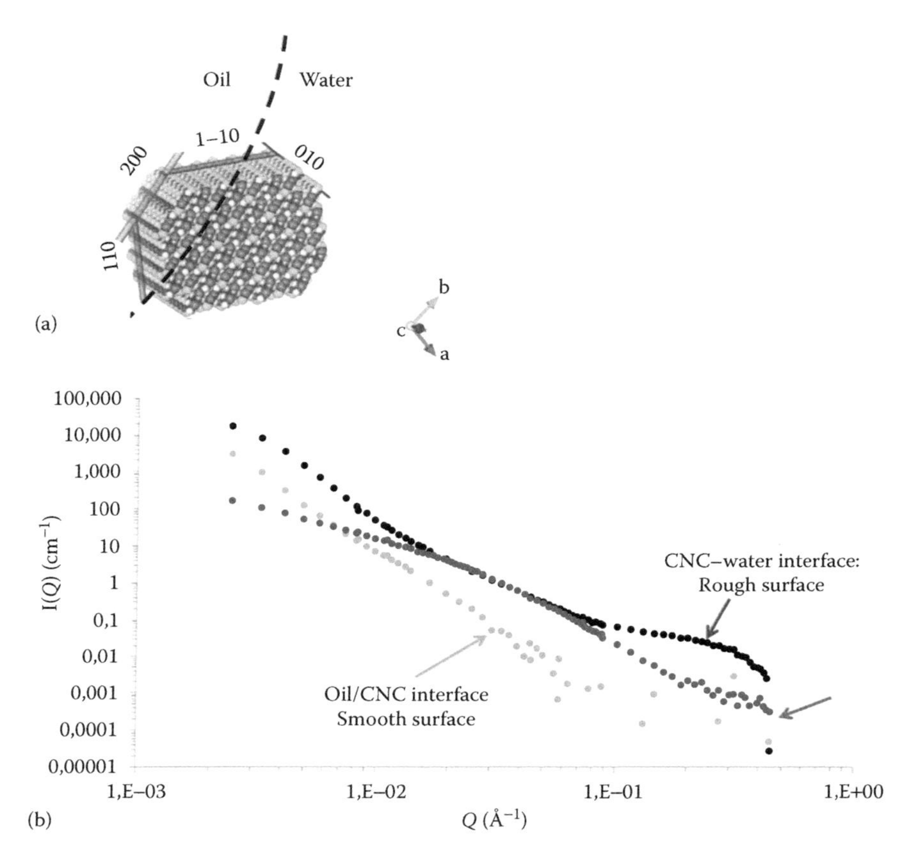

FIGURE 9.8 (a) Schematic representation of the possible stabilisation of the cellulose nanocrystals at the oil/water interface, exposing the hydrophobic edge $(200)_\beta$ or $(110)_\alpha$ to the oil phase (Reprinted with permission from Kalashnikova, I. et al., *Biomacromolecules*, 13, 267–275, 2012. Copyright 2012 American Chemical Society.) and (b) the curves obtained by SANS with CNC contrast-matched to oil (oil / CNC interface) and contrast-matched to water (CNC/water interface), compared to the scattering of CNCs in aqueous suspension at 7.8 g/L in 50 mM NaCl from the reference [45] (intermediate curve, multiplied by a constant factor). These curves illustrate the smooth oil/CNC interface with a continuous decay with a slope of −4 and the rough CNC/water interface with a signal that superimpose with the signal of CNC dispersed in water. (Reprinted with permission from Cherhal, F. et al., *Biomacromolecules*, 17, 496–502, 2016. Copyright 2016 American Chemical Society.)

Differently, when the CNCs are contrast-matched to oil, the scattered intensity superimposed to the scattering curve of CNCs, as obtained in suspensions [45] at intermediate *Q*. There is then evidence that the CNCs are not immersed in oil at the nanometre scale, since the Porod behaviour is observed over the whole *Q*-range, revealing no deformation of the oil surface at all the scales probed. These results strongly suggest that the (200) crystalline plane of the CNCs directly interacts with the interface, without deforming it [41]. Consequently, surface interactions should occur between the CH of the CNCs and the alkyl chain of hexadecane.

9.3.2 Impact of the Shape of Cellulose Nanocrystals

As already reported, cellulose from different biological sources may occur in different fibre-like packing, as dictated by the biosynthesis conditions [2,3] that give rise to different cellulose allomorphs I, II, III_I, III_{II}, IV_I and IV_{II}. Among these allomorphs, native cellulose occurs under of two suballomorphs of cellulose I, termed Iα and Iβ [55]. Furthermore, methods of extraction or treatment can also vary the surface chemistry of nanocellulose. Three sources varying in crystalline suballomorphs of native cellulose, with aspect ratios varying from 13 to 160 and different surface charge densities, have been compared. All of them presented the ability to efficiently stabilise oil-in-water emulsions, irrespective of the crystalline organisation [56] and surface charge density, as long as the charges are screened by using a solution of sufficient ionic strength. However, according to the morphology and the amount of CNCs involved, the coverage of the drop is modified. This surface coverage C is given by the ratio of the theoretical maximum surface area susceptible to be covered by the particles S_p, and the total surface displayed by the oil droplets S_d:

$$C = \frac{S_p}{S_d} = \frac{m_p D}{6h\rho V_{oil}} \tag{9.4}$$

where:

m_p is the mass of CNCs
D is the Sauter mean diameter of the droplets (also called $D_{3,2}$)
h is the thickness of the particle at the interface
ρ is the CNCs density (1.6 g/cm^3)
V_{oil} is the volume of oil included in the emulsion after centrifugation

The experimental drop diameter was compared to the amount of CNCs effectively stabilising the interface in the limited coalescence domain, which corresponds to the low CNCs concentrations. As a result, a minimum of approximately 40% of covered surface is enough to stabilise an emulsion with long *Cladophora* nanocrystals, whereas 60% is needed with medium nanocrystals from bacterial cellulose, and 84% is needed with nanocrystals from cotton. The aspect ratio, corresponding to length-to-width ratio, showed an important impact on the structure of the protecting armour at the interface. In relation to the percolation threshold, when smaller rods are used, more particles are required to get a stable droplet compared with the longer ones. Simultaneously, the longest nanocrystals obtained from *Cladophora* were shown to be of the same order of the drop diameter. Such CNCs were involved in entanglement and bridging phenomena, making some part of the crystals ineffective for interfacial stabilisation but increasing mechanical properties. As a result, one can prepare individual droplets or three-dimensional networks, as well as emulsions of varying surface coverage, thereby modulating the porosity of the interface (Figure 9.9) and viscoelasticity of the emulsion.

Another aspect ratio variation can be obtained by using the long nanofibrilated celluloses. They present alternating crystalline and amorphous domains; however, the crystalline part is identical to the crystalline CNCs. Winuprasith et al. [57] used MFC extracted from mangosteen. They varied microfibrillated cellulose (MFC) number of

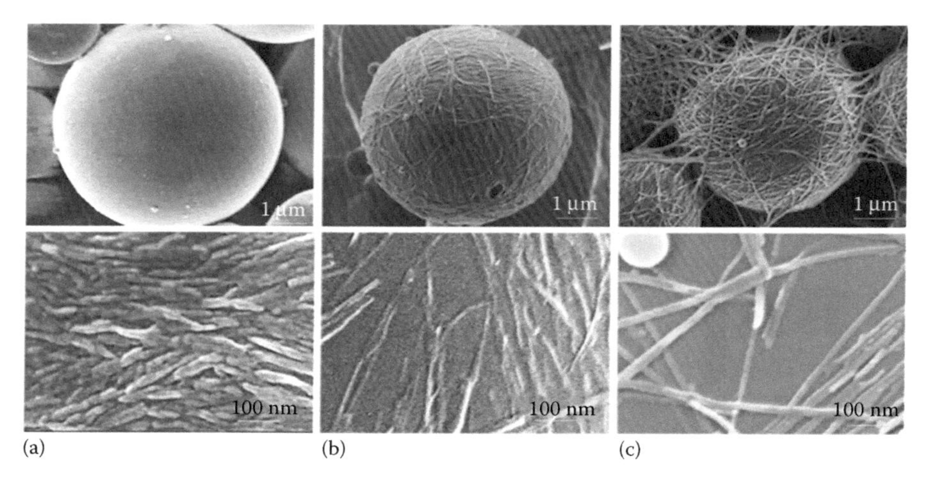

FIGURE 9.9 Scanning electron microscopic images of polymerised styrene-water emulsions stabilised by nanocrystals issued from (a) cotton (200-nm long), (b) bacterial cellulose (900-nm long), and (c) *Cladophora* algae (4-µm long), revealing the coverage variation as a function of nanocrystal aspect ratio. (Kalashnikova, I. et al., *Soft Matt.*, 9, 952–959, 2013. Reproduced by permission of The Royal Society of Chemistry.)

passes through a high-pressure homogenizer and investigated the influence of such degrading process on oil–water MFC-stabilised emulsions. All the emulsions prepared were stable to coalescence. However, the amount of excess non-adsorbing MFC particles present in the continuous aqueous phase increased with increasing MFC concentration from 0.5 to 7 g/L in the aqueous phase, and the mean droplet size increased with increasing MFC concentration [57]. Another study comparing emulsion stabilisation using by CNC and nanofibrillated cellulose (NFC) noted a minor contribution of the type of nanocellulose [58]. It appears that both types of nanocellulose can be used as very efficient stabilising agents to produce oil-in-water surfactant-free emulsions. Nanocellulose post-treatment might also vary the surface charge density such as oxidation promoted by the well-known 2,2,6,6-tetramethylpiperidine-1-oxyl (TEMPO) radical. The TEMPO-mediated oxidation leads to higher charged cellulose, where C6-hydroxyl groups are substituted by C6-carboxyl groups on the surface. Such surface modification has no major impact on emulsion stability, either of CNCs or of NFC [59,60]. As a global evidence, as long as the crystalline part is accessible to the interface, stable emulsions may be performed. Only the characteristics of the emulsion differ with the length of the particles according to the drop mean diameter and the network-type organisation from isolated droplets to totally interconnected system.

9.3.3 High Internal Phase Emulsions

The CNCs from cotton are crystalline rods of approximately 200 nm in length and 6–18 nm in thickness, which allow a good control of the architecture of the drop interface. When a Pickering emulsion is formed, monodisperse droplets with diameters of about 4–10 µm are obtained. It has been observed that the addition of oil phase with a double-cylinder-type homogenizer yields a new, highly stable gel structure,

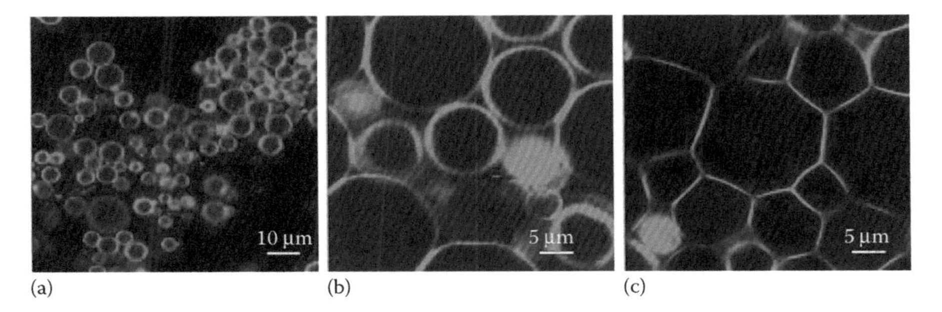

FIGURE 9.10 **(See colour insert.)** Confocal laser scanning microscopic images of emulsions stabilised by cotton CNCs containing increasing amounts of hexadecane stained with Bodipy from (a) the original 10/90 oil–water Pickering emulsion; (b) 65% of internal phase; (c) 85.6% of internal phase. (Reprinted with permission from Capron, I. and Cathala, B., *Biomacromolecules*, 14, 291–296, 2013. Copyright 2013 American Chemical Society.)

with a volume fraction of internal phase greater than 0.85 [61]. This is known as high internal phase emulsion (HIPE). The HIPEs are emulsified systems with an internal phase volume fraction greater than 0.74, which is the maximum packing density of monodispersed hard spheres. A new cellular tridimensional structure is then observed. Results from confocal laser scanning microscopy of the HIPEs (Figure 9.10) revealed that cells of larger dimension were produced. It implies that CNCs are redistributed from spherical armour to a continuous organisation. Oil drops are swellings stabilised by a fixed irreversibly adsorbed CNCs content up to a critical coverage stability of the droplet of 84%, at which coalescence or partition of particles occurs. Consequently, they promote an open structure, with solid particles at the interface resisting to coalescence by steric hindrance. Such emulsions interface proved to be very robust, since they kept their structure upon freeze drying, leading to hierarchical solid structures. As a result, dry architecture might be obtained with a porosity imposed by the emulsion drop diameter, providing a modular cellulosic template [62].

9.4 TOWARDS NEW FORMULATIONS

Based on such development, it is clear that Pickering emulsions with high stability, high resistance to shear and chemical versatility are good candidates to develop more complex structures. Unmodified CNCs proved to produce only oil-in-water emulsions. As described earlier, the surface chemistry is the main parameter that controls the type of emulsion. Consequently, to inverse the emulsion, hydrophilicity should be turned to hydrophobic characteristic [10,14,63]. Water-in-oil emulsions of high stability have been produced by using various hydrophobic cellulose derivatives, such as hypromellose phthalate (HP) [35], silylated cellulose [37] and alkyl modified CNCs esterified through acetic (C2–), hexanoic (C6–) and dodecanoic (C12–) acids [64,65]. Recently, there has been a growing interest in using surfactants, as demonstrated first by the team in Cermav [66] and then by others authors [67–70].

In these studies, they generally adsorb cationic surfactants such as cationic trimethyl-ammonium bromide-type molecule on the anionic CNCs via electrostatic interactions, followed by hydrophobic interactions. Moreover, Salajkova et al. [71] used quaternary ammonium salts bearing C18 alkyl chains for counterion exchange. Such surfactant-modified CNCs lead to stable surfactant-coated CNCs that can generally be dried from solvent.

Whatever be the route chosen to modify CNCs, inverse water-in-oil emulsions may be obtained. Combining both the oil-in-water and water-in-oil emulsions may lead to double emulsions. Thereby, oil-in-water-in-oil emulsions were prepared using only native and modified CNC and NFC [72] or in combination with surfactants [73] (Figure 9.11).

Efforts have dealt with stimuli-responsive cellulosic nanomaterials for sensing, detection and separation. This has been approached by covalent surface modification or adsorption of different biopolymers on the surface. It may involve impregnation of responsive particles into a cellulosic matrix to prepare, such as photo-switchable films [74]. An increasing amount of studies involve such smart modification of the surface of the CNCs for functional emulsion properties. Functional droplets were carried out by using stimuli-responsive N-isopropylacrylamide (NIPAM) modified cellulose [75]. The emulsions stabilised by poly(NIPAM) grafted CNCs gain temperature responsiveness. They break upon heat treatment at a temperature greater than the lower critical solution temperature of poly(NIPAM). Similarly, pH-responsive properties were demonstrated from PDMAEMA-grafted CNCs [76].

Moreover, in order to preserve the 'green' interest of cellulose, strong efforts are currently made on methods to modify the surface of CNCs without using chemicals that are environmentally undesired. The route to physically adsorbed molecules is of particular interest. Hydrosoluble methyl cellulose and hydroxypropyl cellulose were shown to adsorb at the surface of CNCs [77,78]. As CNCs are negatively charged, multilayers can be obtained by electrostatic interactions, covering the droplet with molecules or particles alternatively positively and negatively charged. This was carried out by using positively charged chitosan to reinforce the interface [62] or by water-soluble polyelectrolytes such as poly(allylamine hydrochloride) that can support a next layer of CNCs [60]. In addition, emulsions were prepared using CNCs pre-adsorbed with methyl cellulose, uncharged water-soluble cellulose derivatives. Adding tannic acid to such emulsions allows preparation of stable dry emulsions with high oil content (94 wt.%) that can be dried and re-dispersed readily, without the need for high energy mixing [78]. The authors propose that tannic acid condense on the surface of the shell where coated CNCs are stabilised, preventing from coalescence during a drying process.

In a different manner, covalently linked hybrid nanoparticles involving organic and inorganic materials are produced. Notably, CNC–$CoFe_2O_4$ hybrid nanoparticles synthesised by co-precipitation were assembled at oil–water interface by Nypelo et al. [79]. Ferromagnetic material was produced as solid microbeads obtained by polymerisation of styrene used as dispersed phase and hollow microcapsules after solubilisation of the polystyrene (PS). Nanocellulose might be used for micro- or nanostructuration in composites. With a polymerisable dispersed phase such as styrene monomer in a CNF-stabilised Pickering emulsion,

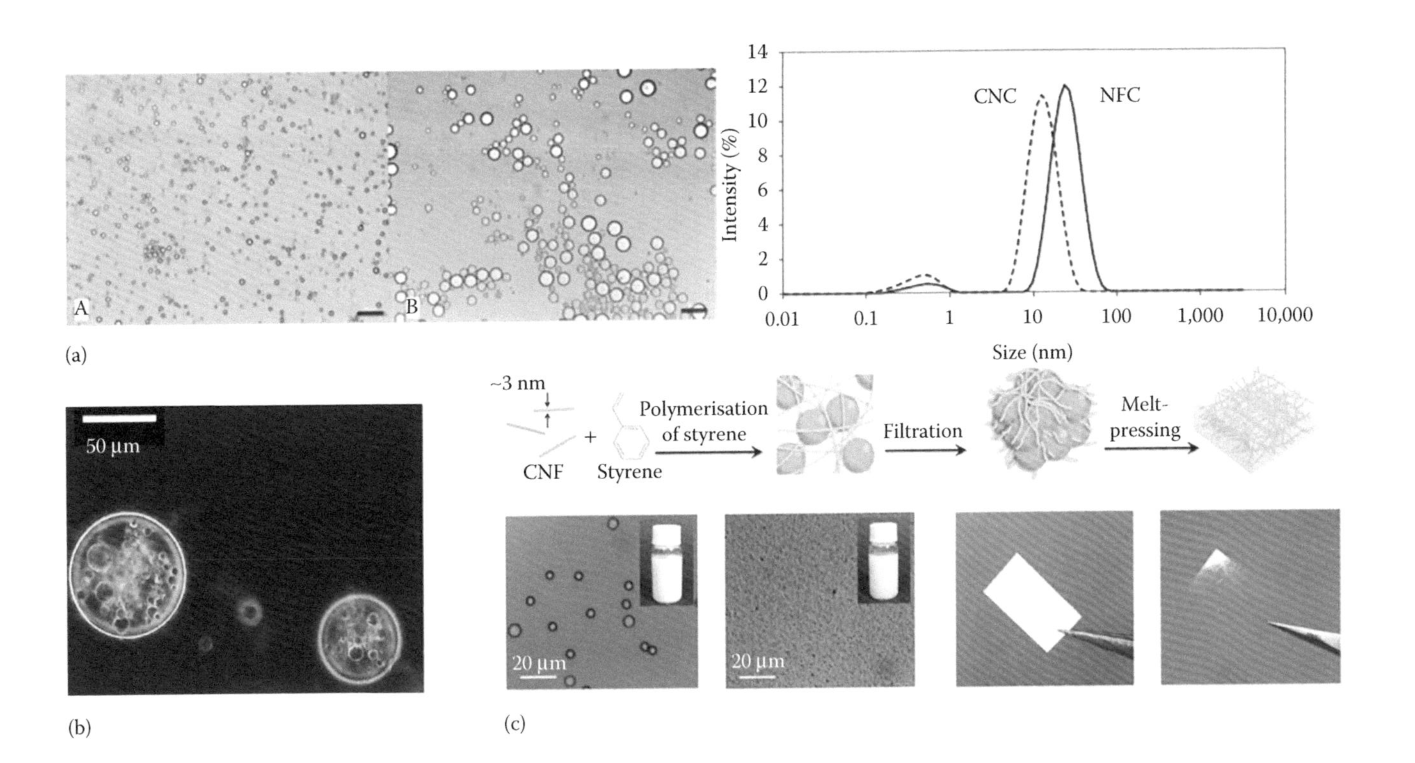

FIGURE 9.11 (a) Optical microscopy of limonene emulsions with 0.5 wt.% of CNC (A) and MFC (B) (scale bar: 50 µm). (Reprinted from *Food Hydrocoll.*, 61, Mikulcova, V. et al., On the preparation and antibacterial activity of emulsions stabilised with nanocellulose particles, 780–792, Copyright 2016, with permission from Elsevier.) (b) Solely cellulose-stabilised oil–water–oil double emulsions. (Reprinted with permission from Cunha, A.G. et al., *Langmuir*, 30, 9327–9335, 2014. Copyright 2014 American Chemical Society.) (c) CNF-stabilised polystyrene-in-water emulsion before and after polymerisation, leading to a transparent PS/CNF composite film. (Reprinted with permission from Fujisawa, S. et al., *Biomacromolecules*, 18, 266–271, 2016. Copyright 2017 American Chemical Society.)

subsequent polymerisation leads to PS/CNF composite. Subsequent melt pressing allows the preparation of a transparent PS/CNF composite film [80]. Such material might be used for in situ capturing and separation of molecules of interest from liquid media. Other hybrids particles composed of CNCs and calcium carbonate ($CaCO_3$) were used for multicore double emulsion elaborated in a microfluidic device. The hybrids served both as emulsion stabilisers and as calcium release reservoir, an in situ gelling agent for alginate microgel. The gelling step is triggered by acidic conditions. This two-step approach allows the preparation of alginate microgels containing several oil microdroplets that might be used as multi-compartmented reservoir [81]. In this chapter, only a few examples are given, but a wide literature is appearing that ambitions to develop emulsions with new functionalities and more complex structures [82].

9.5 CONCLUSION

Nanocelluloses are biobased nanomaterials that have been studied for decades. They are now well controlled in shape and surface chemistry. Furthermore, their recent industrial production had made them available for applications at a large-scale production. As such, there is a growing interest for its various applications. Their ability to stabilise interfaces has allowed to develop a large range of biphasic systems. Increasing environmental awareness is prompting scientists and manufacturers to develop strategies for environmental sustainability. Nanocellulose offers a large range of new opportunities that makes them good candidates to replace petroleum-based surfactants and polymers. Three main characteristics of nanocelluloses that provide a number of benefits for formulations are as follows: (a) irreversible adsorption at the interface, resulting in highly stable structures; (b) a rod-like shape that gives rise to a percolating network, thus increasing cohesion and stability; and (c) a non-toxic, abundant and renewable source. It gives important opportunities for future applications. Among the various applications using surfactants, there are a number of challenges of choice for nanocelluloses in complex formulations involving smart template organisation and functionalities.

REFERENCES

1. Aulin C, Netrval J, Wagberg L, Lindstrom T. Aerogels from nanofibrillated cellulose with tunable oleophobicity. *Soft Matter.* 2010;6(14):3298–3305.
2. Klemm D, Kramer F, Moritz S, Lindstrom T, Ankerfors M, Gray D et al. Nanocelluloses: A new family of nature-based materials. *Angewandte Chemie-International Edition.* 2011;50(24):5438–5466.
3. Habibi Y, Lucia LA, Rojas OJ. Cellulose nanocrystals: Chemistry, self-assembly, and applications. *Chemical Reviews.* 2010;110(6):3479–3500.
4. Lin N, Huang J, Dufresne A. Preparation, properties and applications of polysaccharide nanocrystals in advanced functional nanomaterials: A review. *Nanoscale.* 2012;4(11):3274–3294.
5. Ranby BG. The colloidal properties of cellulose micelles. *Discussions Faraday Society.* 1951;11:158–164.

6. Battista OA. Hydrolysis and crystallization of cellulose. *Industrial & Engineering Chemistry*. 1950;42:502–507.
7. Marchessault RH, Morehead FF, Koch MJ. Some hydrodynamic properties of neutral suspensions of cellulose crystallites as related to size and shape. *Journal of Colloid Science*. 1961;16:327–344.
8. Moon RJ, Martini A, Nairn J, Simonsen J, Youngblood J. Cellulose nanomaterials review: Structure, properties and nanocomposites. *Chemical Society Reviews*. 2011;40(7):3941–3994.
9. Charreau H, Foresti ML, Vazquez A. Nanocellulose patents trends: A comprehensive review on patents on cellulose nanocrystals, microfibrillated and bacterial cellulose. *Recent Patents on Nanotechnology*. 2013;7(1):56–80.
10. Habibi Y. Key advances in the chemical modification of nanocelluloses. *Chemical Society Reviews*. 2014;43:1519–1542.
11. Eichhorn SJ, Dufresne A, Aranguren M, Marcovich NE, Capadona JR, Rowan SJ et al. Review: Current international research into cellulose nanofibres and nanocomposites. *Journal of Materials Science*. 2010;45(1):1–33.
12. Stenius P, Andersen M. *Preparation, Properties and Chemical Modification of Nanosized Cellulose Fibrils*, 2007. pp. 135–154.
13. Missoum K, Belgacem MN, Bras J. Nanofibrillated cellulose surface modification: A review. *Materials*. 2013;6(5):1745–1766.
14. Jasmani L, Eyley S, Schutz C, Van Gorp H, De Feyter S, Thielemans W. One-pot functionalization of cellulose nanocrystals with various cationic groups. *Cellulose*. 2016;23(6):3569–3576.
15. Lagerwall JPF, Schutz C, Salajkova M, Noh J, Park JH, Scalia G et al. Cellulose nanocrystal-based materials: From liquid crystal self-assembly and glass formation to multifunctional thin films. *NPG Asia Materials*. 2014;6:E80.
16. Chen LF, Huang ZH, Liang HW, Gao HL, Yu SH. Three-dimensional heteroatom-doped carbon nanofiber networks derived from bacterial cellulose for supercapacitors. *Advanced Functional Materials*. 2014;24(32):5104–5111.
17. Yan CY, Wang JX, Kang WB, Cui MQ, Wang X, Foo CY et al. Highly stretchable piezoresistive graphene-nanocellulose nanopaper for strain sensors. *Advanced Materials*. 2014;26(13):2022–2027.
18. Liu HZ, Geng BY, Chen YF, Wang HY. Review on the aerogel-type oil sorbents derived from nanocellulose. *ACS Sustainable Chemistry & Engineering*. 2017;5(1):49–66.
19. Roman M. Toxicity of cellulose nanocrystals: A review. *Industrial Biotechnology*. 2015;11(1):25–33.
20. Lin N, Dufresne A. Nanocellulose in biomedicine: Current status and future prospect. *European Polymer Journal*. 2014;59:302–325.
21. Cathala B, Capron I, Bizot H, Kalashnikova I, Buleon A. Brevet WO2012/017160 - Composition sous forme d'émulsion, comprenant une phase hydrophobe dispersée dans une phase aqueuse (émulsions de Pickering). Patent WO2012/0171602010.
22. Ramsden W. Separation of solids in the surface-layers of solutions and 'suspensions'. *Proceedings of the Royal Society*. 1903;72:156.
23. Pickering SU. Emulsions. *Journal of Chemical Society*. 1907;91:2001–2021.
24. Binks BP. Macroporous silica from solid-stabilized emulsion templates. *Advanced Materials*. 2002;14(24):1824–1827.
25. Leal-Calderon F, Schmitt V. Solid-stabilized emulsions. *Current Opinion in Colloid & Interface Science*. 2008;13(4):217–227.
26. Arditty S, Schmitt V, Giermanska-Kahn J, Leal-Calderon F. Materials based on solid-stabilized emulsions. *Journal of Colloid and Interface Science*. 2004;275(2):659–664.

27. Arditty S, Whitby CP, Binks BP, Schmitt V, Leal-Calderon F. Some general features of limited coalescence in solid-stabilized emulsions. *European Physical Journal E.* 2003;11(3):273–281.
28. Destribats M, Gineste S, Laurichesse E, Tanner H, Leal-Calderon F, Heroguez V et al. Pickering emulsions: What are the main parameters determining the emulsion type and interfacial properties? *Langmuir.* 2014;30(31):9313–9326.
29. Ashby NP, Binks BP. Pickering emulsions stabilised by Laponite clay particles. *Physical Chemistry Chemical Physics.* 2000;2(24):5640–5646.
30. Bon SAF, Colver PJ. Pickering miniemulsion polymerization using Laponite clay as a stabilizer. *Langmuir.* 2007;23(16):8316–8322.
31. Wongkongkatep P, Manopwisedjaroen K, Tiposoth P, Archakunakorn S, Pongtharangkul T, Suphantharika M et al. Bacteria interface pickering emulsions stabilized by self-assembled bacteria-chitosan network. *Langmuir.* 2012;28(13):5729–5736.
32. Gao ZM, Zhao JJ, Huang Y, Yao XL, Zhang K, Fang YP et al. Edible pickering emulsion stabilized by protein fibrils. Part 1: Effects of pH and fibrils concentration. *LWT-Food Science and Technology.* 2017;76:1–8.
33. Oza KP, Frank SG. Microcrystalline cellulose stabilized emulsions. *Journal of Dispersion Science and Technology.* 1986;7(5):543–561.
34. Ougiya H, Watanabe K, Morinaga Y, Yoshinaga F. Emulsion-stabilizing effect of bacterial cellulose. *Bioscience Biotechnology and Biochemistry.* 1997;61(9):1541–1545.
35. Wege HA, Kim S, Paunov VN, Zhong QX, Velev OD. Long-term stabilization of foams and emulsions with in-situ formed microparticles from hydrophobic cellulose. *Langmuir.* 2008;24(17):9245–9253.
36. Lee K-Y, Blaker JJ, Murakami R, Heng JYY, Bismarck A. Phase behavior of medium and high internal phase water-in-oil emulsions stabilized solely by hydrophobized bacterial cellulose nanofibrils. *Langmuir.* 2014;30:452–460.
37. Blaker JJ, Lee KY, Li XX, Menner A, Bismarck A. Renewable nanocomposite polymer foams synthesized from Pickering emulsion templates. *Green Chemistry.* 2009;11(9):1321–1326.
38. Kalashnikova I, Bizot H, Cathala B, Capron I. New pickering emulsions stabilized by bacterial cellulose nanocrystals. *Langmuir.* 2011;27(12):7471–7479.
39. Aveyard R, Binks BP, Clint JH. Emulsions stabilised solely by colloidal particles. *Advances in Colloid and Interface Science.* 2003;100:503–546.
40. Arditty S, Whitby CP, Binks BP, Schmitt V, Leal-Calderon F. Some general features of limited coalescence in solid-stabilized emulsions. *European Physical Journal E.* 2003;12(2):355–355.
41. Cherhal F, Cousin F, Capron I. Structural description of the interface of pickering emulsions stabilized by cellulose nanocrystals. *Biomacromolecules.* 2016;17(2):496–502.
42. Kalashnikova I, Bizot H, Cathala B, Capron I. Modulation of cellulose nanocrystals amphiphilic properties to stabilize oil/water interface. *Biomacromolecules.* 2012;13(1):267–275.
43. Araki J, Wada M, Kuga S, Okana T. Influence of surface charge on viscosity behavior of cellulose microcrystal suspension. *Journal of Wood Science.* 1999;45(3):258–261.
44. Boluk Y, Zhao LY, Incani V. Dispersions of nanocrystalline cellulose in aqueous polymer solutions: Structure formation of colloidal rods. *Langmuir.* 2012;28(14): 6114–6123.
45. Cherhal F, Cousin F, Capron I. Influence of charge density and ionic strength on the aggregation process of cellulose nanocrystals in aqueous suspension, as revealed by small-angle neutron scattering. *Langmuir.* 2015;31(20):5596–5602.
46. Peddireddy KR, Capron I, Nicolai T, Benyahia L. Gelation kinetics and network structure of cellulose nanocrystals in aqueous solution. *Biomacromolecules.* 2016;17(10):3298–3304.

47. Jestin J, Simon S, Zupancic L, Barre L. A small angle neutron scattering study of the adsorbed asphaltene layer in water-in-hydrocarbon emulsions: Structural description related to stability. *Langmuir*. 2007;23(21):10471–10478.
48. Inglesby MK, Zeronian SH. Direct dyes as molecular sensors to characterize cellulose substrates. *Cellulose*. 2002;9(1):19–29.
49. Haigler CH, Brown RM, Benziman M. Calcofluor white st alters the in vivo assembly of cellulose microfibrils. *Science*. 1980;210(4472):903–906.
50. Mazeau K. On the external morphology of native cellulose microfibrils. *Carbohydrate Polymers*. 2011;84(1):524–532.
51. Mazeau K, Rivet A. Wetting the (110) and (100) surfaces of Iβ cellulose studied by molecular dynamics. *Biomacromolecules*. 2008;9(4):1352–1354.
52. Lehtio J, Sugiyama J, Gustavsson M, Fransson L, Linder M, Teeri TT. The binding specificity and affinity determinants of family 1 and family 3 cellulose binding modules. *Proceedings of the National Academy of Sciences of the United States of America*. 2003;100(2):484–489.
53. Fink HP, Hofmann D, Philipp B. Some aspects of lateral chain order in cellulosics from X-ray-scattering. *Cellulose*. 1995;2(1):51–70.
54. Sugiyama J, Vuong R, Chanzy H. Electron-diffraction study on the 2 crystalline phases occuring in native cellulose from an algal cell-wall. *Macromolecules*. 1991;24(14):4168–4175.
55. Atalla RH, Vanderhart DL. Native cellulose - A composite of 2 distinct crystalline forms. *Science*. 1984;223(4633):283–285.
56. Kalashnikova I, Bizot H, Bertoncini P, Cathala B, Capron I. Cellulosic nanorods of various aspect ratios for oil in water pickering emulsions. *Soft Matter*. 2013;9:952–959.
57. Winuprasith T, Suphantharika M. Properties and stability of oil-in-water emulsions stabilized by microfibrillated cellulose from mangosteen rind. *Food Hydrocolloids*. 2015;43:690–699.
58. Mikulcova V, Bordes R, Kasparkova V. On the preparation and antibacterial activity of emulsions stabilized with nanocellulose particles. *Food Hydrocolloids*. 2016;61:780–792.
59. Jia YY, Zhai XL, Fu W, Liu Y, Li F, Zhong C. Surfactant-free emulsions stabilized by tempo-oxidized bacterial cellulose. *Carbohydrate Polymers*. 2016;151:907–915.
60. Saidane D, Perrin E, Cherhal F, Guellec F, Capron I. Some modification of cellulose nanocrystals for functional pickering emulsions. *Philosophical Transactions of the Royal Society A-Mathematical Physical and Engineering Sciences*. 2016;374(2072):1–11.
61. Capron I, Cathala B. Surfactant-free high internal phase emulsions stabilized by cellulose nanocrystals. *Biomacromolecules*. 2013;14(2):291–296.
62. Tasset S, Cathala B, Bizot H, Capron I. Versatile cellular foams derived from CNC-stabilized pickering emulsions. *RSC Advances*. 2014;4(2):893–898.
63. Cunha AG, Gandini A. Turning polysaccharides into hydrophobic materials: A critical review. Part 1. Cellulose. *Cellulose*. 2010;17(5):875–889.
64. Lee K-Y, Blaker JJ, Heng JYY, Murakami R, Bismarck A. pH-triggered phase inversion and separation of hydrophobised bacterial cellulose stabilised pickering emulsions. *Reactive & Functional Polymers*. 2014;85:208–213.
65. Lee K-Y, Blaker JJ, Murakami R, Heng JYY, Bismarck A. Phase behavior of medium and high internal phase water-in-oil emulsions stabilized solely by hydrophobized bacterial cellulose nanofibrils. *Langmuir*. 2014;30(2):452–460.
66. Bonini C, Heux L, Cavaille JY, Lindner P, Dewhurst C, Terech P. Rodlike cellulose whiskers coated with surfactant: A small-angle neutron scattering characterization. *Langmuir*. 2002;18(8):3311–3314.
67. Habibi Y, Hoeger I, Kelley SS, Rojas OJ. Development of langmuir-schaeffer cellulose nanocrystal monolayers and their interfacial behaviors. *Langmuir*. 2010;26(2):990–1001.

68. Dhar N, Au D, Berry RC, Tam KC. Interactions of nanocrystalline cellulose with an oppositely charged surfactant in aqueous medium. *Colloids and Surfaces A-Physicochemical and Engineering Aspects*. 2012;415:310–319.
69. Brinatti C, Huang J, Berry RM, Tam KC, Loh W. Structural and energetic studies on the interaction of cationic surfactants and cellulose nanocrystals. *Langmuir*. 2016;32(3):689–698.
70. Prathapan R, Thapa R, Garnier G, Tabor RF. Modulating the zeta potential of cellulose nanocrystals using salts and surfactants. *Colloids and Surfaces A-Physicochemical and Engineering Aspects*. 2016;509:11–18.
71. Salajkova M, Berglund LA, Zhou Q. Hydrophobic cellulose nanocrystals modified with quaternary ammonium salts. *Journal of Materials Chemistry*. 2012;22(37):19798–19805.
72. Cunha AG, Mougel J-B, Cathala B, Berglund LA, Capron I. Preparation of double pickering emulsions stabilized by chemically tailored nanocelluloses. *Langmuir*. 2014;30(31):9327–9335.
73. Carrillo CA, Nypelo T, Rojas OJ. Double emulsions for the compatibilization of hydrophilic nanocellulose with non-polar polymers and validation in the synthesis of composite fibers. *Soft Matter*. 2016;12(10):2721–2728.
74. Gutierrez J, Fernandes SCM, Mondragon I, Tercjak A. Conductive photoswitchable vanadium oxide nanopaper based on bacterial cellulose. *ChemSusChem*. 2012;5(12):2323–2327.
75. Zoppe JO, Venditti RA, Rojas OJ. Pickering emulsions stabilized by cellulose nanocrystals grafted with thermo-responsive polymer brushes. *Journal of Colloid and Interface Science*. 2012;369:202–209.
76. Tang J, Lee MFX, Zhang W, Zhao B, Berry RM, Tam KC. Dual responsive pickering emulsion stabilized by poly 2-(dimethylamino)ethyl methacrylate grafted cellulose nanocrystals. *Biomacromolecules*. 2014;15(8):3052–3060.
77. Hu Z, Patten T, Pelton R, Cranston ED. Synergistic stabilization of emulsions and emulsion gels with water-soluble polymers and cellulose nanocrystals. *ACS Sustainable Chemistry & Engineering*. 2015;3(5):1023–1031.
78. Hu Z, Xu R, Cranston ED, Pelton RH. Stable aqueous foams from cellulose nanocrystals and methyl cellulose. *Biomacromolecules*. 2016;17(12):4095–4099.
79. Nypeloe T, Rodriguez-Abreu C, Kolen'ko YV, Rivas J, Rojas OJ. Microbeads and hollow microcapsules obtained by self-assembly of pickering magneto-responsive cellulose nanocrystals. *ACS Applied Materials & Interfaces*. 2014;6(19):16851–16858.
80. Fujisawa S, Togawa E, Kuroda K. Facile route to transparent, strong, and thermally stable nanocellulose/polymer nanocomposites from an aqueous pickering emulsion. *Biomacromolecules*. 2017;18(1):266–271.
81. Marquis M, Alix V, Capron I, Cuenot S, Zykwinska A. Microfluidic encapsulation of pickering oil microdroplets into alginate microgels for lipophilic compound delivery. *ACS Biomaterials Science & Engineering*. 2016;2(4):535–543.
82. Salas C, Nypeloe T, Rodriguez-Abreu C, Carrillo C, Rojas OJ. Nanocellulose properties and applications in colloids and interfaces. *Current Opinion in Colloid & Interface Science*. 2014;19(5):383–396.

10 Upgrading the Properties of Woven and Non-Woven (Ligno)Cellulosic Fibre Preforms with Nanocellulose

Marta Fortea-Verdejo and Alexander Bismarck

CONTENTS

10.1 Introduction 197
10.2 Surface Fibrillation of (Ligno)Cellulosic Fibres to Create Self-Binding Materials 198
10.3 Nanocellulose Binder for Plant-Based Natural Fibres 201
10.4 Nanocellulose as Binder for Wood Pulp Fibres 205
10.5 Nanocellulose Coated Natural Fibres 209
10.6 Conclusion 213
Acknowledgements 213
References 214

10.1 INTRODUCTION

The use of biodegradable materials, which can be obtained from natural resources, has been a research trend over the past decades. The world's globalisation and manufacturing capability has very much improved the quality of life; however, it has had several negative side effects, such as an increasing amount of waste and a progressive decrease of various limited resources. In order to reduce the amount of plastics we use, we can substitute some by natural fibres as reinforcements and create natural fibre composites. These reinforcements provide an improved performance and have the added value of being fully biodegradable and relatively cheap, as compared with other reinforcements, such as carbon or glass fibres [1]. Other advantages of natural fibres are that they are a CO_2 neutral resource, have high specific properties and are widely available [2].

All these advantages are reflected in the renewed interest that natural fibre composites have gained over the last two decades. Although natural fibre composites are being used in many applications, some key issues still need to be addressed to optimise their performance. Cellulosic fibres are difficult to be dispersed into polymer matrices, resulting in fibre agglomeration and breakage during processing of the composites [3], ultimately resulting in reduced mechanical properties of the composite. Furthermore, the inherent properties of natural fibres, such as their low density and their low thermal stability [4], complicate the manufacturing processes and limit the choices of potential polymer matrices [5]. Numerous efforts have been made to enhance the mechanical performance of natural fibre composites, such as surface modifications of natural fibres and matrix modifications to improve the adhesion between fibres and matrix, limited by the hydrophilic nature of the natural fibres [6–10]. However, these modifications normally require chemical treatments, which generate chemical waste [10,11]. Another way to enhance the properties of natural fibre composites is by processing the natural fibres into fabrics/mats by interlacing their yarns before their use as reinforcement or by using them in the form of non-wovens. Non-wovens from only (lingo) cellulosic fibres can be made by mechanical entanglement of the fibres (and are commercially available), so no polymer is used. However, due to the lack of a binder, their applications are limited [12]. How can these (non)wovens/loose fibres be used? This can be done by using nanocellulose as binder. Owing to the dimensions of nanocellulose it does not need to be used in high amounts to properly surround the cellulosic fibres, but it can be used as a binder for those fibres producing numerous contact points.

The aim of this chapter is to describe how nanocellulose can be used to bind natural fibres and pulp. To do so, nanocellulose can be obtained by a partial fibrillation of the surface of the natural fibres to be bonded, or it can be simply attached onto them. The use of nanocellulose as an additional binder for pulp fibres will also be discussed, and at the end of the chapter, the coating of individual natural fibres with nanocellulose will be presented, since it is important to understand the adhesion between them to further improve the use of nanocellulose as binder.

10.2 SURFACE FIBRILLATION OF (LIGNO)CELLULOSIC FIBRES TO CREATE SELF-BINDING MATERIALS

Nanocellulose can be created by fibrillating the surface of (ligno)cellulosic fibres. The surface fibrillation of (ligno)cellulosic fibres causes the elementary fibrils on the surface of the (ligno)cellulosic fibres to disassemble. Depending on the starting materials, for example, fibres or fabrics, different surface fibrillation techniques could be used. Miao et al. [13] used high-pressure jets (hydroentanglement) to partially fibrillate and generate nanofibres at the surface flax fabric, without damaging the body of the mat. A Rieter Perfojet pilot spun-lace machine equipped with three injectors, the last one working with a high-water pressure of 30 MPa (300 bar), was used. Both sides of the flax fibre fabrics were subjected to this high-pressure (hydroentanglement) treatment. Nanofibrils of between 10 nm and 50 nm in diameter were found to be liberated from the surface of the flax fibres due to this hydroentanglement treatment and forming a continuous network wrapping the flax fibres. The filtration efficiency of hydroentanglement-treated flax fibre fabrics was found improve significantly over neat flax fibre fabrics. For particles smaller than 1 μm, the filtration efficiency of neat flax fibre fabric was found to be 10%, whereas the filtration efficiency of hydroentanglement-treated flax fibre fabric was found to be 55%. The filtration efficiency of hydroentanglement-treated flax fibre fabric is close to that of commercial fabrics for surgical gowns. In addition to this, the abrasion resistance of the hydroentanglement-treated flax fibre fabric was found to improve by a factor of 10 compared with neat flax fibre fabrics. This improvement was attributed to the filling of the macroscale voids between the flax fibres, with the nanocellulose fibre network created by surface fibrillation of the fabric, using high-pressure jets. However, the hydroentanglement treatment reduced the fabric strength, hypothesised to be due to the weakening of the flax fibres within the fabric (Table 10.1).

In a subsequent work, the authors applied the same hydroentanglement process to non-woven needle-punched mats of flax (tradename: Durafibre) and hemp (tradename: DA Therm), respectively [14]. Micro- and nanocellulose fibres (10–50 nm) were liberated from the surface of the non-woven mats to form the superficial network through the fibres. During the hydroentanglement process, the needle-punched

TABLE 10.1
Mechanical Properties of the Selected Cellulosic Materials After a Surface Fibrillation (Hydroentanglement) Process

Authors	Starting Material	Fibrillation Time (h)	Fibrillation Pressure (MPa)	Tensile Strength ($kN \cdot m^{-1}$)	Flexural Strength (MPa)	Flexural Modulus (GPa)
Miao et al. [13][a]	Flax fabric		0	28.2 ± 2.2		
			150	17.6 ± 2.3		
			200	19.0 ± 13		
			250	18.6 ± 1.1		
			300	17.4 ± 7.0		
Miao et al. [14][b]	Flax/hemp non-woven		0	0.002		
			20	3		
Arévalo et al. [15][c]	Loose flax fibres	1.5			50	3
		3			57	4
		4			62	4
		5			70	5.5
		6			85	7.5
		12			75	7

[a] Tensile properties in the weft direction. The original data were given in $N \cdot cm^{-1}$.
[b] The original data were given in $N \cdot gsm^{-1}$, and it was in the form of a figure.
[c] The original data were given in a figure.

mats were pre-wetted at low water pressure (20 bar), followed by high-pressure treatment of 200 bar. The thickness of the non-wovens was reduced due to densification as a result of the hydroentanglement process. In this case, the tensile strength of the hydroentanglement-treated non-wovens increased by more than 50 times (Table 10.1). It is worth mentioning at this point that in needle-punched non-wovens, the fibres are held together only by friction and mechanical fibre entanglement, so the mechanical properties of these non-wovens are worse as compared to flax fibre fabrics (Table 10.1). Although the surface fibrillation of the fabrics led to better performance of the resulting self-binding materials than the partial fibrillation of non-wovens, the fibrillation process itself was more successful in the case of non-wovens.

The network-like structure that nanocellulose forms when the nanofibrils are peeled off the surface of the (lingo)cellulosic fibres is useful as, there is no need for the (ligno)cellulosic fibres to be previously entangled or interlaced. The use of individual fibres (or a web of fibres) as a source for nanocellulose through partial fibrillation of their surface has been reported to create self-binding materials. For example, Arévalo et al. [15] created binder-less all-cellulose fibreboards by using flax fibres as raw material. The choice of these fibres was motivated by their low lignin content, which ensures that the main self-binding mechanism was the hydrogen network formed by the nanocellulose exposed at the fibre surface and not the lignin naturally present in the fibres (even though they could not discard that it may happen). A scheme of the process used by the authors to produce the binder-less fibreboards is shown in Figure 10.1. To partially fibrillate the fibre surface, they used a standard Valley (Hollander) beater to refine 20-mm flax fibres in the presence of a large amount of water (98%). Once the mechanical refining was completed, the resulting pulp was dewatered by filteration and hot-pressed to completely remove the water and consolidate the fibreboards. The authors chose the heating temperature of 140°C to limit any effects of lignin acting as a plasticiser. The refining times the authors chose were 1.5, 3, 4, 5, 6 and 12 h. The authors observed that nanofibrils in the form of branches were created on the surface of the flax fibres, due to the refining process. With increasing refining times, smaller and more dispersed/branched branches were created, with fibre and fibril diameters ranging approximately from 1 μm to 100 nm (or even smaller). The micro/nanocellulose created at the surface provided a network-like structure that increased the fibre surface area and filled the voids between the otherwise-loose flax fibres, thus reducing the porosity of the fibreboards removing the need for a matrix or plasticiser to bind the fibres together.

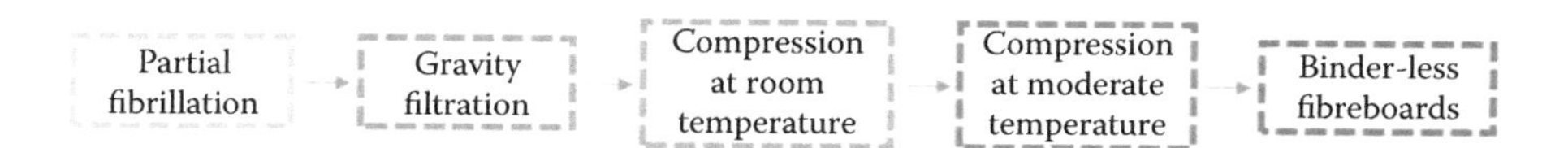

FIGURE 10.1 Schematic manufacturing from the binder-less fibreboards. (Adapted from Arévalo, R. and Peijs, T., *Compos. Part A*, 83, 38–46, 2016.)

The porosity of the fibreboards was reduced from 17% to 11% with an increase of the refining times from 1 to 6 h, respectively. Furthermore, the flexural strength and modulus of the fibreboards increased progressively with the refinement time up to 85 ± 4 MPa and 7.5 ± 0.2 GPa for 6 h refining time (Table 10.1). After 12 h of refining, a slight decrease of the flexural properties was seen, due a possible damage of the flax fibre structure. The authors also observed that an increase of the refinement time led to a lower fracture resistance, due the shortening of the fibres after the treatment and formation of the nanocellulose network. The authors also investigated the initial fibre length before the mechanical fibrillation. They observed that the flexural strength and modulus increased when increasing the initial fibre length from 5 to 10 mm, due to a better ability of the long fibres to transfer stress and enhanced fibre efficiency. However, they found a decrease of the flexural properties, when fibres with an initial length of 20 mm were used, which the authors attribute to (a) the difficulty of the fibrillation process with long fibres, due to entanglements, failing to create a dense network, (b) the increasing probability of out-of-plane fibre orientations due to longer fibres and (c) a less effective packing of the fibres. In another publication [16], the authors also noticed that the fracture resistance for crack initiation and peak cohesive stress for these fibreboards increased with increasing fibre fibrillation time, whereas the steady-state fracture resistance decreased. This process was then further optimised and scaled up, and it is commercialised by Zelfo Technology GmbH® (Section 10.3).

10.3 NANOCELLULOSE BINDER FOR PLANT-BASED NATURAL FIBRES

Instead of surface fibrillation of (ligno)cellulosic fibres, nanocellulose can be added directly onto (ligno)cellulosic fibres, thereby avoiding damage to the main reinforcing fibres [17,18]. Lee et al. [19,20] used bacterial cellulose (BC), a type of nanocellulose synthesised by cellulose-producing bacteria, as binder for loose sisal fibres. BC was used as binder to create sisal non-wovens using a simple papermaking process to create the final non-wovens, without the need of machinery typically to produce non-wovens. The authors soaked short sisal fibres in a suspension with BC overnight, followed by dewatering, cold compression and hot compression at 120°C for 4 h to consolidate the nanocellulose network to form the BC/sisal fibre-based non-wovens. The measured tensile properties and porosity of the non-wovens are summarised in Table 10.2. They found that the addition of only 10 wt.% of BC to the fibre preforms resulted in a decrease of the porosity to 61% and an increase in the tensile strength of the non-woven to 13 $kN{\cdot}m^{-1}$ (Table 10.2). It should be noted that without BC binder, the tensile properties of the non-wovens could not be measured, as the loose fibres are held together only by friction.

Following this concept, Fortea-Verdejo et al. [21] used pulp and different nanocelluloses, namely cellulose nanofibres produced by mechanically refining pulp and BC, to bind loose flax fibres. Two different manufacturing methods were compared in their work: a single-step filtration process (first used by Lee et al. [19,20]) and a layer-by-layer filtration process. A schematic representation of the manufacturing method can be observed in Figure 10.2. The authors found that the use of nanocellulose as binder (from any source) undoubtedly outperformed the use of pulp as

TABLE 10.2
Mechanical Properties of Natural Fibres Bonded by Nanocellulose or Other Materials for Comparison

Authors	Fibres	Type of Binder	Binder (wt.%)	Tensile Strength (MPa)	Tensile Strength ($kN \cdot m^{-1}$)	Tensile Index ($Nm \cdot g^{-1}$)	Flexural Strength (MPa)	Flexural Modulus (GPa)
Lee et al. [19,20]	Sisal	BC	0	–	–	–	–	–
			10		13.1 ± 2.1			
Fortea-Verdejo et al. (layer-by-layer filtration) (unpublished)	Sisal	BC	10	6.2 ± 1.5	10.1 ± 2.5	11.9 ± 3.1	8.5 ± 2.3	0.6 ± 0.2
Fortea-Verdejo et al. [21] (single-layer filtration)	Flax	BC	0	–	–	–	–	–
			10	4.4 ± 0.9	6.5 ± 1.5	7.5 ± 1.9	9.0 ± 1.7	0.8 ± 0:3
			20	6.9 ± 2.0	10.6 ± 1.9	11.9 ± 2.8	12.3 ± 2.2	1.2 ± 0.4
			30	13.3 ± 2.3	25.7 ± 6.1	18.3 ± 3.4	19.0 ± 6	1.0 ± 0.3
Fortea-Verdejo et al. [21] (layer-by-layer filtration)		BC	0		–			
			10	12.7 ± 1.1	17.6 ± 2.5	19.6 ± 2.3	14.6 ± 4.1	1.5 ± 0.6
			20	12.9 ± 2.6	19.4 ± 0.9	16.9 ± 1.7	17.9 ± 1.8	1.8 ± 0.4
			30	19.3 ± 1.8	32.7 ± 2.9	26.7 ± 3.3	21.1 ± 1.7	2.3 ± 0.4
		NFC (wood)	0	–	–	–	–	–
			10	5.0 ± 1.4	7.5 ± 1.4	7.8 ± 1.6	6.9 ± 0.6	0.5 ± 0.1
			20	13.1 ± 2.2	19.1 ± 2.5	19.6 ± 1.7	15.4 ± 1.8	1.3 ± 0.7
			30	17.1 ± 0.9	23.6 ± 1.0	25.4 ± 1.7	18.1 ± 1.9	1.7 ± 0.8
Fages et al. [22][a]	Flax	PP	10		5.4	14		
			20		10.4	26.9		
			30		10.4	27.9		
Fages et al. [23][b]	Flax	PVA	10		1.8	4.6		
			20		1.4	3.6		
			30		1.2	2.6		
Flexform[c]		PP	50	13.8				1.2
Zelfo HG[d]				7			17	
Zelfo HZ[e]				55			95	

[a] The referenced data reported in the table were not reported as tensile strength and index but as load to failure in the longitudinal direction. To ease direct comparison, we calculated the tensile strength and index from the thickness, width and grammage reported in the papers.

[b] The referenced data reported in the table were not reported as tensile strength and index but as load to failure in the longitudinal direction. To ease direct comparison, we calculated the tensile strength and index from the thickness, width and grammage reported in the papers.

[c] http://www.flexformtech.com/Auto/Products/(800 gsm).

[d] http://www.matweb.com/search/datasheet.aspx?matguid=808a24e75ac64ba5aa2953f1173d431d&ckck=1.

[e] http://www.matweb.com/search/datasheet.aspx?matguid=da41b2af2e0742988604a62e219222e9&n=1.

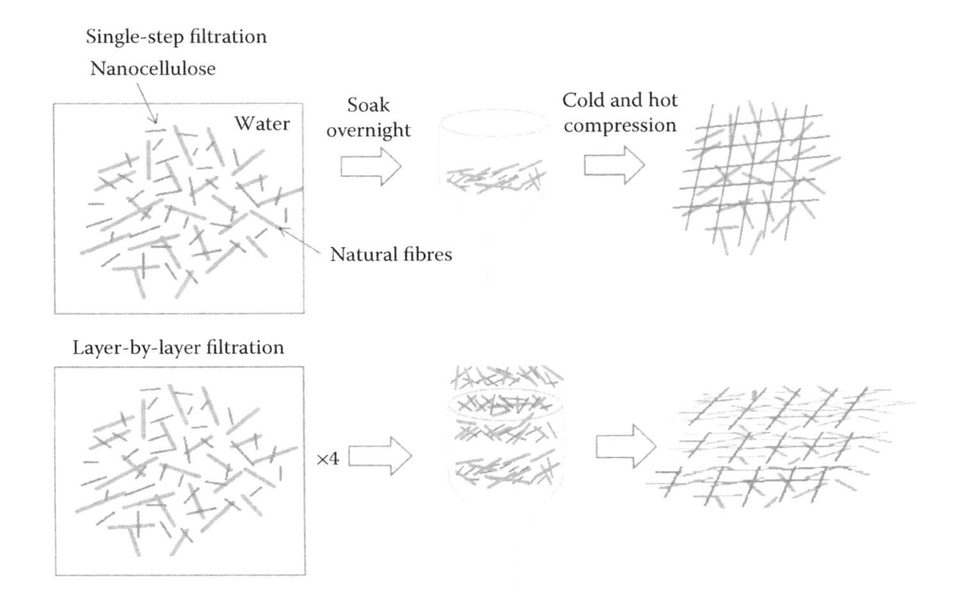

FIGURE 10.2 Representation of the single-step filtration and layer-by-layer filtration as manufacturing methods to create flax non-wovens, as detailed in Reference 21.

"binder", due to the differences in their surface areas. In Figure 10.3, the morphological differences between flax non-wovens containing 10 wt.% BC, CNF and pulp can be observed.

The selected mechanical properties of the non-wovens using different contents of BC and CNF as binders are also summarized in Table 10.2. The BC non-wovens manufactured using a layer-by-layer filtration process possessed better mechanical properties due to the more uniform distribution of the nanocellulose network through thickness of the non-woven preforms achieved during the filtration process (the same trend was reported for nanofibrillated cellulose [NFC]). In addition, an increase in the mechanical properties was observed by increasing the nanocellulose binder content, for both BC and NFC, due to an increase in the number of contact points between the nanocellulose and the natural fibres. The authors also report a slight decrease in the porosity with an increase in the nanocellulose content and by using layer-by-layer filtration. It was concluded that the overall non-wovens preform properties were determined by the nature of the nanocellulose binder and not by the natural fibre type.

In Table 10.2, the properties of non-woven mats made using flax and sisal fibres as main reinforcing fibres are juxtaposed. When comparing the properties of non-wovens produced by the single-step filtration method, one could be under the impression that sisal/BC non-wovens outperform in tension flax/BC non-wovens. However, when comparing the properties of the non-woeven mats produced by the layer-by-layer filtration method (unpublished results), the situation is inverted, for both tensile and flexural properties. These results suggest that the mechanical performance of the non-wovens

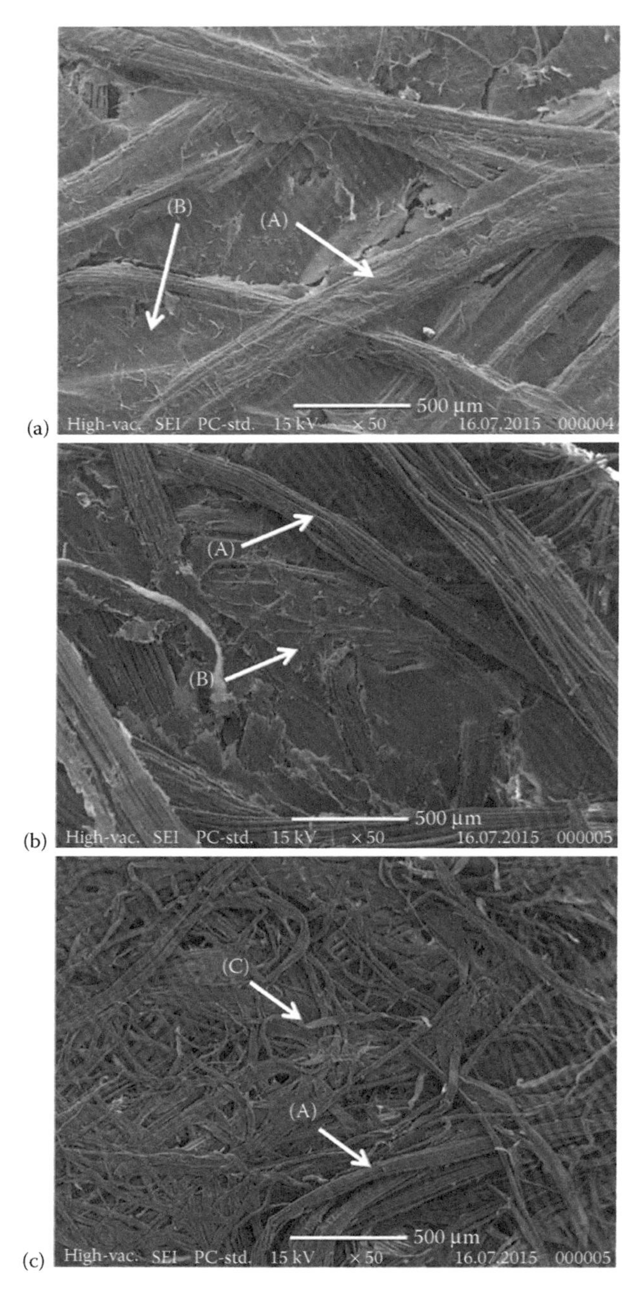

FIGURE 10.3 Non-woven fibre preforms containing 10 wt.% BC (a), 10 wt.% NFC (b) and 10 wt.% pulp (c). Points (A) to flax fibres, (B) to nanocellulose and (C) to pulp fibres. (Reproduced from Fortea-Verdejo, M. et al., *Compos. Part A*. With permission of Pergamon.)

is strongly dominated by the nanocellulose network, no matter which natural fibres are used as reinforcement.

When comparing the use of nanocellulose as a binder with other conventional binders, such as polypropylene or Polyvinyl alcohol, at the same contents (Table 10.2), it was observed that the nanocellulose non-wovens possess comparable mechanical properties or even outperform the non-wovens in which the fibres are held together by polymer binder. Furthermore, BC and NFC non-wovens even outperform commercially available non-wovens (from Flexform®) bonded with as much as 50% polymer (PP) in tension and flexture. A comparison is possible not only with non-wovens bonded with other materials but also with self-bonding materials made only from cellulose. The use of nano- and microcellulose as binder for only cellulosic materials is currently commercialised by Zelfo Technology GmbH®. They patented an energy- and water-efficient technology, which transforms different cellulose raw materials and waste materials into a combination of micro- and nanofibrillated pulp, by mechanical fibrillation, which is then dried following different procedures [24]. The mechanical properties of two of their products, Zelfo HG and Zelfo HZ have been used for comparison (Table 10.2). It can be observed that the mechanical performance of Zelfo HG is similar to that of the flax non-wovens using BC and NFC as a binder [21]. However, Zelfo HZ exceeds by far the mechanical performance of any material whose properties are reported in Table 10.2, demonstrating that these types of materials, made only from cellulosic resources using nanocellulose as a binder, have attractive mechanical properties.

When comparing the mechanical properties of self-binding natural fibre materials produced using of the nanocellulose modified fibres and partial fibrillation of fibres (Tables 10.1 and 10.2), one would expect that the partial fibrillation of fibres would fall behind, due to the damage that occurs to the fibres; however, this was not obvious from the results. The use of fabrics or non-wovens as a source for NFC during the partial fibrillation gives similar results to the ones that use NFC or BC as an external binder. Furthermore, the fibrillation of single fibres and its compaction to create self-binding materials outperform the use of binders (nanocellulose and other polymers) to create non-wovens.

10.4 NANOCELLULOSE AS BINDER FOR WOOD PULP FIBRES

The function or importance of nanocellulose acting as a binder for natural fibres differs when used to bind pulp fibres. Bonds between natural fibres, such as hydrogen bonds, cannot be created; however, they can be formed between pulp fibers [25]. In pulp, as for nanocellulose, an inter-fibre network is formed by hydrogen bonding by a process called hornification [26]. There are several mechanisms contributing to the formation of this network, such as mechanical interlocking, van der Waals forces and hydrogen bonds [27]. These cellulose fibre–fibre bonds have been traditionally utilised in the paper industry and in composites. However, nanocellulose possesses a higher surface area than pulp [28], resulting in stronger interactions with itself and other materials [29]. The effect of combining both pulp fibre–fibre interaction with the extra interaction between nanocellulose fibrils and pulp fibres explored to enhance paper properties. Table 10.3 summarises the tensile properties of different pulp-NFC material combinations.

TABLE 10.3
Tensile Properties of Selected Composites Made from Pulp and Nanocellulose. All the Tensile Strengths with the Different Units Were Calculated, When Possible, from the Information of the Papers in Which It Was Published

Authors	Pulp	Binder	Amount of Binder (wt.%)	Tensile Strength (MPa)	Tensile Strength ($kN \cdot m^{-1}$)	Tensile Index ($Nm \cdot g^{-1}$)
Ghazali et al. [31]	Pulp fibres from brunches	–	0			25
			2.5			30
			5			43
			8.5			55
Eriksen et al. [32]	Kraft pulp	NFC	0			36
			1			37
			2			36
			5			41
			10			49
Taipale et al. [33]	TMP	NFC	0			95
			1			96
			1.5			97
			2			98
			3			99
			4.5			100
			10			113
Rezayati Charani et al. [34][a]	Hardwood Kraft pulp	NFC (kenaf Kraft pulp)	0	29.5 ± 0.8	2.8	47.2 ± 1.7
			2	31.9 ± 0.6	3.3	49.7 ± 0.9
			4	34.9 ± 0.7	3.2	52.9 ± 1.0
			6	40.4 ± 0.6	3.5	56.7 ± 0.2
			10	46.0 ± 0.6	4.2	62.4 ± 0.7
		NFC (softwood Kraft pulp)	2	32.8 ± 0.2	3.1	49.1 ± 0.2
			4	33.2 ± 0.3	3.3	52.0 ± 1.4
			6	38.9 ± 0.4	3.8	55.5 ± 2.1
			10	43.1 ± 0.5	4.1	61.5 ± 0.8
Manninen et al. [35][b]	Softwood pulp	NFC	0		49.2	82
			6		63	105
			20		63.6	106
Kajanto et al. [36][c]	Finish SW/HW pulp		0		3.3	
		NFC1	1–2		3.7	
		NFC2	1–2		4	
Hassan et al. [37]	Bagasse	NFC[d]	0			30
			10			32
			20			35
			30			36
			40			38
			50			39

(Continued)

TABLE 10.3 (*Continued*)
Tensile Properties of Selected Composites Made from Pulp and Nanocellulose. All the Tensile Strengths with the Different Units Were Calculated, When Possible, from the Information of the Papers in Which It Was Published

Authors	Pulp	Binder	Amount of Binder (wt.%)	Tensile Strength (MPa)	Tensile Strength ($kN{\cdot}m^{-1}$)	Tensile Index ($Nm{\cdot}g^{-1}$)
			60			38
			70			39
			80			38
			90			41
			100			40
Sehaqui et al. [38][e]	Wood pulp	NFC (wood)	0	97.5 ± 15	6.8	117 ± 28
			2	141 ± 13	9.9	149 ± 29
			5	141 ± 8	9.9	151 ± 20
			7	146 ± 14	10.2	155 ± 24
			10	160 ± 9	11.2	165 ± 16
			100	235 ± 21	16.4	218 ± 27
Sehaqui et al. [39][f]	Softwood sulphite pulp	NFC	0		2.8	41.1 ± 0.6
			10		4.7	69.4 ± 2.4
			100		10.5	156 ± 10

[a] The tensile strength in $kN{\cdot}m^{-1}$ was calculated from the tensile index by using the grammage reported in the original paper.
[b] The tensile strength in $kN{\cdot}m^{-1}$ was calculated by assuming a thickness of 60 μm from the tensile strength reported in MPa.
[c] NFC1 and NFC2 are obtained after non-specified pre-treatments.
[d] It is not addition, it is percentage.
[e] The tensile strength in $kN{\cdot}m^{-1}$ was calculated by assuming a thickness of 70 μm from the tensile strength reported in MPa.
[f] The tensile strength in $kN{\cdot}m^{-1}$ was calculated by assuming a grammage of 67 from the tensile index.

In order to utilise nanocellulose fibrils partially detached from the surface of natural fibres, the fibres do not need to be in their raw fibre form; they can also be in a pulp form. The partial fibrillation of sisal pulp fibres has been previously reported; to detach nanofibres from its surface; as a reinforcement for composites [30]. On the other hand, Ghazali et al. [31] partially fibrillated pulp from oil palm empty fruit bunches to create self-binding materials. The authors found that when the fibrillation energy used to fibrillate the pulp was 8.5 KWh/mt, the tensile index of the papers obtained was two times higher than when the papers were made from non-fibrillated pulp (Table 10.3). The authors also attributed the increase of the tensile index to the filling of the porosity of the paper with NFC and the additional bonding between the fibres.

As observed for natural fibres, nanocellulose can also be simply attached to the pulp fibres. It should be noted that the concept of further reinforcing pulp with cellulose has already been long used, but instead of using nanocellulose pulp fines were used [33]. The use of nanocellulose as paper additive has been reported by several authors. Eriksen et al. [32] added microfibrillated cellulose (mechanically obtained) to thermomechanical pulp (TMP) handsheets. The addition of small amounts of NFC did not improve the tensile index of the modified papers; however, additions of 5% to 10% of NFC strongly increased the tensile index up to around 49 N m/g. The values are also similar to the ones reported by Ghazali et al. [33], where the fibres were partially fibrillated; however, a comparison is not possible due to the unknown content of NFC created by the fibrillation process. Apart from these results, the authors also observed an increase in the resistance to air flow through the paper. A similar tensile properties were observed by Taipale et al. [33], who used NFC to further reinforce elemental chlorine-free (ECF)-bleached hardwood Kraft pulp, following a similar manufacturing method (a mixing process followed by a filtration process). However, it has to be noted that even though the addition of nanocellulose produced a similar effect, the mechanical performance of the pulp sheets/papers is strongly affected by the pulp used as the starting material (the tensile index of the reference values from Taipale et al. [33] are twice as high as the properties of the reference paper made by Eriksen et al. [32]). Rezayati et al. [34] found the same increase in mechanical properties, when adding 5%–10% NFC to kraft pulp. They compared the use of nanocellulose obtained from different types of pulps and found that both provided a similar reinforcement. Manninen et al. [35] used a Masuko's Supermasscolloider to produce nanocellulose from ECF-bleached chemical birch pulp and added this nanocellulose to bleached chemical softwood pulp from pine. The authors found that the tensile index increased by 30% and the tensile strength by 60% upon the addition of only 6 wt.% NFC (Table 10.3). It is important to note that contrary to other authors [33,38], Eriksen et al. [32] did not found such high increases in paper tensile properties at low NFC loadings. A slight increase was observed upon the addition of further NFC (10 wt.%). These were not the only authors who observed a big increase upon the addition of very small amounts of NFC. Kajanto et al. [36] explored the use of nanocellulose as a 'paper additive' on a pilot paper machine, in order to prove its scalability (Table 10.3). To make it comparable, the authors used chemical pulp from a paper mill (and the recirculating water of the paper mill too), from a mixture of Finish softwood/hardwood pulp (SW/HW) without fillers. As an additive, they used two different types of nanocellulose subjected to different treatments (which they do not specify). They investigated the mechanical performance of the papers upon the addition of the nanocelluloses and concluded that an 8 g/m^2 reduction of the paper basis weight would be possible upon the addition of only 1–2% nanocellulose to obtain the same tensile strength. They also reported an increase in the apparent density and a reduction in the air permeability and opacity. Another important highlight of these work is that the authors reported that no major production problem was encountered in the production with the addition of the nanocellulose (in the dewatering step).

Hassan et al. [37] used NFC obtained from bagasse pulp to self-reinforce the same bagasse pulp. They investigated the whole spectrum of possible nanocellulose contents, from 0 to 100 wt.% NFC within the papers. They did not observe an important increase in the tensile index of the composites when increasing NFC content; the difference between the reference and the 100 wt.% NFC nanopaper was only

10 Nm·g^{-1}, which means an increase of only 30%, much lower than that reported by Taipale et al. [33], Eriksen et al. [32] and Manninen et al. [35] using much lower NFC loadings.

The word nanocomposites has also been used to refer to these materials. Sehaqui et al. [38] used NFC to further reinforce never-dried softwood pulp. They followed a papermaking process, where pulp was disintegrated in water and different contents of NFC were added to the suspension. Afterwards, they used a vacuum filtration process to produce the wet composite, and they further dried it under pressure and temperature. A small addition of NFC (2 wt.%) significantly increased the tensile strength of the composites (Table 10.3). Moreover, a further increasing the NFC loading resulted in further increased tensile strength (and the work of fracture). The authors stated that the NFC network formed through the pulp fibres improved the load transfer between the pulp fibres when the damage started, delaying large-scale damage. The authors later used NFC from never-dried bleached softwood sulphite pulp from spruce to self-reinforce the same pulp [39] and obtained the same trend as before and similar results, keeping in mind that the pulp and NFC used were of different origin (Table 10.3).

It can be seen in Table 10.3 that the NFC contents used as an extra binder for pulp fibres are normally less than 5%–10%, whereas for natural fibres, these are normally greater than 10 wt.% (Table 10.2). Furthermore, it can be stated that when using nanocellulose as additional binder, the main reinforcing fibres (pulp fibres) strongly influence the mechanical performance of the resulting composites/papers. On the other hand, it was shown before that in the case of non-wovens, the reinforcing fibres did not have a big effect on the mechanical performance of the nanocellulose non-wovens. Overall, it can be observed that the tensile strength and the tensile index are generally higher when nanocellulose is used as a binder for pulp fibres than for natural fibres (Tables 10.2 and 10.3). However, the nominal tensile strength is comparable.

10.5 NANOCELLULOSE COATED NATURAL FIBRES

Nanocellulose has not only been used for binding loose fibres together but also to coat them individually, and therefore, improving the natural fibres' surface properties resulting in improved adhesion to polymer matrices. Understanding 'nanocellulose bonding' is a key factor for the further use of nanocellulose as binder.

Before, we showed that nanocellulose could be partially be liberated from natural fibres or simply attached to them, to bind them together. In the case of coating individual fibres, another route was used: nanocellulose can be attached to natural fibres by directly synthesising it in their presence. Juntaro et al. [40,41] used sisal fibres as a substrate during a bacterial fermentation of BC, attaching bacterial nanocellulose directly onto sisal fibres. The authors incubated bacteria (*Acetobacter xylinum*) at 30°C for 4–7 days. After this period of time, they observed a transparent pellicle covering the sisal fibres. They then treated the coated fibres with NaOH to remove the bacteria and other by-products, which did not destroy the nanocellulose layer. The authors justified the non-destruction of the layer by the strong interaction between the BC and the sisal surface, due to the self-affinity of cellulosic materials through hydrogen bonding. However, they observed that the nanocellulose (50–100 nm) was

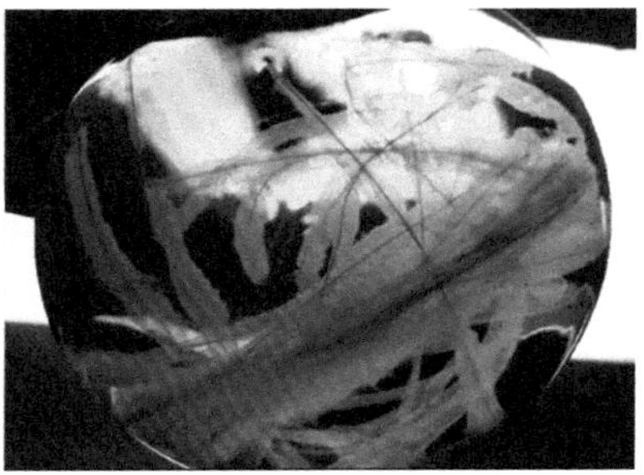

FIGURE 10.4 Differences of sisal fibres before and after 2 days on bacterial culture. (Reprinted with permission from Pommet, M. et al., *Biomacromolecules*, 9, 1643–1651, 2008. Copyright 2008 American Chemical Society.)

not covering the entire fibre surface, which they attributed to the presence of waxes or other organic compounds on the fibre surface. To improve the BC deposition this problem, they washed the fibre surface with acetone, using a Soxhlet extraction. After removal of the waxy substances from the fibres, the amount of nanocellulose attached to the fibres was still relatively low (5–6 wt.%) but the entire surface of the natural fibres was covered by a layer of nanocellulose. In order to investigate the effectivity of the nanocellulose coating, the authors measured single-fibre tensile strength and the interfacial shear strength (IFFS) between the BC coated fibres and a poly(L-lactic) acid (PLLA) matrix. They found that the tensile strength and modulus slightly decreased after the BC coating however, the IFFS slightly improved for the untreated fibres, with BC attached to their surface. The authors continued their work by further depositing BC onto sisal and hemp fibres, changing the manufacturing method and fibre pre-treatments [42]. Before the introduction of a 3-day-old broth of a culture of *Acetobacter xylinum* BPR2001, they introduced the natural fibres in to a culture medium and autoclaved them at 121°C for 20 min. After innoculation with bacteria, the authors left the fermentation to occur for a week in an environmental chamber at 30°C (Figure 10.4).

In order to scale-up the production, the authors used an agitated 5 L fermentator, into which they introduced fibre mats or natural fibres (1 cm long) directly to the culture medium and introduced again a 3-day-old broth of a culture of bacteria, to be fermented for a week at 30°C. After the fermentation, the authors removed the fibres and purified them with NaOH at 80°C for 20 min. In small scale, they discovered that the bacteria strain used grew preferably on the fibre surface rather than in the culture medium. When culturing at a large scale, they found that the BC produced was preferably produced was the support cassette (holding a fibremat) rather than the fibres themselves, which the authors explained by the possibility of too anaerobic conditions in the cassette. So, instead of fibre mats, they used loose fibres. They then realised that the fibres agglomerated in the cassette, so BC was only growing on the fibre surface and around the agglomerates, covering the fibres partially. The coated fibres were then difficult to be isolated, and no tests were performed on them (Figure 10.5).

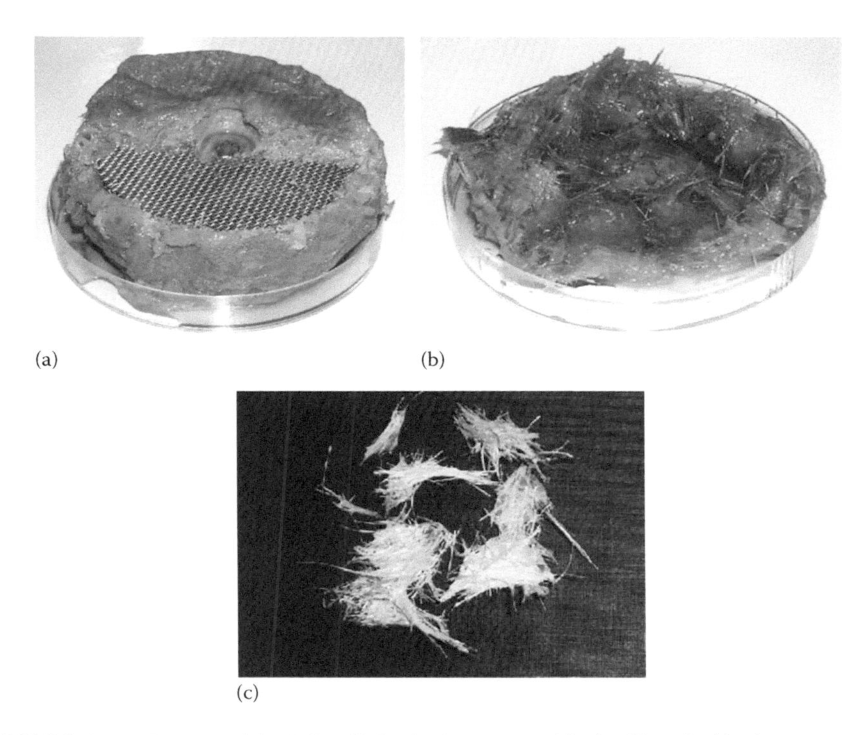

(a) (b) (c)

FIGURE 10.5 Images of the BC pellicles in the cassette (a), the fibres inside the cassette (b) and BC covering sisal fibres (c). (Reprinted with permission from Pommet, M. et al., *Biomacromolecules*, 9, 1643–1651, 2008. Copyright 2008 American Chemical Society.)

On the other hand, nanocellulose can also be coated onto natural fibres by simply immersing them on a nanocellulose suspension. For example, Dai et al. [43] obtained nanocellulose by oxidation hydrolysis of short hemp fibres (0.5–1 cm) and used it as a 'coupling agent' for the same hemp fibres. They modified the hemp fibres before the nanocellulose incorporation. The modified fibres were then soaked in a 2% nanocellulose suspension for 10 min and then dried in an oven. The successful coating of the fibres was obvious as diameter increase (from 45.1 ± 9.7 to 51.4 ± 7.1 μm). X-ray photoelectron spectroscopy (XPS) was showed that the fibre was nearly fully covered by nanocellulose. They observed an increase in the tensile properties of the fibres coated with nanocellulose. The tensile modulus, strength and strain increased from 29.83 ± 7.95 GPa, 735.29 ± 7.65 MPa and 2.47% ± 9.98%, respectively, for modified fibres up to 38.51 ± 8.44 GPa, 1203.85 ± 9.25 MPa and 3.84% ± 5.92%, respectively, after the nanocellulose coating/modification.

Lee et al. [44,45] deposited BC onto sisal fibres to form either densely coated hairy fibres. In order to produce the BC coating, they soaked sisal fibres in a 0.1% freeze-dried BC dispersion for 3 days. Then, they removed the fibres and dried them in two different ways: by drying under vacuum at 80°C overnight to create a dense BC coating densely coated neat sisal (DCNS) fibres or by pressing the fibres between filter papers under a weight of 3 kg for 10 s to partially dry them and then totally

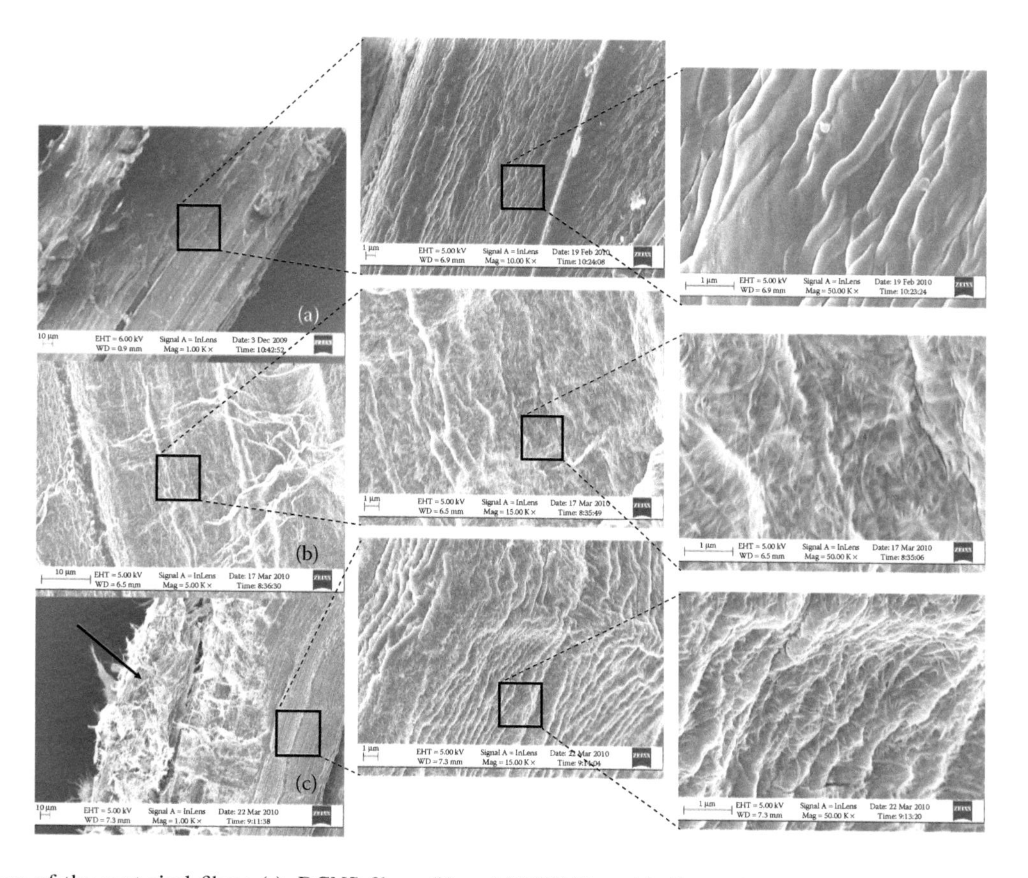

FIGURE 10.6 Images of the neat sisal fibres (a), DCNS fibres (b) and HNSF fibres (c). (Reproduced from Lee, K.Y. et al., *Compos. Part A.*, 43, 2065–2074, 2012. With permission of Pergamon.)

drying them in an air oven at 40°C to obtain hairy fibres of neat sisal fibres (HNSF). The different morphology of coatings can be observed in Figure 10.6. The authors investigated effect of these two types of BC coatings on the surface area of the fibres and on the single-fibre tensile properties. As expected, the surface area increased after coating the fibres with BC, using both coating techniques. The DCNS fibres and the HNSF fibres possessed a surface area of 0.77 ± 0.03 m^2/g and 0.49 ± 0.03 m^2/g, respectively, whereas the unmodified sisal fibres possessed only a surface area of 0.10 ± 0.01 m^2/g. The dense BC coating reduced the tensile strength and modulus of the sisal fibres by nearly half, whereas the 'hairy coating' reduced it only slightly. These fibres were then used as reinforcement for PLLA and PLLA/BC, outperforming the unmodified sisal fibres.

10.6 CONCLUSION

The use of nanocellulose as a binder for natural fibres is motivated by the need to overcome adhesion problems between fibres, n fibres and matrices and the need of enhance the properties of natural fibre mats and to create fully degradable materials. There are two main ways of using nanocellulose as a binder for natural fibres: creating the nanocellulose by partial fibrillation of the surface of the fibres to be bonded or by adding it onto natural fibres.

The partial fibrillation of the surface of the natural fibres partially liberates the nanofibrils from the surface (in the form of branches), which promotes fibre/fibre bonding during mat or fiberboard formation. It is important not to damage the fibrous structure/body of the fibres or mat, in order to achieve enhanced final properties of the mat/fibreboard. On the other hand, nanocellulose (from different sources) can be attached onto natural fibres. The natural fibres to be bonded can be soaked in a nanocellulose suspension and then dried. The nanofibrils then form a three-dimensional network structure by wrapping or linking the loose fibres or mats, enhancing selected properties of the mats or non-wovens. A special case is pulp, where nanocellulose acts as an additional binder, since the pulp fibres form already bond to themselves. The improvement of the mechanical properties upon the addition of nanocellulose to pulp fibres strongly depends on the properties of the fibres, which is not the case for natural fibres, where the properties depend on the network structure formed by the nanocellulose within the fibre mat. It is worth mentioning that for both natural and pulp fibres, the use of nanocellulose as a binder has been scaled up and even commercialised (Zelfo Technology GmbH®).

In addition, it is important to understand the effect of the nanocellulose adhesion to the natural fibres (coating of single fibres) to further improve the nanocellulose binding. By soaking single fibres in nanocellulose suspensions, the fibres can be coated, and depending on the way it is done, different types of coatings can be created. Furthermore, BC gives a great opportunity for in situ coating, where nanocellulose is directly synthesised on the surface of the natural fibres in the culture medium.

ACKNOWLEDGEMENTS

The authors would like to thank Universität Wien, Wien, Austria, for funding MFV.

REFERENCES

1. Celino A, Freour S, Jacquemin F, Casari P. The hygroscopic behavior of plant fibers: A review. *Frontiers in Chemistry*. 2013;1:43.
2. Summerscales J, Dissanayake NPJ, Virk AS, Hall W. A review of bast fibres and their composites. Part 1-Fibres as reinforcements. *Composites Part A*. 2010;41(10):1329–1335.
3. John MJ, Thomas S. Biofibres and biocomposites. *Carbohydrate Polymers*. 2008;71(3):343–364.
4. Wielage B, Lampke T, Marx G, Nestler K, Starke D. Thermogravimetric and differential scanning calorimetric analysis of natural fibres and polypropylene. *Thermochimica Acta*. 1999;337(1):169–177.
5. Kalia S, Dufresne A, Cherian BM, Kaith BS, Avérous L, Njuguna J et al. Cellulose-based bio- and nanocomposites: A review. *International Journal of Polymer Science*. 2011;2011:1–35.
6. George J, Sreekala MS, Thomas S. A review on interface modification and characterization of natural fiber reinforced plastic composites. *Polymer Engineering & Science*. 2001;41(9):1471–1485.
7. Tserki V, Zafeiropoulos NE, Simon F, Panayiotou C. A study of the effect of acetylation and propionylation surface treatments on natural fibres. *Composites Part A*. 2005;36(8):1110–1118.
8. Kim JT, Netravali AN. Mercerization of sisal fibers: Effect of tension on mechanical properties of sisal fiber and fiber-reinforced composites. *Composites Part A*. 2010;41(9):1245–1252.
9. Sangthong S, Pongprayoon T, Yanumet N. Mechanical property improvement of unsaturated polyester composite reinforced with admicellar-treated sisal fibers. *Composites Part A*. 2009;40(6–7):687–694.
10. Lee KY, Delille A, Bismarck A. Greener surface treatments of natural fibres for the production of renewable composite materials. In: Kalia, S. (Ed.). *Cellulose Fibers: Bio- and Nano-Polymer Composites*, Berlin, Germany; 2011. pp. 155–178.
11. Rowell RM, Stout HP. Jute and kenaf. *International Fiber Science and Technology*. 2007;16:405–452.
12. Anandjiwala RD, Boguslavsky L. Development of needle-punched nonwoven fabrics from flax fibers for air filtration applications. *Textile Research Journal*. 2008;78(7):614–624.
13. Miao MH, Shan MJ, Finn N, Schutz JA, Wood R, Ahern M. Improvement of filtration efficiency by fibre surface nanofibrillation. *Journal of the Textile Institute*. 2012;103(7):719–723.
14. Miao MH, Pierlot AP, Millington K, Gordon SG, Best A, Clarke M. Biodegradable mulch fabric by surface fibrillation and entanglement of plant fibers. *Journal of the Textile Institute*. 2013;83(18):1906–1917.
15. Arévalo R, Peijs T. Binderless all-cellulose fibreboard from microfibrillated lignocellulosic natural fibres. *Composites Part A*. 2016;83:38–46.
16. Goutianos S, Arévalo R, Sørensen BF, Peijs T. Effect of processing conditions on fracture resistance and cohesive laws of binderfree all-cellulose composites. *Applied Composite Materials*. 2014;21(6):805–825.
17. Hampton JM, Jones DO, Shenoy SL. Nonwoven filtration medium useful for air or liquid filtration, comprises a fibrous base media comprising synthetic and/or natural fibers, and microfibrillated cellulose fibers. Donaldson Co Inc; Hampton JM; Jones DO; Shenoy SL Patent WO 2014164127 A1.

18. Cartier N, Dufour M, Mavrikos F, Merlet S, Vincent A. Wet-laid nonwoven for e.g. healthcare, medical, surgical, personal care, and textiles comprises long fibers consisting of synthetic and/or natural fibers and nanofibrillar cellulose which impregnates nonwoven for its entire thickness. Ahlstroem Corp A; Ahlstroem OY A.
19. Lee K-Y, Ho KKC, Schlufter K, Bismarck A. Hierarchical composites reinforced with robust short sisal fibre preforms utilising bacterial cellulose as binder. *Composites Science and Technology*. 2012;72(13):1479–1486.
20. Lee K-Y, Shamsuddin SR, Fortea-Verdejo M, Bismarck A. Manufacturing of robust natural fiber preforms utilizing bacterial cellulose as binder. *Journal of Visualized Experiments*. 2014;87:e51432. doi:10.3791/51432.
21. Fortea-Verdejo M, Lee K-Y, Zimmermann T, Bismarck A. Upgrading flax nonwovens: Nanocellulose as binder to produce rigid and robust flax fibre preforms. *Composites Part A*. 2016;83:63–71.
22. Fages E, Girones S, Sanchez-Nacher L, Garcia-Sanoguera D, Balart R. Use of wet-laid techniques to form flax-polypropylene nonwovens as base substrates for eco-friendly composites by using hot-press molding. *Polymer Composites*. 2012;33(2):253–261.
23. Fages E, Cano MA, Girones S, Boronat T, Fenollar O, Balart R. The use of wet-laid techniques to obtain flax nonwovens with different thermoplastic binding fibers for technical insulation applications. *Textile Research Journal*. 2013;83(4):426–437.
24. Svoboda MA, Lang RW, Bramsteidl R, Ernegg M, Stadlbauer W. Zelfo. An engineering material fully based on renewable resources. *Molecular Crystals and Liquid Crystals*. 2000;353:47–58.
25. Gardner DJ, Oporto GS, Mills R, Samir MASA. Adhesion and surface issues in cellulose and nanocellulose. *Journal of Adhesion Science and Technology*. 2008;22(5–6):545–567.
26. Fernandes DJMB, Gil MH, Castro JAAM. Hornification—its origin and interpretation in wood pulps. *Wood Science and Technology*. 2004;37(6):489–494.
27. Schmied FJ, Teichert C, Kappel L, Hirn U, Bauer W, Schennach R. What holds paper together: Nanometre scale exploration of bonding between paper fibres. *Scientific Reports-UK*. 2013;3:2432.
28. Aulin C, Lindström T. *Biopolymer Coatings for Paper and Paperboard. Biopolymers – New Materials for Sustainable Films and Coatings*, New York: John Wiley & Sons; 2011. pp. 255–276.
29. Klemm D, Schumann D, Kramer F, Hessler N, Koth D, Sultanova B. Nanocellulose materials – Different cellulose, different functionality. *Macromolecular Symposia*. 2009;280:60–71.
30. Zhong LX, Fu SY, Zhou XS, Zhan HY. Effect of surface microfibrillation of sisal fibre on the mechanical properties of sisal/aramid fibre hybrid composites. *Composites Part A*. 2011;42(3):244–252.
31. Ghazali A, Zukeri MRHM, Daud WRW, Azhari B, Ibrahim R, Mohamed IA et al. Augmentation of EFB fiber web by nano-scale fibrous elements. *Advanced Materials Research*. 2014;832:494–499.
32. Eriksen O, Syverud K, Gregersen O. The use of microfibrillated cellulose produced from kraft pulp as strength enhancer in TMP paper. *Nordic Pulp & Paper Research Journal*. 2008;23(3):299–304.
33. Taipale T, Österberg M, Nykänen A, Ruokolainen J, Laine J. Effect of microfibrillated cellulose and fines on the drainage of kraft pulp suspension and paper strength. *Cellulose*. 2010;17(5):1005–1020.

34. Rezayati CP, Dehghani-Firouzabadi M, Afra E, Blademo A, Naderi A, Lindström T. Production of microfibrillated cellulose from unbleached kraft pulp of Kenaf and Scotch Pine and its effect on the properties of hardwood kraft: Microfibrillated cellulose paper. *Cellulose*. 2013;20(5):2559–2567.
35. Manninen M, Kajanto I, Happonen J, Paltakari J. The effect of microfibrillated cellulose addition on drying shrinkage and dimensional stability of wood-free paper. *Nordic Pulp and Paper Research Journal*. 2011;26(3):297.
36. Kajanto I, Kosonen M. The potential use of micro- and nano-fibrillated cellulose as a reinforcing element in paper. *Journal of Science & Technology for Forest Products and Processes*. 2012;2(6):42–48.
37. Hassan EA, Hassan ML, Oksman K. Improving bagasse pulp paper sheet properties with microfibrillated cellulose isolated from xylanase-treated bagasse. *Wood Fiber Science*. 2011;43(1):76–82.
38. Sehaqui H, Allais M, Zhou Q, Berglund LA. Wood cellulose biocomposites with fibrous structures at micro- and nanoscale. *Composites Science and Technology*. 2011;71(3):382–387.
39. Sehaqui H, Zhou Q, Berglund LA. Nanofibrillated cellulose for enhancement of strength in high-density paper structures. *Nordic Pulp and Paper Research Journal*. 2013;28(2):182–189.
40. Juntaro J, Pommet M, Kalinka G, Mantalaris A, Shaffer MSP, Bismarck A. Creating hierarchical structures in renewable composites by attaching bacterial cellulose onto sisal fibers. *Advanced Materials*. 2008;20(16):3122–3126.
41. Bismarck A, Juntaro J, Mantalaris A, Pommet M, Shaffer MSP. Material for manufacturing articles useful in e.g. packaging comprises cellulose produced by a microorganism, and associated with a support selected from a polymer and/or a fiber. Imperial Innovations Ltd. Patent WO2008020187 A1.
42. Pommet M, Juntaro J, Heng JYY, Mantalaris A, Lee AF, Wilson K et al. Surface modification of natural fibers using bacteria: Depositing bacterial cellulose onto natural fibers to create hierarchical fiber reinforced nanocomposites. *Biomacromolecules*. 2008;9(6):1643–1651.
43. Dai D, Fan M, Collins P. Fabrication of nanocelluloses from hemp fibers and their application for the reinforcement of hemp fibers. *Industrial Crops and Products*. 2013;44:192–199.
44. Lee KY, Bharadia P, Bismarck A. Nanocellulose surface coated support material. Patent 2011, US 9193130 B2.
45. Lee KY, Bharadia P, Blaker JJ, Bismarck A. Short sisal fibre reinforced bacterial cellulose polylactide nanocomposites using hairy sisal fibres as reinforcement. *Composites Part A*. 2012;43(11):2065–2074.

11 Cellulose-Based Aerogels

Jian Yu and Jun Zhang

CONTENTS

11.1 Introduction 217
11.2 Preparation of Cellulose-Based Aerogels 219
11.2.1 Ce-II Aerogels 219
11.2.2 Ce-I Aerogels 221
11.2.3 Ce-D Aerogels 224
11.3 Cellulose-Based Composite Aerogels 225
11.3.1 Before Formation of Cellulose Gels 225
11.3.2 After Formation of Cellulose Gels 229
11.3.3 After Formation of Cellulose Aerogels 231
11.4 Functional Cellulose-Based Aerogels 235
11.4.1 Optical Properties 235
11.4.2 Flame Retardancy 236
11.4.3 Antibacterial Activity 237
11.4.4 Magnetic Properties 238
11.4.5 Hydrophobicity 239
11.4.6 Absorption 242
11.4.7 Oxidation Resistance 247
11.4.8 Thermal Conductivity 247
11.4.9 Sound Absorption 247
11.4.10 Electrical Properties 249
11.4.11 Catalytic Activity 250
11.4.12 Biomedical Features (Biocompatibility) 250
11.5 Cellulose-Based Carbon Aerogels 251
11.6 Outlook 252
Acknowledgements 252
References 253

11.1 INTRODUCTION

Cellulose, the most abundant renewable biomass on earth, has plentiful hydroxyl groups in the molecules, allowing for the formation of strong intra- and intermolecular hydrogen bonding. The effect of these interactions results in a physical cross-linking network required for the cellulose gels and further for aerogels. The combination of high porosity and surface area, superior processability and mechanical properties and excellent biocompatibility and biodegradability makes cellulose-based aerogel a valuable biomass as a new-generation aerogel following the inorganic oxide and the synthetic polymer aerogels [1,2].

The first systematic preparation of cellulose aerogels was reported in 1971 by Weatherwax and Caulfield [3]. They obtained cellulose aerogels with high specific surface areas of about 200 m^2/g by supercritical CO_2 drying (SD) after solvent exchanging wood pulp with ethanol. However, cellulose aerogels did not attract extensive research until about 30 years later. In 2001, Tan et al. [4] used cellulose acetate and cellulose acetate butyrate as starting materials and diisocyanate as cross-linking agent to obtain cellulose-derived aerogels with high-impact strength, which exhibited specific surface areas about 300 m^2/g and densities in the range of 0.1–0.4 g/cm^3. In 2004, Jin et al. [5] reported the preparation of regenerated cellulose aerogels containing cellulose II crystalline structure (Ce-II aerogels) by using dissolution–regeneration process for the first time. In 2006, nanocellulose isolated from plant sources or bacterial cellulose (BC) was also introduced to prepare aerogels of cellulose I (Ce-I aerogels) [6,7].

The cellulose-based aerogels can be categorised into neat cellulose and cellulose-derived aerogels (Ce-D aerogels) according to the type of starting cellulose. The former includes Ce-II aerogels prepared via dissolution–regeneration method and Ce-I aerogels prepared without dissolving in solvents. Both neat cellulose and Ce-D aerogels can further be incorporated into other components by compounding and/or chemical modification to obtain functional cellulose composite aerogels. In addition, as organic materials with high carbon contents, the above-mentioned cellulose-based aerogels have attracted increasing attention as organic precursors of carbon aerogels. Here, we present the review, beginning with the preparation of neat cellulose aerogels, Ce-D aerogels and cellulose composite aerogels and then discussing the functionalities of these materials, followed by the introduction of cellulose-based carbon aerogels.

In general, gelation and drying are two key steps in the preparation of cellulose-based aerogels. Dissolving cellulose in non-derivatised cellulose solvent results in homogeneous solution, which interacts with non-solvent of cellulose (as coagulation bath) to induce gelation. Physical gel is formed via intermolecular hydrogen bonding among cellulose molecules during gelation. As for nanocellulose, gelation occurs in the aqueous dispersion via intermolecular hydrogen bonding between hydroxyl groups on the surface of cellulose nanofibres and/or entanglement of the nanofibres. Sometimes, chemical cross-linking is introduced to improve the integrity and strength of gels and corresponding aerogels. As the most commonly used drying method, both freeze-drying (FD) and SD have a main shortcoming of high production cost, which hinders the wide application of aerogels. Recently, evaporative drying (ED) under ambient pressure or vacuum has been tried by researchers as a cost-effective method to remove the liquid solvent from gels, without collapsing the network structure.

It is worth noting that aerogel should be a porous material containing nanopores and nanosized structure and obtained by using SD method to maintain the network structure in the gel precursor, according to the definition [8]. Some similar terms for aerogels have been used in the literature. Cellulose foams or sponges mainly consist of large pores with micron size. The freeze-dried porous material is usually called a cryogel, whose morphology has been determined by frozen solvent crystals. The materials obtained by ambient/vacuum drying were often named xerogels, and these are known as ambient dried aerogels. Ce-II aerogels obtained by using SD have been termed aerocellulose since 2006 [9]. Herein, we use the term aerogel for above-mentioned materials, regardless of what has been used in the cited work, and do not strictly distinguish their differences.

11.2 PREPARATION OF CELLULOSE-BASED AEROGELS

11.2.1 Ce-II Aerogels

Ce-II aerogels, commonly referred to as regenerated cellulose aerogels, are obtained by the dissolution and subsequent regeneration/coagulation of cellulose. Therefore, cellulose I crystalline structure in the native cellulose has been transformed to cellulose II in these aerogels. Many direct solvents of cellulose have been used to prepare Ce-II aerogels, including those with low solubility of cellulose, because high cellulose concentration was not necessary for making highly porous aerogels. In 2004, Jin et al. [5] first reported the preparation of Ce-II aerogels by using calcium thiocyanate ($Ca(SCN)_2$) hydrate melt as the cellulose solvent. From then onwards, the solvent systems such as aqueous sodium hydroxide solution, ionic liquids and *N*-methylmorpholine-*N*-oxide (NMMO) monohydrate, along with calcium thiocyanate hydrate melt, have been widely used in the preparation of Ce-II aerogels, as shown in Table 11.1. Until 2016, some new solvents have been applied to fabricate Ce-II aerogels for the first time [10,11].

TABLE 11.1
Ce-II Aerogels Prepared by Various Solvent Systems and Drying Methods

Solvent Systems	Supercritical CO_2 Drying	Freeze-Drying from Hydrogels	Freeze-Drying from Tert-Butanol Gels	Ambient/Vacuum Drying
Aqueous metal hydroxide solution systems				
NaOH	[12–20]	[16,18,21–23]	[24]	
NaOH-urea or LiOH-urea	[25–31]	[30,32–48]	[30,49,50]	
NaOH-thiourea	[51]	[52–55]		
NaOH-PEG			[56–66]	
Ionic liquid systems				
AmimCl	[67–74]	[75]	[76]	
BmimCl or BmimCl-cosolvent	[70,77,78]	[79–87]		
EmimAc or EmimAc-cosolvent	[70,77,88–90]			[91]
HmimCl	[92]			
Other systems				
DMAc-LiCl or DMSO-LiCl		[93]	[94,95[a]]	
NMMO	[9,17,96–99]			
$Ca(SCN)_2$ hydrate	[88,100–104]	[5,102,104,105]	[5]	[102]
$ZnCl_2$ hydrate	[11]			
Trifluoroacetic acid	[10]			

Note: DMAc: *N,N*-dimethylacetamide; DMSO: dimethyl sulphoxide; HmimCl: 1-hexyl-3-methyl-1H-imidazolium chloride; LiCl: lithium chloride; NMMO: *N*-methylmorpholine-*N*-oxide; PEG: polyethylene glycol.

[a] Obtained from 1,1,2,2,3,3,4-heptafluorocyclopentane gel.

Cellulose gelation was induced by changing the temperature of cellulose solution or by using antisolvent as coagulation bath, and cellulose was subsequently regenerated in gel form. The constituents and temperature of the coagulation bath were the main parameters to modulate gelation kinetics of cellulose [9,11,13,18,20,23,30,53,72], which were very important in the formation of aerogel structure. The regeneration of cellulose from its solutions was a diffusion-controlled process and is generally described by the Fick's law [17,18,77]. Ionic liquids, such as 1-ethyl-3-methylimidazolium acetate (EmimAc) and 1-butyl-3-methylimidazolium chloride (BmimCl), had diffusion coefficient of the order of 10^{-11} to 10^{-10} m^2/s at ambient conditions [77,106,107], which were several times lower than those of sodium hydroxide (NaOH) and NMMO, owing to the larger molecular sizes of the former [77]. However, all the solvents exhibited similar activation energy of diffusion, which were calculated to be 15–20 kJ/mol [77].

The porous structure of aerogels can be tailored either by cellulose gelation [9,11,13,18,23,30,53,72] or by additive addition [17,20,25,88,102]. The gelation time was shortened for the cellulose solution with higher cellulose concentration, owing to more entanglement and aggregation of chains to increase the physical cross-linking density [17,20,48,53,104]. The strengthened networks facilitated to homogeneous morphology in aerogels, with fine skeletons and pores [5,30,72,84,91]. On the other hand, adding zinc oxide (ZnO) in the solvent systems of aqueous NaOH solution improved the dissolution of cellulose and significantly delayed the gelation of cellulose solutions [20,25]. The addition of paraffin wax or poly(methylmethacrylate) as porogen resulted in an open and dual porosity and resulted in the formation of macropores (with sizes >100 μm) and nanopores [88]. Foaming cellulose hydrogels by dissolved CO_2 under high pressure produced micro-sized porous cellulose materials [71]. Employing cellulose solution–glyceryl trioctanoate emulsion stabilised by surfactant as template, aerogels containing macropores in the range between 50 μm and 120 μm were obtained by ambient pressure drying. Although with obviously volume shrinkage, the ambient pressure-dried aerogels retained high porosity of 80% and specific surface area of about 107 m^2/g [102].

During the gelation of cellulose, more crystalline regions were formed in a cyclic freezing-thawing process. Therefore, the three-dimensional networks in gels were strengthened by these physical cross-linking sites to prevent the porous structure from collapsing upon drying, resulting in the enhancement of specific surface area [60,68,69]. After dissolution–regeneration process, Ce-II aerogels had lower degree of crystallinity than the native cellulose [16,29,38,95]. Even an amorphous aerogel was reported by using trifluoroacetic acid as the solvent [10]. The degradation of cellulose generally occurred during dissolution, leading to a decrease in the degree of polymerisation [10,84,98].

Nevertheless, one major advantage of dissolution–regeneration process has been in conveniently shaping the cellulose-based aerogels into any arbitrary geometry, from common monolith to beads [9,13,25,67,86] and fibres [101]. In addition, due to the presence of large amount of water, the aqueous NaOH systems had much lower viscosity, facilitating the homogeneous dispersion of nanoparticles (NPs) in the cellulose solution. Consequently, they were among the most used solvent systems to prepare cellulose nanocomposite aerogels (as will be shown in Section 11.3).

As a renewable biomass, lignocellulose is the main source of cellulose. However, the separation of pure cellulose from natural plants requires consumption of a lot of chemicals and energy, which causes the significant loss of lignin and hemicellulose present in lignocellulose. By virtue of the high solubility of lignocellulose, ionic liquids, such as 1-allyl-3-methylimidazolium chloride (AmimCl) [68–70] and 1-ethyl-3-methylimidazolium chloride (EmimCl) [78,86], were used to prepare lignocellulose aerogels, without extraction of cellulose. The strength of lignocellulose hydrogels was enhanced by using a cyclic freeze–thaw method due to the improvement of physical cross-link formation during cellulose gelation [69]. Similarly, waste newspaper consisting of high content of lignin and hemicellulose was dissolved in AmimCl, without any pre-treatment, to prepare aerogels [76]. An additional advantage of ionic liquids was the high yield of their recovery by a simple vacuum distillation and their repeated reuse to dissolve cellulose [71,75,91]. As for the high viscosity of ionic liquid systems, dimethyl sulphoxide (DMSO) was added as a co-solvent to overcome the drawback, thus facilitating the dissolution of cellulose [89].

Cracking tends to occur during the FD of hydrogels, and the resulting aerogels were often partially broken [16,30,104]. Exchanging water in hydrogels with organic solvents having low surface tension helped obtain aerogels with homogeneous morphology after FD [5,16,30]. Freeze-drying from tert-butanol gels of cellulose [30,50,62,63,76,85] as well as Zeorora (1,1,2,2,3,3,4-heptafluorocyclopentane) gels [95] were widely used by researchers. Rein and Cohen [91] used hydrophobic hexamethyldisiloxane in the solvent exchange. In this procedure, the cellulose was simultaneously hydrophobised to obtain wet gels that had shrinkage less than 20% after drying in a vacuum oven. The gels frozen with slow rate were prone to form aerogels consisting of two-dimensional sheet-like structure, instead of three-dimensional fibrillar-like network [5,46,84,108,109].

In both FD and SD, the obtained aerogels generally exhibited hierarchical structure containing micropores, mesopores and macropores, with size ranging from nanometres to microns [42,60,62,66,69,77,90,98,102]. A non-negligible problem that existed in Ce-II aerogels is the residues of cellulose solvents, which were difficult to remove fully due to the limitation of solvent diffusion in the gels [105]. The residual solvents in aerogels should be taken into consideration in the applications of aerogels.

11.2.2 Ce-I Aerogels

Nanocellulose fibres have typical cross-sectional dimensions of 5–20 nm and lengths up to several microns. They are robust materials with high cellulose-I crystal contents, as well as ultrahigh elastic modulus of about 140 GPa and very low thermal expansion coefficient (as low as 10^{-7}/K) [110], and hence are widely used for the preparation of Ce-I aerogels. Depending on the size and production process, the nanocellulose extracted from plant sources are referred to as cellulose nanofibrils (CNFs) or nanofibrillated cellulose, microfibrillated cellulose [7,109,111–185] and cellulose nanocrystals (CNCs) or cellulose nanowhiskers [108,117,119,162,186–197]. Cellulose nanofibril is prepared by mechanical disintegration with chemical or enzymatic pre-treatments. The entanglement of CNFs due to high aspect ratios, combined with intramolecular hydrogen bonding among the CNFs, facilitated to improve the strength of gels. Therefore, Ce-I aerogels with ultralow density (<1 mg/cm^3) were reported from dilute CNF dispersions via FD [130,138,155]. Distinct from CNFs, CNCs were fabricated by strong acid hydrolysis treatment, resulting in well-defined rod-like NPs with high crystallinity and better mechanical properties. In aqueous dispersion, the rigid CNCs cannot entangle with each other to form interconnected networks. However, ultrasound treatment can help form hydrogen bonds between CNCs, leading to a stable hydrogel [197].

Among the production process of nanocellulose, 2,2,6,6-tetramethylpiperidine-1-oxy (TEMPO)-mediated oxidation is of particular interest, owing to its mild processing conditions, high efficiency and low energy consumption to defibrillating cellulose [115–117,121–124,131,133,136,137,139,144,149,151,152,157,158,170,173,176, 179–183]. The obtained TEMPO-oxidised CNF (TOCNF) has abundant surface carboxyl groups about 2 mmol/g, which could be further reacted to introduce new functional groups. With the increase of the surface charge of TOCNF by controlling the degrees of oxidation and protonation, the obtained aerogels had low density and large specific surface area [133,157,158], high thermal stability [133] and excellent mechanical properties [157]. Layer-by-layer (LBL) assembly of functional polymers and NPs was carried out on covalent cross-linked aerogels (CL-aerogels) from negatively charged TOCNF. The aerogels were obtained with several different functionalities and meanwhile retained high porosity and strength [140].

Bacterial cellulose, which is the by-product of the metabolism of certain bacteria, is also among the most common types of nanocellulose. Unlike the plant-based nanocellulose, BC presents advantages such as high purity and degree of polymerisation, free of lignin and hemicellulose; it requires no harsh chemical and/or mechanical treatments during its production [141,148,192,198–234]. As for the aerogels prepared from aqueous nanocellulose dispersion, the microstructure gradually converted from three-dimensional open porous networks of nanofibres into two-dimensional sheet-like skeletons interconnected with each other with the increase of cellulose content (Figure 11.1) [117,122,123,138,155,159,162,164,194]. The types of nanocellulose also influenced the structure and properties of the resulted aerogels. For example, the aspect ratio and surface groups of nanocellulose had effects on the total shrinkage and hence the morphology of the samples, and the nanocellulose with higher aspect ratio resulted in aerogels with better ductility and flexibility to bear large deformation without rupture [117,119].

The BC aerogels with density of 8.8 mg/cm^3 were fully rewetted with water, without collapsing of the porous structure [223]. Sehaqui et al. [164] also reported that the stable hydrogels were recovered by soaking CNF aerogels with densities greater than 40 mg/cm^3 in water, without structural disintegration, whereas Jin et al. [160] suggested that CNF aerogels with density of 20 mg/cm^3 swelled up in water and then disintegration into pieces. Chemical cross-linked CNF (CL-CNF)

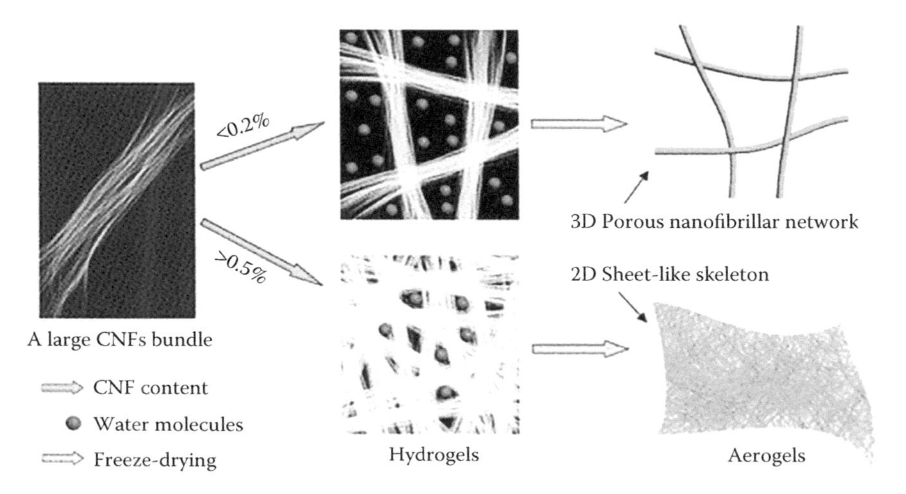

FIGURE 11.1 Scheme of the effect of CNF content in hydrogels on the microstructure of aerogels. (From Chen, W.S. et al., *Soft Matter*, 7, 10360–10368, 2011. Reproduced with permission from The Royal Society of Chemistry.)

dispersed in water led to aerogels possessing high wet strength and stability in water [109,111,112,116,119,134,142,144,156,159,191]. The CL-aerogels can be compressed to large strains, and they can recover their shape only slightly in air. However, they showed high and repeatable shape recovery when they were immersed in water [109,191].

After the gelation of nanocellulose dispersions, FD of hydrogels has been commonly used to prepare Ce-I aerogels. An alternative FD method was freezing tert-butanol gels obtained via solvent exchanging, followed by sublimation [121,128,130,139,141,150,158,170,175,190,206–208,214,217,231]. Meanwhile, SD was also used after solvent exchanging water in hydrogels with water-miscible organic solvents [121,127,135,141,146,149,153,161,163,180,188,191,192,197, 198,205,223,226,229,234] and ambient/vacuum drying after solvent exchanging with octane with low surface tension [184]. Dry ice and its mixtures with organic solvents [137,144,166,167,173,176,179,182,193,195] and liquid propane [7,161,163,168] were used for cooling, along with extensively used liquid nitrogen. Under vacuum drying, some studies suggested that the insignificant shrinkage produced CNF aerogels with low density of about 30 mg/cm^3 and high porosity of about 98% [7,160,161,172]. However, the shrinkage was high with respect to liquid propane drying [7,161], and sheet-like structure was formed due to relatively low cooling rate [7,161,172]. Therefore, the morphology of Ce-I aerogels could be transformed from three-dimensional nanofibrous network to two-dimensional sheet-like structure by adjusting the drying method and the cooling agents [7,141,163,179,190,192].

On the other hand, organic solvents, such as DMSO [187,188] and tert-butanol [113], added to the aqueous nanocellulose dispersion were able to control the microstructure and physical and mechanical properties of Ce-I aerogels. The unidirectional freezing of aqueous cellulose dispersions induces the unidirectional growth of ice crystals to form ice template, resulting in highly ordered porous channels after the sublimation of ice crystals [162,176]. The CL-CNF aerogel beads were fabricated by using spray-FD method [156,177,178] or water-in-oil emulsions [173]. The obtained microspheres had diameter sizes from several tens to several hundreds of microns and densities as low as 1.8 mg/cm^3. Highly porous inorganic titanium oxide microspheres could be further obtained by using the CNF aerogel beads as template [178].

The main problem in the preparation of Ce-I aerogels was the limited range of porosity due to the initial nanocellulose concentration, which must be low, so as to obtain good dispersion. As for the CNF aerogels prepared from FD, Srinivasa et al. [235,236] proposed a model for their compressive response, based on two-dimensional random Voronoi structure, and evaluated the model with experimental data by combining uni-axial and bi-axial compression testing results. In addition, porous Ce-I aerogels were also prepared by FD of the aqueous dispersion of macroscopic cellulose fibres [152,237–242]. Moreover, cellulose pulps were solvent-exchanged with low-surface-tension solvents or mixed with surfactants at high velocity to obtain Ce-I aerogels after vacuum drying [243–246].

11.2.3 Ce-D Aerogels

A variety of cellulose derivatives can be obtained by derivatisation of cellulose [90,247–249], especially homogeneous derivatisation in cellulose solvents [90,247,248], by taking advantage of the abundant hydroxyl groups in the cellulose. Cellulose esters (such as cellulose acetate [4,250–254], cellulose stearoyl ester [248], cellulose carbamate[249], nitrocelluloses [255] and cellulose phosphate [247]) and cellulose ethers (including tritylcellulose [90], hydroxyl propyl methylcellulose [256], carboxymethyl cellulose [195,257] and hydroxyl ethyl cellulose [258]), have been used to prepare Ce-D aerogels. Some cellulose derivatives with low degrees of substitution (DS < 1) still need cellulose solvents for dissolution, and then they form gels via hydrogen bonding between the remaining hydroxyl groups on cellulose chains [90,247–249]. On the other hand, cellulose derivatives having high DS require additional chemical cross-linking agent to induce gelation [4,250,251,254,255,257,259].

Pour et al. [90] synthesised tritylcellulose with bulky hydrophobic trityl groups in DMAc-LiCl homogeneously. With a DS as low as 0.72, Ce-D aerogels with porosities as high as 91% were obtained via ambient drying. However, their specific surface areas were at least one order of magnitude smaller than that obtained via SD. Patel et al. [256] utilised the surface activity of water-soluble hydroxyl propyl methylcellulose (HPMC) to form stable aqueous foams, as shown in Figure 11.2. After further FD, Ce-D aerogels with pore sizes larger than 200 μm were obtained. The presence of propyl and methyl groups enhanced the hydrophobicity of cellulose, and the gelation of polymer at the air–water interfaces stabilised the foams of cellulose derivatives.

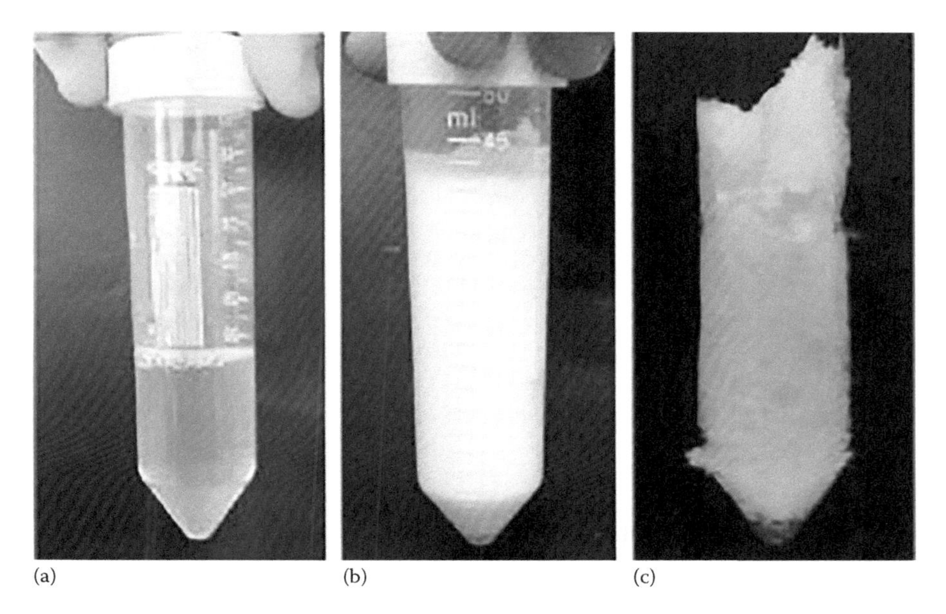

FIGURE 11.2 (a) Hydroxyl propyl methylcellulose (HPMC) solution 1 wt.%, (b) aqueous foam of HPMC and (c) porous aerogel obtained by removal of water by freeze-drying. (Patel, A.R. et al., *RSC Adv.*, 3, 22900–22903, 2013. Reproduced with permission from The Royal Society of Chemistry.)

11.3 CELLULOSE-BASED COMPOSITE AEROGELS

Cellulose aerogels have been modified via chemical reactions or physical blending in the different stages of aerogel preparation to obtain cellulose-based composite aerogels. The incorporation routes of additional components have been classified into three categories: before formation of cellulose gels, after formation of cellulose gels, and after formation of cellulose aerogels.

11.3.1 Before Formation of Cellulose Gels

Before the gelation of cellulose solution or cellulose dispersion, the added inorganic components or their reactive precursors could act as nucleants in the freezing of gels to influence the nucleation and growth of solvent crystals. Subsequent sublimation of solvent crystals produces composite aerogels with smaller sizes and more homogeneous pore size distribution [21,22,33,75,152,179,260,261]. On the other hand, the network structure formed during gelation of cellulose effectively prevented the aggregation of NPs, hence increasing their stability and feasibility [35,36,51,92,131,132,148,149,153,260].

The components added via incorporation or in situ formation could shorten the gelation times [20,24,33,52,81,182] and subsequently influence the degrees of crystallinity [21–23,33,40,87,257], improve mechanical properties [20–22,35,40,52,63,75,81,92, 100,114,127,131,132,144,152,157,164,179,189,193,241,258,261] and enhance thermal stability [22,33,35,52,63,81,258]. It was reported that there were strong hydrogen bonding, electrostatic and/or ionic interaction between these components and cellulose [21–23,33,40,52,75,81,83,87,92,114,132,144,164,179,182]. However, the decrease in thermal stability of cellulose-based composite aerogels was also reported when compared with neat cellulose aerogels [179,257].

With the addition of inorganic nanosized fillers and the increase of their loading, the morphology of aerogels can be transformed between two-dimensional sheet-like structure and three-dimensional porous network structure [33,63,149,179]. Adding porogen to the cellulose solutions led to the formation of large pores, with about 100 μm in the resulted aerogels [41,75,88,102], even in the samples by ambient drying [102]. Adding surfactant Tween® could increase the porosity of aerogels due to the formation of more number of large pores [21–23].

Generally, cross-linking agents were often added before gelation of cellulose to obtain mechanically robust aerogels, which had high stability for various applications. The cross-linking agents include Kymene™ used in aqueous CNF dispersions [109,142,147,156,171,177,242], glutaraldehyde in CNF–polyvinyl alcohol (PVA) systems [137,144,176,179], epichlorohydrin in cellulose in the NaOH–urea solution [29,41–43] and others [48,115,116,131,136,159,173,257,259]. Other components such as porogens [21,22,41,75,88,102], quantum dots (QD) [92] and drugs [148] were also incorporated.

Nanocellulose was surface modified via chemical reactions in aqueous dispersion to introduce functional groups, such as ammonium [142], aldehyde [259] and amino [116,142,154,185,246] and carboxyl [246] groups to increase the affinity towards specific chemicals. Silylation treatment [114,135] and in situ polymerisation of styrene and acrylic monomers [171] on the nanocellulose surface rendered the obtained aerogels hydrophobic, whereas the reaction with borates resulted in flame-retardant aerogels [136]. Various functionalities endowed by chemical modification and physical blending to cellulose-based aerogels before gelation of cellulose included absorption properties [63,115,116,132,142,154,182,185,246,259, 261], antibacterial activities [75], bioactive properties [159,259], catalytic properties [87,124,132], electrical conductivity [33,35,36,40,121,146,149,168,186,260,261], flame retardancy [131,136], low thermal conductivity [100,153], high thermal conductivity [34], hydrophobic [153,171,188], insulation properties [22] and magnetic properties [51,79]. Furthermore, multifunctional cellulose-based aerogels were achieved in some studies the via incorporation of two different components, which simultaneously modulated the morphology of cellulose-based aerogels [21–23,41,75,78,119,131,132,136,142,144,159,171,238,259,260].

In addition, TiO_2 or TiO_2–SiO_2 sols were mixed with TOCNF before gelation to prepare cellulose composite aerogels, which could be further carbonised to remove cellulose to produce inorganic aerogels [183], and Ag(I) ions were subsequently added in aqueous TOCNF dispersion to synthesise Ag NPs via hydrothermal reduction by cellulose in the final aerogels [182]. The cellulose composite aerogels prepared by adding additional components before gelation are summarised in Table 11.2.

TABLE 11.2
Preparation of Cellulose Composite Aerogels by Adding Additional Components before Formation of Cellulose Gels

Cellulose Solvent or Nanocellulose Type	Added Component	Drying Method	Year	Reference
BmimCl	Lignin + xylan	SD	2009	[78]
NaOH	Lignin	SD	2010	[24]
NaOH–urea	GO	FD	2012	[52]
BmimCl–DMSO	Agar	FD	2012	[83]
NaOH–urea	CNT	FD	2013	[40]
NaOH–urea	rGO	FD	2013	[33]
BmimCl	Fe_3O_4 NP	FD	2013	[79]
HmimCl	QD	SD	2013	[92]
BmimCl	PANi NP	FD	2014	[87]
NaOH	Sodium silicate	SD	2014	[20]
NaOH–thiourea	Fe_3O_4 NP	SD	2014	[51]
AmimCl	Chitosan P + sodium sulphate	FD	2014	[75]
NaOH	Modified MMT + Tween® 80	FD	2015	[22]
$Ca(SCN)_2$ hydrate	Silica aerogel P	SD	2015	[100]
BmimCl	rGO	FD	2015	[81]
EmimAc–DMSO	PMMA P	SD	2015	[88]
$Ca(SCN)_2$ hydrate–LiCl	Paraffin P			
NaOH–urea	CNT	FD	2015	[35]
NaOH–urea	CNT	FD	2015	[36]
NaOH–urea	Epichlorohydrin CL	FD	2015	[43]
NaOH–urea	Graphene NP	FD	2016	[34]
NaOH–urea	Chitosan	FD	2016	[38]
NaOH–urea	$CaCO_3$ + epichlorohydrin CL	FD	2016	[41]
NaOH–urea	Epichlorohydrin CL	FD	2016	[42]
NaOH	Modified MMT + Tween® 80	FD	2016	[21]
$Ca(SCN)_2$ hydrate	Glyceryl trioctanoate + surfactant	SD, FD, ED	2016	[102]
NaOH	Modified MMT + Tween® 80	FD	2016	[23]
NaOH–PEG	Graphene	FD	2016	[63]
NaOH–urea	Epichlorohydrin CL	SD	2016	[29]
NaOH–urea	*N,N*-methylenebisacrylamide CL	FD	2016	[48]
CNF	Starch	FD	2008	[165]
CNC	MMT, MMT + PVA	FD	2009	[193]
CNF	Starch	FD	2010	[167]
CNF	Xyloglucan	FD	2010	[164]
CNF	Starch	FD	2011	[166]
CNF	Kymene™ CL	FD	2012	[109]
CNF	Hydroxyapatite	FD	2012	[157]
CNF	Cu(I)	FD	2012	[124]
CNF	PPy	FD, SD	2012	[121]

(Continued)

TABLE 11.2 (*Continued*)
Preparation of Cellulose Composite Aerogels by Adding Additional Components before Formation of Cellulose Gels

Cellulose Solvent or Nanocellulose Type	Added Component	Drying Method	Year	Reference
CNF	Soy protein	FD	2013	[143]
CNF	PVA, PVA + GO NP, PVA + GO NP + glutaraldehyde CL	FD	2013	[144]
CNF	GO	SD	2013	[146]
CNF, BC	Beclomethasone dipropionate NP	FD	2013	[148]
CNF	CNT	SD	2013	[149]
CNF	PVA + glutaraldehyde CL PVA + MWCNT + glutaraldehyde CL	FD	2013	[179]
CNF	CNT	FD	2013	[168]
CNF	$AgNO_3$	FD	2013	[182]
CNF	Kymene™ CL	FD	2014	[156]
CNC	MTMS sol	SD	2014	[188]
Pulp, CNF	Zeolite	FD	2014	[152]
CNF	MTMS sol	SD	2014	[153]
CNF	PVA + glutaraldehyde CL	FD	2014	[137]
CNF	EPTMAC + Kymene™ CL	FD	2014	[142]
CNF	MTMS sol	FD	2014	[114]
CNF	$NaIO_4$ + collagen CL	FD	2014	[259]
CNF	b-PEI	FD	2015	[154]
CNF	MTMS sol	FD	2015	[135]
CNF	PEDS sol	SD	2015	[127]
CNF	GO + sepiolite NP	FD	2015	[131]
CNF	PVA + glutaraldehyde CL	FD	2015	[176]
CNF	rGO, CNT	FD	2015	[260]
CNF	rGO	FD	2015	[261]
CNC	PPy nanofibres PPy-coated CNT Manganese dioxide NP	FD	2015	[186]
CNF	b-PEI CL	FD	2015	[115]
Recycled cellulose fibres	CMC + MMT + APP	FD	2015	[238]
Cellulose fibres	Kymene™ CL	FD	2015	[242]
CNF, CNC	PVA, PVA + $Na_2B_4O_7$ CL	FD	2015	[119]
CNF	b-PEI CL	FD	2015	[116]
CNF	Dendrimer poly(amidoamine)	FD	2015	[185]
CNC	MMT + PVA	FD	2016	[189]
CNF	GO + Fe(III)	FD	2016	[132]
CNF	Kymene™	FD	2016	[147]

(*Continued*)

TABLE 11.2 (*Continued*)
Preparation of Cellulose Composite Aerogels by Adding Additional Components before Formation of Cellulose Gels

Cellulose Solvent or Nanocellulose Type	Added Component	Drying Method	Year	Reference
CNF	Clay + Boric acid CL	FD	2016	[136]
CNF	Gelatin + chitosan + genipin CL	FD	2016	[159]
Cellulose fibres	CNF CNC	FD	2016	[241]
CNF	BA M + St M + Kymene™ CL	FD	2016	[171]
CNF	PVA + BTCA CL	FD	2016	[173]
CNF	Kymene™ CL	FD	2016	[177]
Pulp	AA M + acrylamide M	ED	2016	[246]
Hydroxyethyl cellulose	Alumina	FD	2015	[258]
Carboxymethyl cellulose	$FeCl_3$ CL	FD	2015	[257]

Note: AA: acrylic acid; APP: ammonium polyphosphate; b-PEI: branched polyethyleneimine; BA: n-butyl acrylate; BTCA: 1,2,3,4,-butanetetracarboxylic acid; CL: cross-linking agent; CNT: carbon nanotube; EPTMAC: 2,3-epoxyprcpyltrimethylammonium chloride; GO: graphene oxide; QD: quantum dots; Kymene™: polyamide-epichlorohydrin resin; M: monomer for in situ polymerisation; MMT: montmorillonite; MTMS: methyltrimethoxysilane; MWCNT: multiwall CNT; PANi: polyaniline; PEDS: polyethoxydisiloxane; PEI: polyethyleneimine; PMMA: polymethyl methacrylate; PPy: polypyrrole; PVA: polyvinyl alcohol; rGO: reduced graphene oxide; St: styrene. See also the Note of Table 11.1.

11.3.2 After Formation of Cellulose Gels

Three-dimensional network skeletons were formed after gelation of cellulose, which could be employed as templates for liquid-phase reaction to prepare cellulose composite aerogels. The porous skeletons and abundant surface hydroxyl groups of wet gels facilitated to guide and control the growth of in situ synthesised NPs [26–28,31,32,39,56–58,61,64,65,99,103,183,214,230,234,252,253]. Reducing Ag(I) ions by cellulose led to the formation of Ag NPs [26,42,79]. The confinement effect of micro- or nanosized structure in gels prevented the aggregation of NPs resulting from their high surface energy and chemical activity. Cellulose composite aerogels were reported with high transparency after incorporation of in situ synthesised inorganic NPs [26,31]. Therefore, the gel skeletons had no significant change and retained the interconnected three-dimensional porous structure [26,31,32,42,57,58,99,103,126,183,200,230].

The added components showed strong interactions with cellulose [42,43,59,61,64, 201,253], strengthened effectively the skeletons of aerogels [27,28,32,39,42,43, 65,89,99,198,206], decreased obviously the shrinkage and collapse during drying [27,31,65,89,91] and improved significantly the thermal stability of porous structure [31,32,58,64,198,206,230]. Immersing BC gels into solutions of some specific compounds led to the precursors for various heteroatoms-doped [210,212,221] and TiO_2-loaded [214] carbon nanofibre aerogels.

In situ formation of inorganic NPs in the cellulose gels has been mainly used to modify cellulose aerogels to introduce absorption properties [79,126,183,201], antibacterial activities [42,58,61], bioactive properties [43,223], catalytic properties [57,59,79,103,183,252], hydrophobicity [91,201,226], cross-linking [140], electrical conductivity [28], flame retardancy [32], magnetic properties [39,56,65,200,230] and low thermal conductivity [54,89]. As shown in Table 11.3, aqueous NaOH solutions were the major solvent systems to prepare Ce-II gels, and BC was the main nanocellulose used in Ce-I gels. The cellulose composite aerogels prepared by adding additional components after gelation are summarised in Table 11.3.

TABLE 11.3
Preparation of Cellulose Composite Aerogels by Adding Additional Components after Formation of Cellulose Gels

Cellulose Solvent or Nanocellulose Type	Added Component	Drying Method	Year	Reference
LiOH–urea	$AgNO_3$ $HAuCl_4$ $PtCl_4$	SD	2009	[26]
EmimAc	TMCS + hexamethyldisiloxane	ED	2010	[91]
NMMO	TEOS + 3-chloropropyl trimethoxysilane	SD	2011	[99]
LiOH–urea	$FeCl_3$ + $CoCl_2$	FD	2012	[39]
NaOH–urea	TEOS	SD	2012	[31]
NaOH–thiourea	TEOS	FD	2013	[54]
LiOH–urea	Sodium silicate	SD	2013	[27]
BmimCl	$AgNO_3$	FD	2013	[79]
NaOH–urea	$FeCl_3$ + Py M	SD	2014	[28]
EmimAc–DMSO	Silica sol	SD	2015	[89]
NaOH	Epichlorohydrin CL	SD	2015	[19]
NaOH–PEG	$CoFe_2O_4$ NP	FD	2015	[56]
NaOH–PEG	ZnO NP	FD	2015	[64]
NaOH–urea	$MgCl_2$	FD	2015	[32]
NaOH–PEG	TNBT sol	FD	2015	[57]
NaOH–PEG	TiO_2 NP	FD	2015	[59]
NaOH–PEG	$FeSO_4$ + $FeCl_3$	FD	2015	[65]
$Ca(SCN)_2$	$AgNO_3$	SD	2016	[103]
NaOH–urea	$AgNO_3$	FD	2016	[42]
NaOH–urea	Gelatin	FD	2016	[43]
NaOH–PEG	Py M + $AgNO_3$	FD	2016	[61]
NaOH–PEG	$AgNO_3$	FD	2016	[58]
CA	$Pt(NH_3)_4Cl_2$	FD	2008	[252]
CA	$AgNO_3$	FD	2008	[253]

(*Continued*)

TABLE 11.3 (*Continued*)
Preparation of Cellulose Composite Aerogels by Adding Additional Components after Formation of Cellulose Gels

Cellulose Solvent or Nanocellulose Type	Added Component	Drying Method	Year	Reference
BC	Dexpanthenol L-ascorbic acid	SD	2010	[223]
CNF	AEAPDMS	FD	2011	[126]
BC	Alkyl ketene dimer	SD	2012	[226]
CNF	TTIP sol, TTIP + TEOS sol	FD	2013	[183]
CNF	BTCA CL	FD	2013	[140]
BC	Ni NP	FD	2013	[200]
BC	TNBT sol	FD	2013	[214]
BC	Biocompatible polymers	SD	2014	[198]
BC	GO rGO	FD	2014	[201]
BC	Sodium silicate	FD	2014	[206]
BC	Organic dye containing heteroatom	FD	2014	[210]
BC	H_3PO_4 $NH_4H_2PO_4$ $H_3BO_3 + H_3PO_4$	FD	2014	[212]
BC	Urea	FD	2014	[232]
BC	Lignin + resorcinol + formaldehyde	SD	2015	[234]
BC	$NH_4H_2PO_4$	FD	2016	[221]
BC	$NiCl_2$	FD	2016	[230]

Note: AEAPDMS: *N*-(2-aminoethyl)-3-aminopropylmethyldimethoxysilane; CA: cellulose acetate; Py: pyrrole; TEOS: tetraethyl orthosilicate; TMCS: trimethylchlorosilane; TNBT: tetra-n-butyl titanium; TTIP: titanium tetraisopropoxide. See also the Notes of Tables 11.1 & 11.2.

11.3.3 After Formation of Cellulose Aerogels

The modifications carried out on cellulose aerogels have been focussed on the surface hydrophobisation of cellulose skeletons. The highly porous structure made the network well accessible to reactants to increase the homogeneity of reactions, whereas the resulted composite aerogels retained integrated configuration and network microstructure. The methods of surface modification can be divided into two groups: gas phase and liquid phase.

The gas-phase methods include chemical vapor deposition (CVD) [44,45,48,60,66,76,122,130,137,144,155,160,161,169,173,175,176,192,205,242,243], atomic layer deposition (ALD) [163,172,262], CCl_4 plasma treatment [54,55], gas-phase esterification [128,150,190,192] and gas-phase polymerisation [196]. The unreacted reagents and by-products are removed easily under reduced pressure after

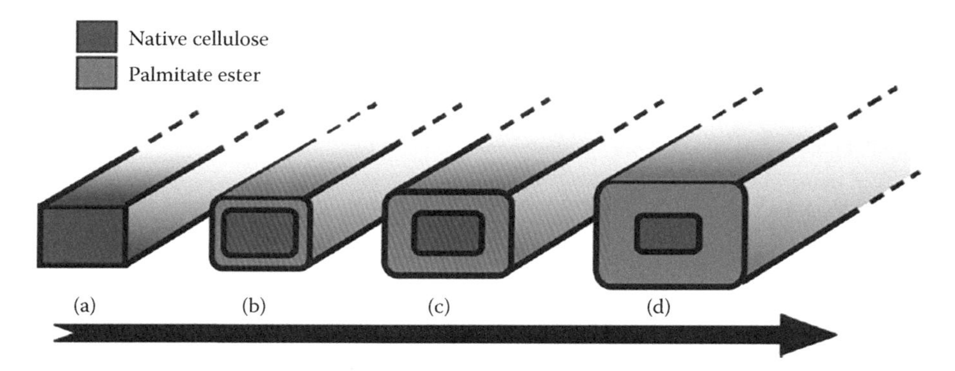

FIGURE 11.3 Schematic representation for the progress of the gas-phase esterification of cellulose nanofibres with palmitoyl chloride, with (a) DS = 0, (b) DS = 0.25, (c) DS = 0.75 and (d) DS = 1.5. (Reprinted with permission from Berlioz, S. et al., *Biomacromolecules*, 10, 2144–2151, 2009. Copyright 2009 American Chemical Society.)

reactions. Other additional benefits include both the internal and external surfaces of aerogels being fully accessible [44,48] and no competition or side reactions with solvents. Moreover, the gaseous reactants diffuse towards the underlying layers to progress the reaction after the hydroxyl groups on the surface are fully reacted, resulting in fully derivatised cellulose [128,150,190,192], as shown in Figure 11.3. However, bi-functional esterifying reagents such as cyclic anhydride and diacyl chloride limit the derivatisation to the surface layers [128]. The SD process renders aerogels with high specific surface areas and three-dimensional fibrillar network structure, which increase the accessibility to gas-phase modification and consequently improve the DS with respect to the FD process [150,192].

The liquid-phase method includes modification of cellulose-based aerogels in solutions or with liquid reactants [7,10,41,51,82,85,118,125,135,140,147,177,178,181,206–208,215,217,219,222,228,231]. After the modifications, FD and SD are used to obtain porous composite aerogels [10,41,51,85,135,177,178,206–208,215,217,219,222,228], as well as ambient/vacuum drying [7,147,216]. The LBL method is a specific technique for the surface modification in liquid solution, which assembles functional materials onto aerogels via ionic interaction to form homogeneous coating with controllable thickness [140]. Compared with gas-phase reaction, liquid-phase reaction is favoured to obtain homogeneous surface modification due to the uniform reaction environment created by rapid diffusion of the reactants to the pores by capillary effects before reactions.

Surface modification rendered the composite aerogels with absorption [82,85,125], antibacterial activities [217], antioxidant activity [181], catalytic properties [161,219], electrical conductivity [7,140], low thermal conductivity [45,135,147], magnetic properties [41,118,215,216], hydrophobicity [44,45,48,51,54,60,76,122,130,137,144,155,160,161,169,172,173,175,176,190,205,207,231,242,243], oleophobicity [155,160], optical properties [10,140,228] and temperature sensitivity [177]. Inorganic NP aerogels can be formed by using Ce-I aerogels as templates after further carbonisation [163,178]. The cellulose composite aerogels prepared by adding additional components after the formation of cellulose aerogels are summarised in Table 11.4.

TABLE 11.4
Preparation of Cellulose Composite Aerogels by Adding Additional Components after Formation of Cellulose Aerogels

Aerogel or Nanocellulose Type	Adding Method	Added Component	Year	Reference
Ce-II + silica	CCl_4 plasma	Chlorination	2013	[54]
Ce-II	CVD	MTMS	2013	[44]
Ce-II	CCl_4 plasma	Chlorination	2013	[55]
Ce-II	CVD	MTMS	2014	[45]
Ce-II + Fe_3O_4	Solution	TTIP	2014	[51]
Ce-II	Solution	Disodium iminodiacetate + epichlorohydrin CL	2014	[85]
CL-Ce-II	Solution	$Cu(NO_3)_2$	2014	[85]
Ce-II	Solution oxidation	Dialdehyde groups	2015	[82]
Ce-II	CVD	MTCS	2015	[66]
Ce-II	CVD	TMCS	2015	[76]
Ce-II	Solution	QD	2016	[10]
CL-Ce-II	Solution	$FeCl_3$ + $FeCl_2$	2016	[41]
CL-Ce-II + Fe_3O_4	Solution	HTMOS	2016	[41]
CL-Ce-II	CVD	MTCS	2016	[48]
CNF	Solution	PANi	2008	[7]
BC	CVD	Palmitoyl chloride	2009	[192]
CNC				
CNF	CVD	PFOTS	2010	[155]
BC	Solution	$FeSO_4$ + $CoCl_2$	2010	[215]
CNF	CVD	FTCS	2011	[160]
CNC	Gas	AA M	2011	[196]
CNF	CVD	TTIP	2011	[161]
CNF	ALD	Titanium tetrachloride Diethyl zinc Trimethyl aluminium	2011	[163]
CNF	ALD	TTIP	2011	[172]
CNF	CVD	OTCS	2012	[145]
CL-CNF	LBL	PAH/HA PEI/PEDOT:PSS PEI/ADS2000P PEI/CNT	2013	[140]
CL-CNF + PVA + GO NP	CVD	FTCS FTEOS	2013	[144]
BC	Solution	TEOS	2013	[207]
BC + silica	Solution	MTMS	2013	[207]
CNC	CVD	Palmitoyl chloride	2013	[190]
CNF	CVD	Palmitoyl chloride	2013	[150]
CNF	Solution reaction	$CoFe_2O_4$ NP	2013	[118]

(Continued)

TABLE 11.4 (*Continued*)
Preparation of Cellulose Composite Aerogels by Adding Additional Components after Formation of Cellulose Aerogels

Aerogel or Nanocellulose Type	Adding Method	Added Component	Year	Reference
CNC	ALD	Al_2O_3	2014	[262]
Wood pulp	CVD	MTCS	2014	[243]
CNF	CVD	TEOS	2014	[122]
BC + silica	Solution	MTMS	2014	[206]
CL-CNF + PVA	CVD	MTCS	2014	[137]
BC	Solution	Zinc acetate dehydrate	2014	[217]
BC	Solution	$FeCl_3$	2014	[219]
BC	Solution	ANi M	2014	[222]
CNF	CVD	MTCS	2015	[130]
Silylated CNF	Solution	TEOS sol	2015	[135]
CNF	CVD	TMCS	2015	[169]
CL-cellulose fibres	CVD	MTMS	2015	[242]
CNF	Solution	TTIP	2015	[178]
CL-CNF + PVA	CVD	MTCS	2015	[176]
CNF	Solution polymerisation	Methacrylic acid M + maleic acid M	2015	[125]
CNF, CNC	CVD	Acyl chloride Cyclic acid anhydride	2015	[128]
BC	Solution	TMCS	2015	[231]
Ce-II	CVD	TMCS	2015	[60]
BC	Solution	$Ca(NO_3)_2$ + TEOS	2016	[208]
CL-CNF	Solution	TEOS + TMCS	2016	[147]
BC	Solution	$FeCl_2 + CoCl_2$	2016	[216]
BC	Solution	Au NP	2016	[228]
CL-CNF	CVD	MTCS	2016	[173]
CNF	CVD	MTCS	2016	[175]
CL-CNF	Solution polymerisation	*N*-isopropyl acrylamide	2016	[177]
BC	CVD	FTCS	2016	[205]
CNF	Solid-state polymerisation	Eumelanin	2016	[181]

Note: ADS2000P: poly(2-(3-thienyl)ethoxy-4-butyl-sulphonate); ANi: aniline; FTEOS: 4-(trifluoromethyl-tetrafluorophenyl) triethoxysilane; FTCS: tridecafluoro-1,1,2,2-tetrahydrooctyl trichlorosilane; HA: hyaluronic acid; HTMOS: hexadecyltrimethoxysilane; MTCS: methyltrichlorosilane; OTCS: octyltrichlorosilane; PAH: polyallylamine hydrochloride; PEDOT: poly(3,4-ethylenedioxythiophene); PFOTS: 1,1,2,2-perfluorodecyl trichlorosilane; PSS: poly(styrene sulphonate). See also the Notes of Tables 11.2 & 11.3.

11.4 FUNCTIONAL CELLULOSE-BASED AEROGELS

11.4.1 Optical Properties

Cellulose does not absorb visible light. Therefore, it is possible to obtain transparent cellulose-based aerogels by manipulating the structure-forming units smaller than the wavelength of visible light. The porous structure and solid skeleton of cellulose-based aerogel are tailorable by controlling the concentration of cellulose solution/dispersion and conditions to form wet gel and its drying. In combination with loaded NPs, it is able to reduce light scattering in cellulose-based aerogels and endow the materials with high transmittance. Compared with FD, SD is an ideal method to prepare transparent cellulose-based aerogels, because it can better preserve the initial three-dimensional porous web-like structure in the wet gel [10,26,30,31,73,74,92,180].

Transparent neat Ce-II aerogels were obtained by using alkaline-urea [30], AmimCl [73] and TFA [10] as cellulose solvents. Using a high concentration, the aqueous AmimCl solution in the coagulation process was particularly facile and effective to prepare Ce-II aerogels with homogeneous nanosized porous morphology from cellulose solution in AmimCl. Monolithic aerogels with 2 mm thickness had a high transmittance at 800 nm up to 80.0% [73], as shown in Figure 11.4. Moreover, functional cellulose composite aerogels also displayed high transmittance by in situ incorporation of inorganic NPs into aforementioned aerogels [26,31,74].

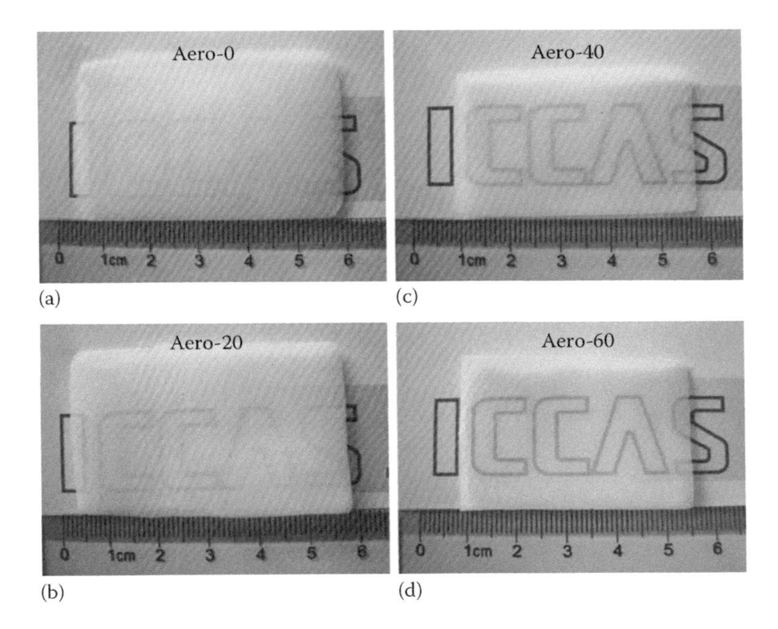

FIGURE 11.4 Neat cellulose aerogels (two-millimetre thickness) made from cellulose-AmimCl solution via coagulation in different baths: (a) deionised water (Aero-0), (b) 20% AmimCl aqueous solution (Aero-20), (c) 40% AmimCl aqueous solution (Aero-40) and (d) 60% AmimCl aqueous solution (Aero-60). (Reprinted with permission from Mi, Q.Y. et al., *ACS Sustain. Chem. Eng.*, 4, 656–660, 2016. Copyright 2016 American Chemical Society.)

Kobayashi et al. [180] used SD to maintain the nematic liquid-crystalline order formed by the cellulose nanofibres dispersed in water, resulting in tough and transparent neat Ce-I aerogels. The 1 mm thick aerogels had 80%–90% transmittance in the visible light range. Hayase et al. [153] prepared polymethylsilsesquioxane (PMSQ) aerogels derived from methyltrimethoxysilane (MTMS), and the CNF-enforced composite aerogels with 10-mm thickness had transmittance up to 75%. Toivonen et al. [184] used ambient drying to prepare transparent CNF aerogel membranes after solvent exchanging water in CNF hydrogels with octane finally. The aerogel membrane with 25-μm thickness had high transmittance (higher than 90%) at 800 nm, and high specific surface area of 208 m^2/g; however, the porosity was only about 60%. For photoluminescent (PL) Ce-II gels prepared by immersing the gels in QD solutions, SD preserved their PL properties to a large extent. The QDs aggregated slightly in the obtained Ce-II aerogels to exhibit PL spectra that can be tuned from visible [10,92] to near-infrared [10] wavelengths. Tian et al. [228] incorporated gold nanorods as plasmonic nanostructure into highly open porous BC aerogels to fabricate plasmonic cellulose composite aerogels. The obtained optically active biofoams exhibited excellent surface-enhanced Raman scattering activity, which facilitated ultrasensitive chemical detecting, and high efficient photothermal heating suitable for steam generation application. They also had especially optical-controllable enzymatic activity.

11.4.2 Flame Retardancy

Despite the renewable and environmentally friendly advantages of cellulose, its intrinsic flammability has hindered the applications of cellulose-based aerogels in many important areas, such as lightweight construction and domestic devices. Therefore, the flame-retardant modification of cellulose-based aerogels is a crucial task for expanding its applications. One of the challenges was to homogeneously disperse flame-retardant agents in the aerogels and simultaneously avoid the collapse of three-dimensional nanoporous structures. The flame-retardant agents can be added into aqueous dispersions of nanocellulose [131,136] or pulp [238] before gelation. Forming flame-retardant components in situ was another way to prepare fire-resistant cellulose composite aerogels [32,74,136].

Montmorillonite (MMT), an inorganic filler, and ammonium polyphosphate (APP), a halogen-free intumescent flame-retardant agent, played a synergistic effect to improve flame retardancy and thermal stability of aerogels prepared from recycled cellulose fibres [238]. By optimising the contents of well-distributed graphene oxide and sepiolite nanorods fillers, CNF-based composite aerogels reached a limiting oxygen index as high as 34 and exhibited high combustion resistance [131]. In situ reaction of boric acid or borate with CNF formed cross-linked complexes, which had high amount of graphitised char, improving ignition resistance and reducing flammability of CNF composite aerogels [136]. In Ce-II gels, a large amount of metal hydroxide NPs such as magnesium hydroxide [32] and aluminium hydroxide (AH) [74] can be synthesised in situ from the added precursor. Therefore, the heat release rate [74] and combustion velocity [32] were reduced significantly, resulting in self-extinguished composite aerogels (Figure 11.5).

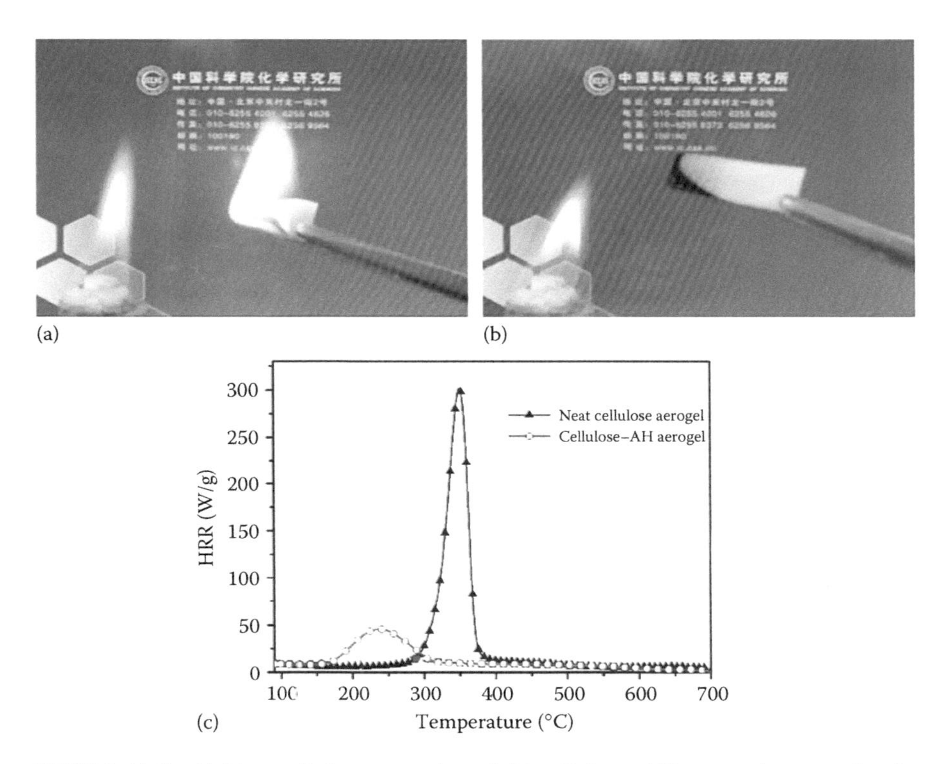

FIGURE 11.5 (a) Neat cellulose aerogels and (b) cellulose–AH composite aerogels after ignition by an alcohol burner and (c) their heat release rate curves. (Reprinted with permission from Yuan, B. et al., *Sci. China Chem.*, 59, 1335–1341, 2016. Copyright 2016 Science China Press and Springer-Verlag Berlin Heidelberg.)

11.4.3 Antibacterial Activity

Cellulose aerogels obviously had no antibacterial function [58]. Ag NP, well known for its bacteriostatic and bactericidal effects, was widely used to improve antibacterial activity of cellulose-based aerogels. In general, Ce-II hydrogels were immersed in the aqueous $AgNO_3$ solution to uniformly immobilise Ag(I) ions, followed by reduction to form Ag NPs [42,58,61,79]. Polypyrrole (PPy), a typical antibacterial polymer, played a synergistic effect with Ag NPs on antibacterial activity [61]. Xiong et al. [79] combined Ag NPs and magnetic Fe_3O_4 NPs to prepare CNF-Fe_3O_4-Ag composite aerogels with high antibacterial performance (Figure 11.6), which could be easily separated from the medium by using a magnetic field. Both the highly porous structure of Ce-II aerogels and antibacterial activity of Ag NPs were confirmed by in vivo tests to contribute for controlling wound infection, preventing wound from purulent infection and promoting wound recovery [42]. ZnO NPs [217] and chitosan particles [75] were also used to induce antibacterial function to cellulose-based aerogels. These antibacterial composite aerogels showed good activity to model bacteria, such as *Escherichia coli* (ATCC25922, model gram-negative bacteria) [42,58,61,75,217], *Staphylococcus*

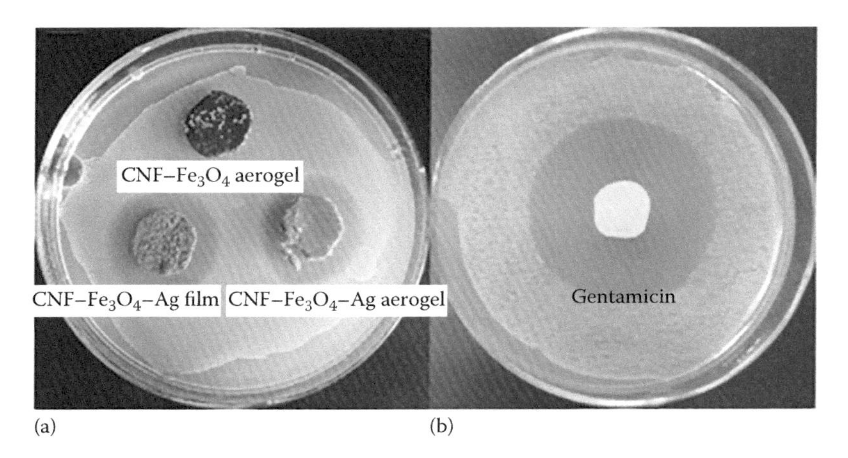

FIGURE 11.6 (a) Inhibition zones of the CNF–Fe_3O_4 aerogel, CNF–Fe_3O_4–Ag aerogel and the CNF–Fe_3O_4–Ag film against *S. aureus* and (b) inhibition zone of the positive control (gentamicin disc). (Xiong, R. et al., *J. Mater. Chem. A*, 1, 14910–14918, 2013. Reproduced by permission of The Royal Society of Chemistry.)

aureus (ATCC6538, gram-positive bacteria) [42,58,61,75,79] and *Listeria monocytogenes* (NICPBP 54002, intracellular bacteria) [61].

11.4.4 Magnetic Properties

$CoFe_2O_4$ and Fe_3O_4 NPs were the most frequently used magnetic particles for cellulose-based aerogels (Table 11.5). By immersing cellulose-based hydrogels [39,56,65] or aerogels [41,118,215,216] in the aqueous solution of Fe(II)/Fe(III) or

TABLE 11.5
Magnetic Cellulose Composite Aerogels

Reference	Magnetic Particles	Adding Stage	Content (wt.%)	Particle Size (nm)	Year
[215]	$CoFe_2O_4$	BC aerogel	71–94	40–120	2010
[39]	$CoFe_2O_4$	Ce-II hydrogel	2.3–10.4	–	2012
[118]	$CoFe_2O_4$	CNF aerogel	34–75	30	2013
[200]	Ni	BC hydrogel	–	3.2–140	2013
[79]	Fe_3O_4	Cellulose solution	9.5	17.1	2013
[51]	Fe_3O_4	Cellulose solution	50	7–10	2014
[56]	$CoFe_2O_4$	Ce-II hydrogel	16.5	98.5	2015
[65]	Fe_3O_4	Ce-II hydrogel	14.1–34.1	8.1–11.7	2015
[41]	Fe_3O_4	Ce-II aerogel	–	20	2016
[216]	$CoFe_2O_4$	BC aerogel	87.6	9–13	2016
[230]	Ni	BC hydrogel	78–83.8	3–200	2016

Fe(II)/Co(II), magnetic NPs were formed via in situ coprecipitation. Magnetic cellulose aerogels were also prepared by direct addition of Fe_3O_4 NPs into cellulose solution [51,79] or in situ formation of Ni NPs [200,230] in cellulose hydrogels, as shown in Table 11.5.

The magnetic properties of cellulose composite aerogels were strongly dependent on the properties of NPs in terms of size, size distribution and content [200,230]. If the particle size was below a critical value, the magnetic NPs were in superparamagnetic state, resulting in a composite aerogel in the absence of hysteresis and coercivity in magnetic field [39,65,79]. However, a saturation magnetisation higher than 50 emu/g was generally obtained [65,79,118,215,216], thereby ensuring strong magnetic responsiveness of the material. In particular, composite aerogels containing magnetic particles with wide size distribution enabled a temperature-dependent transition from ferromagnetic state to superferromagnetic state [200,230].

11.4.5 Hydrophobicity

The abundant hydroxyl groups made contribution to strong hydrophilicity, hygroscopicity and swelling in water of cellulose materials, which significantly affected the performance of cellulose in many applications. The most used hydrophobisation method for cellulose-based aerogels was to consume surface hydroxyls of aerogels via gas-phase reaction [44,45,48,60,76,130,137,144,145,155,160,161,169,172,175,176,190, 205,242,243], as shown in Table 11.6. However, this method was not applicable to the large-scale production and was unable to form a uniform hydrophobisation in the obtained materials. It has been previously suggested that only the external surface of aerogels and the internal surface of micropores had taken part in the gas-phase reaction, whereas the surface of the nanopores was inaccessible to the reactant molecules [243]. Recently, some researchers have hydrophobised the three-dimensional network skeleton structure in liquid phases, in which the uniformly dispersed reactants can be rapidly diffused into the pores of gels or aerogels [41,51,147,181,206,207,231], or the aqueous nanocellulose dispersion and the precursors solution of modifiers can be coagulated together to form composite gels [114,153,171,188]. The morphology of aerogels is kept almost intact during hydrophobisation [45,153,176,207]. The rough surface of aerogels benefitted to improve their water contact angles [41,155,160,161,248].

Both chlorosilane and alkoxysilane containing hydrolysable groups are generally used to modify the surface chemistry of cellulose-based aerogels. Silanols are formed by hydrolysis; they then react with hydroxyl groups of cellulose, resulting in strong hydrophobic and oleophilic aerogels with water contact angle higher than 130° [41,44,45,48,60,76,114,130,137,145,147,153,169,175,176,188,206,207,231,242,243]. Water droplets easily rolled off the silanised surface of the aerogels due to low adhesion [76,137], and the hydrophobicity was often improved by increasing the silane dosage.

Fluorinated silane, such as 1,1,2,2-perfluorodecyl trichlorosilane (PFOTS) and tridecafluoro-1,1,2,2-tetrahydrooctyl trichlorosilane (FTCS), imparted not only superhydrophobicity [144,155,160,205] but also superoleophobicity to aerogels [155,160]. Moreover, the modified surface had high adhesion and high contact angles

TABLE 11.6
Hydrophobic Cellulose-Based Aerogels

Aerogel Type	Hydrophobic Modifier	Hydrophobisation Method	Water Contact Angle	Year	Reference
CNF	PFOTS	Gas phase	≤166 (castor oil) ≤144 (hexadecane)	2010	[155]
CNF	FTCS	Gas phase	160 153 (paraffin oil) 158 (mineral oil)	2011	[160]
CNF	TTIP	Gas phase	140 Hydrophilic (after UV)	2011	[161]
CNF	TTIP	Gas phase	>90	2011	[172]
Cellulose stearoyl esters	–	–	124 (DS = 0.07)	2011	[248]
BC	Alkyl ketene dimer	Gas phase or liquid phase	–	2012	[226]
CNF	OTCS	Gas phase	150	2012	[145]
BC + silica	MTMS	Liquid phase	133	2013	[207]
CNF + PVA + GO	FTCS FTEOS	Gas phase Gas phase	139.2 143.6	2013	[144]
CNF + PVA	MTCS	Gas phase	141.8	2013	[176]
Ce-II + silica	CCl_4	Gas phase	50–130	2013	[54]
Ce-II	CCl_4	Gas phase	60–102	2013	[55]
CNC	Palmitoyl chloride	Gas phase	Dispersible in toluene	2013	[190]
Ce-II	MTMS	Gas phase	145 (external) 143 (internal)	2013	[44]
Pulp	MTCS	Gas phase	150	2014	[243]
CNF	MTMS	Liquid phase	136	2014	[114]
CNC	MTMS	Liquid phase	125 (10% CNC) 115 (20% CNC)	2014	[188]
BC + silica	MTMS	Liquid phase	145	2014	[206]
CL-CNF + PVA	MTCS	Gas phase	150	2014	[137]
Ce-II + Fe_3O_4 NP	TTIP	Liquid phase	Absorb no water	2014	[51]
Ce-II	MTMS	Gas phase	135.2	2014	[45]
CNF	MTMS	Liquid phase	145.4–154.3	2014	[153]
Ce-II	TMCS	Gas phase	136	2015	[60]
BC	TMCS	Liquid phase	≤146.5	2015	[231]
Ce-II	TMCS	Gas phase	136	2015	[76]
CNF	MTCS	Gas phase	139	2015	[130]
CL-cellulose fibres	MTMS	Gas phase	153.8 (external) 150.8 (internal)	2015	[242]
CL–CNF + PVA	MTCS	Gas phase	141.8	2015	[176]
CNF	TMCS	Gas phase	135	2015	[169]

(Continued)

TABLE 11.6 (*Continued*)
Hydrophobic Cellulose-Based Aerogels

Aerogel Type	Hydrophobic Modifier	Hydrophobisation Method	Water Contact Angle	Year	Reference
Tritylcellulose	–	–	138 (DS = 0.72) 97 (DS = 0.38)	2015	[90]
CNF	MTCS	Gas phase	151–155	2016	[175]
CNF	DHI	Liquid phase	94.2	2016	[181]
CL-CNF	BA M + St M + EDMA M	Liquid phase	149	2016	[171]
CL-Ce-II	MTCS	Gas phase	141 (external) 128.4 (internal)	2016	[48]
CL-CNF	TEOS + TMCS	Liquid phase	152.1	2016	[147]
CL-Ce-II + Fe_3O_4 NP	HTMOS	Liquid phase	156	2016	[41]
BC	FTCS	Gas phase	156	2016	[205]

Note: DHI: 5,6-dihydroxyindole; EDMA: ethylene dimethylacrylate; UV: ultraviolet. See also the Notes of Tables 11.2–11.4.

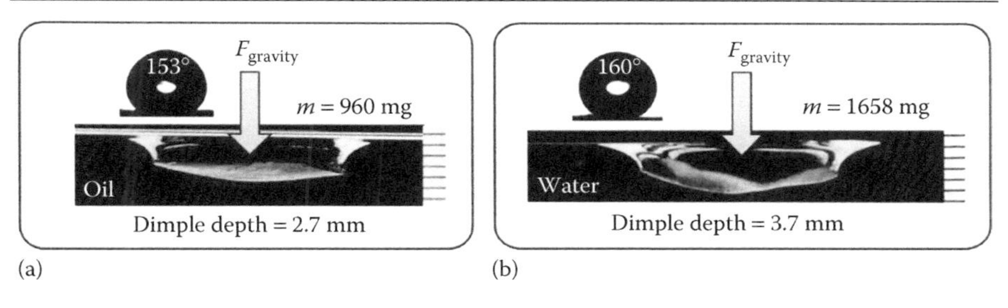

FIGURE 11.7 Contact angle measurement and load-carrying experiment of the aerogel on (a) paraffin oil and (b) water, respectively. The side-view photograph of the aerogel load carrier on paraffin oil and water shows the dimple at maximum supportable weight. (Reprinted with permission from Jin, H. et al., *Langmuir*, 27, 1930–1934, 2011. Copyright 2011 American Chemical Society.)

to both water and oil droplets (Figure 11.7) [160]. Therefore, superoleophobic surfaces can alleviate the contamination of oily substances to superhydrophobic effect of the surface.

TiO_2-coated CNF aerogels were also hydrophobic, with high adhesion pinning to water droplets, which did not roll off even by turning the surface upside down. It was worth noting that the coated surface switched from hydrophobic state to water-superabsorbent state under ultraviolet radiation and recovered gradually to original wetting properties in the dark [161]. CCl_4 plasma was an alternative effective method to conduct hydrophobic modification of cellulose aerogels via chlorination of surface [54,55]. The water contact angles of aerogels related to discharge duration and power of cold plasma in the preparation.

11.4.6 Absorption

Cellulose has an amphiphilic character due to the presence of both hydrophilic hydroxyls and hydrophobic pyranose rings. With a combination of high porosity of aerogel and amphiphilicity, the neat cellulose aerogels can absorb large amount of water and saline [10,45,109,122,133,138,151,156,161,191] and also oils and organic solvents [49,51,63,82,122,133,145,191,201]. After hydrophobisation, the cellulose-based aerogels showed larger oil absorption capacity and selectivity from oil–water mixture, as well as higher uptake rate for oil–water separation. The absorption capacities for oils and solvents were related to the densities of these absorbed compounds. In general, chloroform has been the one with highest absorbed capacity among the tested solvents, owing to its high density. Most hydrophobised aerogels displayed good recoverability, maintaining their absorption capacity for multiple sorption–desorption cycles. Moreover, the open porous structure of aerogels offered the possibility in the construction of devices to separate oil–water mixture continuously and effectively [41,201].

Heavy metal ions are also the main pollutants in water, which cause serious environmental and public health problems. Absorption process is quite convenient and effective for the removal of heavy metal ions from water. Grafting functional groups, which are effective for metal binding, such as carboxyl [125,137,246] and amino groups [116,142,185,246], can enhance the absorption capacity and rates. Modified cellulose-based aerogels were used for CO_2 absorption [126], removal of urea from water [82] and immobilisation of amino acids, peptides and proteins [85]. The CNF aerogels cross-linked with branched polyethyleneimine (b-PEI) showed selective colorimetric response to fluoride anions, changing colour from yellow to deep reddish orange (Figure 11.8). It can provide naked-eye sensing of fluoride over other commonly competing anions, such as chloride, acetate and dihydrogen phosphate [115]. Modified cellulose-based aerogels used for absorption are summarised in Table 11.7.

FIGURE 11.8 (a) CNF aerogels cross-linked with b-PEI in DMSO, and different solutions: (b) tetrabutyl ammonium (TBA) fluoride 0.05 M, (c) TBA fluoride 0.5 M, (d) TBA chloride 0.5 M, (e) TBA acetate 0.5 M and (f) TBA phosphate 0.5 M. Temperature: 55°C, contact time: 30 min. (Melone, L. et al., *RSC Adv.*, 5, 83197–83205, 2015. Reproduced by permission of The Royal Society of Chemistry.)

TABLE 11.7
Cellulose-Based Aerogels Used for Absorption after Modification

Aerogel	Modifier	Aerogel Density (mg/cm³)	Absorption Capacity (g/g)	Absorbate	Recyclability	Year	Reference
CNF	TTIP	20–30	20–40	Benzene, chloroform (maximum), dodecane, hexadecane, hexane (minimum), octane, octanol, paraffin oil, petroleum, mineral oil, toluene	Good	2011	[172]
CNF	AEAPDMS	26	0.062	CO_2	Good	2011	[126]
CNF	OTCS	8	<45	Hexadecane	–	2012	[145]
Ce-II	MTMS	40	13.9–24.4	Crude oil	Limited	2013	[44]
HPMC	–	–	98	Sunflower oil	–	2013	[256]
CNF	–	1.1	0.0037–0.0042	Methylene blue (minimum), toluidine blue (maximum)	–	2014	[117]
CNF	MTMS	11	49–102	Acetone, chloroform (maximum), dichloromethane, dodecane, ethanol, mineral oil, motor oil (minimum), silicone oil, toluene	Good	2014	[114]
BC	rGO		135–150	Cyclohexane (maximum), DMF (minimum)	–	2014	[201]
CNF	TEOS	2.7	139–356	Acetone, chloroform (maximum), cyclohexane, decane, DMF, DMSO, ethylene glycol (minimum), hexadecane, hexane, octane, pump oil, soybean oil, toluene	Good	2014	[122]
CNF + PVA	MTCS	13	44.56–95.25	Chloroform (maximum), corn oil, crude oil, diesel oil, gasoline, hexane (minimum), pump oil, toluene		2014	[137]
Ce-II + Fe_3O_4	TTIP		<28	Paraffin oil	Good	2014	[51]
Ce-II	MTMS	40	~18	Cooking oil, motor oil	–	2014	[45]
CL-CNC	–	5.6	72–160	Ethanol, dodecane (minimum), DMSO, water (maximum)	Good	2014	[191]
CL-CNF	EPTMAC		0.018	Cr(VI)	Good	2014	[142]

(Continued)

TABLE 11.7 (*Continued*)
Cellulose-Based Aerogels Used for Absorption after Modification

Aerogel	Modifier	Aerogel Density (mg/cm^3)	Absorption Capacity (g/g)	Absorbate	Recyclability	Year	Reference
CL-Ce-II	Cu(II) immobilised		0.005–0.367	Angiotensin I, l-carnosine, haemoglobin (maximum), l-histidine (minimum), lysozyme, myoglobin		2014	[85]
CNF	TMCS	3.12	52	Cooking oil	–	2015	[169]
CL-cellulose fibres	MTMS	7	95	Motor oil	–	2015	[242]
CNF	b-PEI	–	0.205–1.630	Amoxicillin, p-nitrophenol (maximum), 2,4,5-trichlorophenol (minimum)	–	2015	[116]
CNF	β-CD	10	0.0167	p-chlorophenol	Good	2015	[112]
Ce-II	TMCS	44 (maximum)–150 (minimum)	6.9–19.2	Motor oil	–	2015	[60]
BC	TMCS	6.7	86–185	Acetone, chlorobenzene, chloroform (maximum), dichloromethane, gasoline, n-hexene (minimum), paraffin, plant oil, toluene	Good	2015	[231]
Ce-II	TMCS	29	12–22	Benzene, chloroform (maximum), ethanol (minimum), ethyl acetate, methanol, methylbenzene, motor oil, waste engine oil, vegetable oil	Bad	2015	[76]
Ce-II	–	44 (maximum)–88 (minimum)	9.7–22.4	Soybean oil	–	2015	[82]
Ce-II	Dialdehyde modification	88	0.007	Urea	–	2015	[82]
CNF	MTCS	0.84	296–669	Acetone, benzene (minimum), t-butanol, diesel oil, ethanol, ethylene glycol, methanol, motor oil (maximum), trichloromethane	Limited	2015	[130]
Ce-II	MTCS	53 (maximum)–92 (minimum)	13.5–20.6	Waste motor oil	–	2015	[66]

(Continued)

TABLE 11.7 (*Continued*)
Cellulose-Based Aerogels Used for Absorption after Modification

Aerogel	Modifier	Aerogel Density (mg/cm³)	Absorption Capacity (g/g)	Absorbate	Recyclability	Year	Reference
CNF	rGO	5.6	44–265	Acetone, chloroform (maximum), colza oil, cyclohexane, DMF, DMSO, ethanol, ethyl acetate, toluene, MMA, oleic acid, petroleum ether (minimum), pump oil, St, THF, water	Good	2015	[261]
CNF	b-PEI		0.057–0.155	Cd(II) (maximum), Co(II) (minimum), Cu(II), Ni(II)	–	2015	[116]
CNF	PAMAM	10	0.377	Cr(VI)	–	2015	[185]
Ce-II	Fe_3O_4		0.001–0.010	Cr(VI)	–	2015	[65]
CNF	Methacrylic acid M + maleic acid M		0.117–0.165	Cd(II), Ni(II) (minimum), Pb(II) (maximum), Zn(II)	Good	2015	[125]
CNF	b-PEI with NO_2 groups		–	F anion	Good	2015	[115]
CNF	Eumelanin	–	–	Methylene blue	–	2016	[181]
Ce-II	–	–	–	Rhodamine		2016	[80]
CNF	GO + Fe(III) (minimum), GO (maximum)		0.127–0.142	Methyl blue	Limited	2016	[132]
CNF	–	–	101–145	Toluene	–	2016	[133]
Ce-II	Graphene	–	20.4	Waste machine oil	–	2016	[63]
CL-Ce-II + Fe_3O_4	HTMOS	–	Separating oil/water emulsion	–	Good	2016	[41]

(*Continued*)

TABLE 11.7 (*Continued*)
Cellulose-Based Aerogels Used for Absorption after Modification

Aerogel	Modifier	Aerogel Density (mg/cm^3)	Absorption Capacity (g/g)	Absorbate	Recyclability	Year	Reference
CL-CNF	BA M + St M + EDMA M	23.2	23–46.6	Chloroform (maximum), DMF, DMSO, hexane (minimum), Isopar™ M fluid, methanol, toluene	Limited	2016	[171]
CL-CNF	MTCS	5.98–16.54	54–140	Acetone, chloroform (maximum), crude oil, DMF, ethanol, hexane (minimum), toluene	–	2016	[173]
CNF	MTCS	4.9	53–93	Acetone (minimum), acetylacetone, benzene, chloroform, ethanol, ethyl acetate, n-hexane, machine oil, methylbenzene, oleic acid (maximum)	Good	2016	[175]
CL-Ce-II	MTCS	31	30–59	Chloroform, colza oil, cyclohexane (minimum), ethyl acetate, hexane, methylbenzene, petroleum ether, pump oil (maximum)	Good	2016	[48]
Pulp	AA M + acrylamide M	–	0.172	Ni(II)	Good	2016	[246]

Note: CD: cyclodextrin; DMF: N,N-dimethylformamide; MMA: methyl methacrylate; THF: tetrahydrofuran. See also the Notes of Tables 11.1–11.4, 11.6

11.4.7 Oxidation Resistance

Panzella et al. [181] coated TOCNF aerogels with melanin (Mel-TOCNF) via ammonia-induced solid-state polymerisation of 5,6-dihydroxyindole (DHI) and 5,6-dihydroxyindole-2-carboxylic acid (DHICA). DHI Mel-TOCNF aerogel can turn deep-violet 2,2-diphenyl-1-picrylhydrazyl, a stable free radical, into a yellow hydrazine in an antioxidant assay, indicating a high antioxidant activity. By comparison, DHICA Mel-TOCNF aerogel had low activity to scavenging free radicals. Propensity of CNC towards oxidation limited its application as nanofiller in polymers. Smith et al. [262] deposited thin layers of Al_2O_3 onto the CNC aerogels by using the ALD process to obtain protection against oxidation. Therefore, ALD-coated aerogels were suitable to reinforce mechanical properties of polymers as nanofillers owing to their significantly increased oxidation resistance. The more the completion of the layers, the better the antioxidant properties of CNC obtained.

11.4.8 Thermal Conductivity

Aerogels exhibit extremely low thermal conductivity (<0.050 W/m/K) owing to the ultralow density and highly nanoporous structure. The former decreases the contribution of the solid phase to the thermal conductivity, whereas the latter decreases the contribution of the gas phase when the pore size is smaller than the mean free path of gas molecules in air (ca. 70 nm). Generally, low density and high porosity facilitated to reduce the thermal conductivity of cellulose-based aerogels [10,31,55,68,117,127,179,180]. Optimising the conditions used for gel formation and drying and/or incorporating nanofillers can prepare cellulose-based aerogels with thermal conductivity below that of air (0.025 W/m/K) [117,127,131,135,147,152,153,180,241]. The heat insulation properties obviously decreased with the increase in moisture uptake [54,55]. Consequently, hydrophobisation reduced the influence of humidity on the thermal insulation performance and decreased the thermal conductivity of cellulose-based aerogels [45,55,135]. Baillis et al. [263] developed a heat transfer model for cellulose-based aerogels and analysed the contribution of conduction from the gas and solid phases and radiation. It was suggested that the presence of macropores would have detrimental effect on thermal insulation. Table 11.8 summarises the values of thermal conductivity reported in cellulose-based aerogel systems.

11.4.9 Sound Absorption

Chen et al. [117] showed that the sound absorption ratio of nanocellulose aerogels (density ~ 5 mg/cm^3) improved rapidly with the increase in frequency. It was lower than 20% at frequencies less than 1,000 Hz, whereas it reached a high value (up to 57.1%) at 4,000 Hz. Meanwhile, the aerogels obtained from CNFs displayed higher sound absorption ability than those obtained from CNCs with low aspect ratios. The results of Lu et al. [68] revealed that lignocellulose with density of 95 mg/cm^3 had sound absorption ratio higher than 76% in the frequency range between 1,000 Hz and 2,000 Hz, which was desirable for applications of sound absorption and noise reduction.

TABLE 11.8
Thermal Conductivities of Cellulose-Based Aerogels

Aerogel Type	Drying Method	Thermal Conductivity (W/m/K)	Density (g/cm^3)	Year	Reference
CL-CA	SD	0.029	0.25	2006	[250]
Ce-II	SD	0.025	0.14	2012	[31]
Ce-II + silica	SD	0.025–0.045	0.34–0.58		
Ce-II	FD	0.038	0.218	2012	[39]
Ce-II + $CoFe_2O_4$	FD	0.035–0.049	0.254–0.390		
Regenerated lignocellulose	SD	0.030–0.054	0.095–0.143	2012	[68]
CL-CNF + PVA	FD	0.040	0.020–0.031	2013	[179]
CL-CNF + PVA + MWCNT	FD	0.029–0.031	0.027		
BC	FD	0.0295	0.007	2013	[207]
BC + silica	FD	0.0313–0.0368	0.020–0.201		
CNF + PVA + GO	FD	0.045	0.035	2013	[144]
Ce-II + silica	FD	0.0257–0.031	0.233–0.45	2013	[54]
Ce-II	FD	0.029–0.032	0.2–0.4	2013	[55]
Ce-II with 11% hygroscopicity	FD	0.20	0.2		
PMSQ–CNF	SD	0.0162–0.0243	0.020–0.186	2014	[153]
CNF	SD	0.018	0.017	2014	[180]
Ce-II	FD	0.032	0.040	2014	[45]
Silylated Ce-II	FD	0.029	–		
CNF	FD	0.014–0.016	0.005	2014	[117]
BC	FD	0.0295	0.007	2014	[206]
BC + silica	FD	0.0308–0.0369	0.011–0.369		
CNF	SD	0.0339–0.0386	0.0074–0.0183	2015	[135]
CNF + silica	SD	0.0176–0.0201	0.122–0.129		
Silylated CNF	SD	0.0313–0.0363	0.0098–0.0306		
Silylated CNF + silica	SD	0.0138–0.0175	0.130–0.146		
Ce-II	SD	0.044–0.055	0.041–0.113	2015	[100]
Ce-II + silica	SD	0.040–0.052	0.04–0.15		
Ce-II fibres	SD	0.04–0.075	0.009–0.137	2015	[101]
Ce-II	FD	0.056	0.22	2015	[32]
Ce-II + $Mg(OH)_2$	FD	0.063–0.081	0.24–0.40		
Ce-II	FD	0.029	0.233	2015	[53]
CNF + silica	SD	0.0153	0.127	2015	[127]
Ce-II	SD	0.033	0.125	2015	[89]
Ce-II + silica	SD	0.026–0.028	0.155–0.225		
CNF	FD	0.018	0.0056	2015	[131]
CNF + GO + sepiolite	FD	0.015	0.0075		
Ce-II	FD	0.0359–0.0363	–	2015	[22]
Ce-II + MMT	FD	0.0326–0.0357	–		
Cellulosic microfibre, CNF	FD	0.028	–	2015	[152]

(Continued)

TABLE 11.8 (*Continued*)
Thermal Conductivities of Cellulose-Based Aerogels

Aerogel Type	Drying Method	Thermal Conductivity (W/m/K)	Density (g/cm³)	Year	Reference
Cellulosic microfibre + nanozeolites	FD	0.029–0.031	–		
NFC + nanozeolites	FD	0.018–0.026	–		
Cellulose fibres	FD	0.028	0.40	2016	[241]
Cellulose fibres + CNC	FD	0.025	0.40		
Cellulose fibres + CNF	FD	0.023	0.40		
Ce-II	SD	0.024–0.033	0.072–0.220	2016	[10]
CNF + silica	ED	0.0226	0.11	2016	[147]
Ce-II + graphene (impregnated with PEG)	FD	1.35	–	2016	[34]
CNC + PVA + MMT	FD	0.041–0.044	0.055–0.108	2016	[189]
BC + FTCS	SD	0.027	–	2016	[205]
Ce-II	SD	0.033	0.024	2016	[73]

11.4.10 Electrical Properties

Electrically conductive substances, including inorganic reduced graphene oxide (rGO) [33,146,260,261] and carbon nanotube (CNT) [35,36,40,149,168,260], organic PPy [28,121,186] and polyaniline (PANi) [7], can be incorporated into cellulose aerogels to improve conductivity. The obtained conductivity generally reached a value greater than 10^{-2} S/cm [7,28,33,35,40,168]. Moreover, the high porosity of aerogels resulted in a threshold lower than 0.1% volume fraction [7,36,40]. The electrical resistivity of conductive cellulose composite aerogels responded quickly to compressive strain and gas molecule adsorption, providing potential for applications of these materials as pressure-responsive sensor [168,261] or chemical sensor for gases [35,40] (Figure 11.9). The electrically conductive cellulose–CNT composite aerogels showed highly efficient electromagnetic interference (EMI) shielding in the microwave frequency range of 8.2–12.4 GHz. It was revealed that the contribution of absorption of electromagnetic energy to the EMI shielding was much higher than that of the reflection from the material surface [36].

Electrically conductive cellulose-based aerogels were also promising candidates for electrodes used in supercapacitors. The highly open porous structure provided fast transport channels for electrons and ions, resulting in ultralow internal resistance. In the interconnected three-dimensional network, the incorporated active NPs offered large accessible surface area and hence high charge storage. By using cellulose-based aerogels, especially nanocellulose-based aerogels, as electrodes, the obtained highly flexible and all-solid-state supercapacitor had high capacitance, power density and energy density and excellent cyclic stability [33,146,149,186,260].

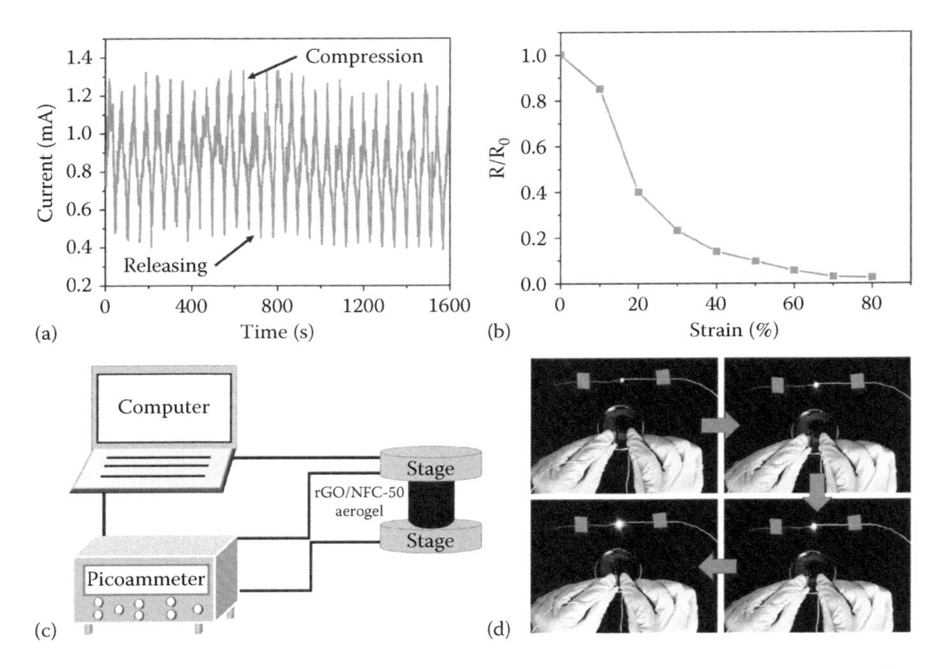

FIGURE 11.9 Strain-sensitive conductivity of CNF-rGO aerogel. (a) The variation of electric current with cyclic compression in a closed circuit. (b) Plot of resistance variation with compressive strain. (c) Set-up for the in situ measurement of the electrical resistance during cyclic compression. (d) CNF-rGO aerogel is used as the conductive bulk for the construction of a circuit. The brightness of the light-emitting diode (LED) lamp gradually becomes stronger upon compression. (Yao, X.L. et al., *Nanoscale*, 7, 3959–3964, 2015. Reproduced by permission of The Royal Society of Chemistry.)

11.4.11 Catalytic Activity

Although without catalytic activity, cellulose-based aerogels have high porosity, favourable to mass transport and effective contact with the reactants, and high specific surface area, facilitating the loading and dispersion of catalysts. Moreover, the network structure of aerogels guided the in situ growth of NPs with catalytic activity and prevented their aggregation. Therefore, cellulose composite aerogels displayed high conversion rate in the catalytic reaction [59,87,103]. On the other hand, the aerogels showed excellent recyclability after simple isolation after reactions [57,79,103,124,132]. The cellulose composite aerogels with catalytic activity are included in Table 11.9.

11.4.12 Biomedical Features (Biocompatibility)

Cellulose is a biocompatible biomass. Porous cellulose-based aerogels were suitable for cell-culture scaffold application. Aerogels prepared from cellulose phosphate with very low degree of phosphorylation ($\leq$0.20) exhibited good hemocompatibility, and their biomineralising properties were improved by the presence of nitrogen

TABLE 11.9
Cellulose Composite Aerogels with Catalytic Activity

Aerogel Type	Catalytic Component Type	Size (nm)	Content (wt.%)	Reaction	Year	Reference
CNF	TiO_2	7[a]	13.3	Photocatalytic degradation of methylene blue	2011	[161]
CNF	Cu(I)	–	10 μmol	Catalytic cycloaddition of Huisgen [3 + 2]	2012	[124]
CNF	Ag	21.1	4.72	Catalytic reduction of 4-nitrophenol	2013	[79]
Ce-II	PANi	150	20	Photocatalytic degradation of methylene blue	2014	[87]
Ce-II	TiO_2	3.69	17.7	Photocatalytic degradation of indigo carmine dye	2015	[57]
Ce-II	TiO_2	2–5	35.7	Photocatalytic degradation of Rhodamine B and methyl orange	2015	[59]
CNF + GO	Fe(III)	–	0.05–0.30	Catalytic Fenton's oxidation	2016	[132]
Ce-II	Ag	10–30	–	Catalytic reduction of 4-nitrophenol	2016	[103]

[a] Layer thickness.

functional groups [247]. The CL-CNF–gelatin–chitosan aerogels, which had tailorable three-dimensional porous structure and mechanical properties, provided favourable environment for cell interactions and extracellular matrix production. They were potentially used as scaffolds for soft tissue regeneration, such as cartilage [159]. The porous structure of biocompatible cellulose-based aerogels was essential for cell attachment, proliferation and differentiation [156,158]. After immobilising gelatin, the composite aerogels can promote cell growth, resulting in enhanced biocompatibility and excellent matrix for cell culture [43,259]. In vitro studies showed that the incorporation of electrically conductive PPy in Ce-II aerogel facilitated the adhesion and proliferation of neural cells on the surface, making it a promising material for nerve regeneration [28]. Cellulose-based aerogels can be used as drug delivery carriers [29,148,223], and incorporating amino groups was favourable for the enhancement of drug loading capacity [154]. On the other hand, grafting temperature-sensitive poly(*N*-isopropylacrylamide) can be used to prepare aerogel microspheres for controlled drug release, with good temperature response [177].

11.5 CELLULOSE-BASED CARBON AEROGELS

Carbonisation of cellulose-based aerogels at high temperatures can produce carbon aerogels. Owing to the strong correlations between crystallinities of the raw materials and the derived carbon materials, researchers have mostly focussed on highly crystalline BC aerogels [141,203,204,209–214,218–222,224,225,227,232–234,264] and CNF aerogels [141,174,186] as carbon precursors. Other Ce-I aerogels prepared from pulp were also used [141,237,239,240,244]. On the other hand, Ce-II aerogels [12,14–16,37,47,49,62], Ce-D aerogels [249,251,252,254] and waste cigarette filters [265] were also selected as precursor materials. Some biomasses, such as cotton, sugarcane and straw, and cellulose suspensions [264,266–272] were transformed directly to highly porous carbon aerogels through hydrothermal treatment. Carbonising cellulose with compounds containing heteroatoms was a facile method to obtain carbon aerogels doped with nitrogen [210,212,221,222,232,237,266,267,270], phosphorus [212,221], boron [212] and sulphur [47,210] heteroatoms. Further heating of carbon aerogels under an ammonia atmosphere also yielded nitrogen-doped carbon materials [209,213,233,251]. In addition, porous carbon materials can be further loaded with inorganic NPs to increase the functionalisation of carbon materials [12,14,214,218, 220,224,227,252,254,264]. Moreover, the carbon aerogels featured excellent chemical and thermal stability, flame retardance and fire resistance [62,202,244,270,271], which greatly facilitated their potential in versatile application fields. Upon activation, the carbonised cellulose-based carbon showed high adsorption capacity and selectivity to dye [264,270], heavy metal ions [49,266], organic solvents and oils [129,174,202,239, 240,244,269,271] and so on. Consequently, they can serve as promising absorbents for the removal of pollutants from water resources. The high electrical conductivity possessed by carbon materials made them suitable for applications as catalysts in the oxygen reduction reaction [14,16,210,227,232,233,254,267], electrodes for supercapacitors [37,204,210–213,220–222,234,237] and Li-ion batteries [47,214,218,219,224,225], pressure sensors [202] and EMI shielding material [265,268].

11.6 OUTLOOK

Cellulose-based aerogels have gained rapidly growing technological and scientific interest owing to their sustainable and environmentally friendly origins since the beginning of the new century. Their unique morphologies and properties highlighted in this review can be tailored to render materials suitable for a wide range of applications. The highly porous materials also provide an interesting platform upon which to develop high-performance functional materials with unique properties via a template route. Use of three-dimensional cellulose aerogel as the template can open new applications to construct flexible three-dimensional devices with high surface area [120], scaffolding materials for tissue engineering [208], inorganic nanomaterials [163,183] and polymer nanocomposites [176,228].

On the other hand, it is still highly challenging to fabricate large-scale cellulose-based aerogels by facile and economic process. Reinforcing skeletal structure of network, combined with introducing large pores and hydrophobicity, makes it possible to obtain aerogels via cost-effective ambient/vacuum drying method. Another method is to use raw materials with low quality, such as straws, grasses, barks and waste papers, without isolating cellulose, to decrease the cost of aerogels. Cellulose solvent system plays a crucial role in the sustainability of cellulose-based aerogels. It is worth noting that the cellulose solvent systems that can be recycled efficiently by simple distillation have great advantages over others, which deserve more attention.

ACKNOWLEDGEMENTS

Some of our work reported in this chapter was supported in part by grants from the National Natural Science Foundation of China (Grant Nos. 51425307 and 51273206).

REFERENCES

1. Garcia-Gonzalez CA, Alnaief M, Smirnova I. Polysaccharide-based aerogels-Promising biodegradable carriers for drug delivery systems. *Carbohydr Polym* 2011;86(4):1425–1438.
2. Ma SR, Mi QY, Yu J, He JS, Zhang J. Aerogel materials based on cellulose. *Prog Chem* 2014;26(5):796–809.
3. Weatherw RC, Caulfiel DF. Cellulose aerogels–improved method for preparing a highly expanded form of dry cellulose. *Tappi* 1971;54(6):985–986.
4. Tan CB, Fung BM, Newman JK, Vu C. Organic aerogels with very high impact strength. *Adv Mater* 2001;13(9):644–646.
5. Jin H, Nishiyama Y, Wada M, Kuga S. Nanofibrillar cellulose aerogels. *Colloid Surf A-Physicochem Eng Asp* 2004;240(1–3):63–67.
6. Maeda H, Nakajima M, Hagwara T, Sawaguchi T, Yano S. Preparation and properties of bacterial cellulose aerogel. *Kobunshi Ronbunshu* 2006;63(2):135–137.
7. Paakko M, Vapaavuori J, Silvennoinen R, Kosonen H, Ankerfors M, Lindstrom T, Berglund LA, Ikkala O. Long and entangled native cellulose I nanofibers allow flexible aerogels and hierarchically porous templates for functionalities. *Soft Matter* 2008;4(12):2492–2499.
8. Husing N, Schubert U. Aerogels-airy materials: Chemistry, structure, and properties. *Angew Chem Int Ed* 1998;37:22–45.
9. Innerlohinger J, Weber HK, Kraft G. Aerocellulose: Aerogels and aerogel-like materials made from cellulose. *Macromol Symp* 2006;244(1):126–135.
10. Ayadi F, Martin-Garcia B, Colombo M, Polovitsyn A, Scarpellini A, Ceseracciu L et al. Mechanically flexible and optically transparent three-dimensional nanofibrous amorphous aerocellulose. *Carbohydr Polym* 2016;149:217–223.
11. Schestakow M, Karadagli I, Ratke L. Cellulose aerogels prepared from an aqueous zinc chloride salt hydrate melt. *Carbohydr Polym* 2016;137:642–649.
12. Guilminot E, Gavillon R, Chatenet M, Berthon-Fabry S, Rigacci A, Budtova T. New nanostructured carbons based on porous cellulose: Elaboration, pyrolysis and use as platinum nanoparticles substrate for oxygen reduction electrocatalysis. *J Power Sources* 2008;185(2):717–726.
13. Sescousse R, Gavillon R, Budtova T. Wet and dry highly porous cellulose beads from cellulose-NaOH-water solutions: Influence of the preparation conditions on beads shape and encapsulation of inorganic particles. *J Mater Sci* 2011;46(3):759–765.
14. Rooke J, Passos CdM, Chatenet M, Sescousse R, Budtova T, Berthon-Fabry S, Mosdale R, Maillard F. Synthesis and properties of platinum nanocatalyst supported on cellulose-based carbon aerogel for applications in PEMFCs. *J Electrochem Soc* 2011;158(7):B779–B789.
15. Dassanayake RS, Gunathilake C, Jackson T, Jaroniec M, Abidi N. Preparation and adsorption properties of aerocellulose-derived activated carbon monoliths. *Cellulose* 2016;23(2):1363–1374.
16. Zu GQ, Shen J, Zou LP, Wang F, Wang XD, Zhang YW, Yao X. Nanocellulose-derived highly porous carbon aerogels for supercapacitors. *Carbon* 2016;99:203–211.
17. Gavillon R, Budtova T. Aerocellulose: New highly porous cellulose prepared from cellulose-NaOH aqueous solutions. *Biomacromolecules* 2008;9(1):269–277.
18. Sescousse R, Budtova T. Influence of processing parameters on regeneration kinetics and morphology of porous cellulose from cellulose-NaOH-water solutions. *Cellulose* 2009;16(3):417–426.
19. Kovarskii AL, Sorokina ON, Shapiro AB, Nikitin LN. A spin-probe study of the porous structure of cellulose treated with supercritical carbon dioxide. *Polym Sci Ser A* 2015;57(3):309–314.

20. Demilecamps A, Reichenauer G, Rigacci A, Budtova T. Cellulose-silica composite aerogels from "one-pot" synthesis. *Cellulose* 2014;21(4):2625–2636.
21. Ahmadzadeh S, Keramat J, Nasirpour A, Hamdami N, Behzad T, Aranda L, Vilasi M, Desobry S. Structural and mechanical properties of clay nanocomposite foams based on cellulose for the food-packaging industry. *J Appl Polym Sci* 2016;133(2):10.
22. Ahmadzadeh S, Nasirpour A, Keramat J, Hamdami N, Behzad T, Desobry S. Nanoporous cellulose nanocomposite foams as high insulated food packaging materials. *Colloid Surf A-Physicochem Eng Asp* 2015;468:201–210.
23. Ahmadzadeh S, Desobry S, Keramat J, Nasirpour A. Crystalline structure and morphological properties of porous cellulose/clay composites: The effect of water and ethanol as coagulants. *Carbohydr Polym* 2016;141:211–219.
24. Sescousse R, Smacchia A, Budtova T. Influence of lignin on cellulose-NaOH-water mixtures properties and on Aerocellulose morphology. *Cellulose* 2010;17(6):1137–1146.
25. Mohamed SMK, Ganesan K, Milow B, Ratke L. The effect of zinc oxide (ZnO) addition on the physical and morphological properties of cellulose aerogel beads. *RSC Adv* 2015;5(109):90193–90201.
26. Cai J, Kimura S, Wada M, Kuga S. Nanoporous cellulose as metal nanoparticles support. *Biomacromolecules* 2009;10(1):87–94.
27. Liu SL, Yu TF, Hu NN, Liu R, Liu XY. High strength cellulose aerogels prepared by spatially confined synthesis of silica in bioscaffolds. *Colloid Surf A-Physicochem Eng Asp* 2013;439:159–166.
28. Shi ZQ, Gao HC, Feng J, Ding BB, Cao XD, Kuga S et al. In situ synthesis of robust conductive cellulose/polypyrrole composite aerogels and their potential application in nerve regeneration. *Angew Chem Int Ed* 2014;53(21):5380–5384.
29. Shen XP, Shamshina JL, Berton P, Bandomir J, Wang H, Gurau G, Rogers RD. Comparison of hydrogels prepared with ionic-liquid-isolated vs commercial chitin and cellulose. *ACS Sustain Chem Eng* 2016;4(2):471–480.
30. Cai J, Kimura S, Wada M, Kuga S, Zhang L. Cellulose aerogels from aqueous alkali hydroxide-urea solution. *ChemSusChem* 2008;1(1–2):149–154.
31. Cai J, Liu SL, Feng J, Kimura S, Wada M, Kuga S et al. Cellulose-silica nanocomposite aerogels by in situ formation of silica in cellulose gel. *Angew Chem Int Ed* 2012;51(9):2076–2079.
32. Han YY, Zhang XX, Wu XD, Lu CH. Flame retardant, heat insulating cellulose aerogels from waste cotton fabrics by in situ formation of magnesium hydroxide nanoparticles in cellulose gel nanostructures. *ACS Sustain Chem Eng* 2015;3(8):1853–1859.
33. Ouyang WZ, Sun JH, Memon J, Wang C, Geng JX, Huang Y. Scalable preparation of three-dimensional porous structures of reduced graphene oxide/cellulose composites and their application in supercapacitors. *Carbon* 2013;62:501–509.
34. Yang J, Zhang EW, Li XF, Zhang YT, Qu J, Yu ZZ. Cellulose/graphene aerogel supported phase change composites with high thermal conductivity and good shape stability for thermal energy storage. *Carbon* 2016;98:50–57.
35. Qi HS, Liu JW, Pionteck J, Potschke P, Madera E. Carbon nanotube-cellulose composite aerogels for vapour sensing. *Sens Actuator B-Chem* 2015;213:20–26.
36. Huang HD, Liu CY, Zhou D, Jiang X, Zhong GJ, Yan DX, Li ZM. Cellulose composite aerogel for highly efficient electromagnetic interference shielding. *J Mater Chem A* 2015;3(9):4983–4991.
37. Hao P, Zhao ZH, Tian J, Li HD, Sang YH, Yu GW, Cai H, Liu H, Wong CP, Umar A. Hierarchical porous carbon aerogel derived from bagasse for high performance supercapacitor electrode. *Nanoscale* 2014;6(20):12120–12129.
38. Peng HL, Wu JN, Wang YX, Wang H, Liu ZY, Shi YL, Guo X. A facile approach for preparation of underwater superoleophobicity cellulose/chitosan composite aerogel for oil/water separation. *Appl Phys A-Mater Sci Process* 2016;122(5):7.

39. Liu SL, Yan QF, Tao DD, Yu TF, Liu XY. Highly flexible magnetic composite aerogels prepared by using cellulose nanofibril networks as templates. *Carbohydr Polym* 2012;89(2):551–557.
40. Qi HS, Mader E, Liu JW. Electrically conductive aerogels composed of cellulose and carbon nanotubes. *J Mater Chem A* 2013;1(34):9714–9720.
41. Peng HL, Wang H, Wu JN, Meng GH, Wang YX, Shi YL, Liu Z, Guo X. Preparation of superhydrophobic magnetic cellulose sponge for removing oil from water. *Ind Eng Chem Res* 2016;55(3):832–838.
42. Ye DD, Zhong ZB, Xu H, Chang CY, Yang ZX, Wang YF, Ye Q, Zhang L. Construction of cellulose/nanosilver sponge materials and their antibacterial activities for infected wounds healing. *Cellulose* 2016;23(1):749–763.
43. Pei Y, Ye DD, Zhao Q, Wang XY, Zhang C, Huang WH, Zhang N, Liu S, Zhang L. Effectively promoting wound healing with cellulose/gelatin sponges constructed directly from a cellulose solution. *J Mat Chem B* 2015;3(38):7518–7528.
44. Nguyen ST, Feng JD, Le NT, Le ATT, Hoang N, Tan VBC, Duong HM. Cellulose aerogel from paper waste for crude oil spill cleaning. *Ind Eng Chem Res* 2013;52(51):18386–18391.
45. Nguyen ST, Feng JD, Ng SK, Wong JPW, Tan VBC, Duong HM. Advanced thermal insulation and absorption properties of recycled cellulose aerogels. *Colloid Surf A-Physicochem Eng Asp* 2014;445:128–134.
46. Liu CY, Zhong GJ, Huang HD, Li ZM. Phase assembly-induced transition of three dimensional nanofibril- to sheet-networks in porous cellulose with tunable properties. *Cellulose* 2014;21(1):383–394.
47. Yun YS, Song MY, Kim NR, Jin HJ. Sulfur-enriched, hierarchically nanoporous carbonaceous materials for sodium-ion storage. *Synth Met* 2015;210:357–362.
48. Liao Q, Su XP, Zhu WJ, Hua W, Qian ZQ, Liu L, Yao J. Flexible and durable cellulose aerogels for highly effective oil/water separation. *RSC Adv* 2016;6(68):63773–63781.
49. Wang H, Gong Y, Wang Y. Cellulose-based hydrophobic carbon aerogels as versatile and superior adsorbents for sewage treatment. *RSC Adv* 2014;4(86):45753–45759.
50. Isobe N, Nishiyama Y, Kimura S, Wada M, Kuga S. Origin of hydrophilicity of cellulose hydrogel from aqueous LiOH/urea solvent coagulated with alkyl alcohols. *Cellulose* 2014;21(2):1043–1050.
51. Chin SF, Romainor ANB, Pang SC. Fabrication of hydrophobic and magnetic cellulose aerogel with high oil absorption capacity. *Mater Lett* 2014;115:241–243.
52. Zhang J, Cao YW, Feng JC, Wu PY. Graphene-oxide-sheet-induced gelation of cellulose and promoted mechanical properties of composite aerogels. *J Phys Chem C* 2012;116(14):8063–8068.
53. Shi J, Lu L, Guo W, Liu M, Cao Y. On preparation, structure and performance of high porosity bulk cellulose aerogel. *Plast Rubber Compos* 2015;44(1):26–32.
54. Shi JJ, Lu LB, Guo WT, Zhang JY, Cao Y. Heat insulation performance, mechanics and hydrophobic modification of cellulose-SiO_2 composite aerogels. *Carbohydr Polym* 2013;98(1):282–289.
55. Shi JJ, Lu LB, Guo WT, Sun YJ, Cao Y. An environment-friendly thermal insulation material from cellulose and plasma modification. *J Appl Polym Sci* 2013;130(5):3652–3658.
56. Wan CC, Li J. Synthesis of well-dispersed magnetic $CoFe_2O_4$ nanoparticles in cellulose aerogels via a facile oxidative co-precipitation method. *Carbohydr Polym* 2015;134:144–150.
57. Jiao Y, Wan CC, Li J. Room-temperature embedment of anatase titania nanoparticles into porous cellulose aerogels. *Appl Phys A-Mater Sci Process* 2015;120(1):341–347.
58. Wan C, Jiao Y, Sun Q, Li J. Preparation, characterization, and antibacterial properties of silver nanoparticles embedded into cellulose aerogels. *Polym Compos* 2016:1137–1142.

59. Wan CC, Lu Y, Jin CD, Sun QF, Li J. A facile low-temperature hydrothermal method to prepare anatase titania/cellulose aerogels with strong photocatalytic activities for rhodamine B and methyl orange degradations. *J Nanomater* 2015;2015:8.
60. Wan C, Lu Y, Jiao Y, Cao J, Sun Q, Li J. Cellulose aerogels from cellulose-NaOH/PEG solution and comparison with different cellulose contents. *Mater Sci Technol* 2015;31(9):1096–1102.
61. Wan CC, Li J. Cellulose aerogels functionalized with polypyrrole and silver nanoparticles: In-situ synthesis, characterization and antibacterial activity. *Carbohydr Polym* 2016;146:362–367.
62. Wan CC, Lu Y, Jiao Y, Jin CD, Sun QF, Li J. Fabrication of hydrophobic, electrically conductive and flame-resistant carbon aerogels by pyrolysis of regenerated cellulose aerogels. *Carbohydr Polym* 2015;118:115–118.
63. Wan CC, Li J. Incorporation of graphene nanosheets into cellulose aerogels: Enhanced mechanical, thermal, and oil adsorption properties. *Appl Phys A-Mater Sci Process* 2016;122(2):7.
64. Wan CC, Li J. Embedding ZnO nanorods into porous cellulose aerogels via a facile one-step low-temperature hydrothermal method. *Mater Des* 2015;83:620–625.
65. Wan CC, Li J. Facile synthesis of well-dispersed superparamagnetic gamma-Fe_2O_3 nanoparticles encapsulated in three-dimensional architectures of cellulose aerogels and their applications for Cr(VI) removal from contaminated water. *ACS Sustain Chem Eng* 2015;3(9):2142–2152.
66. Wan C, Lu Y, Cao J, Sun Q, Li J. Preparation, characterization and oil adsorption properties of cellulose aerogels from four kinds of plant materials via a NaOH/PEG aqueous solution. *Fiber Polym* 2015;16(2):302–307.
67. Voon LK, Pang SC, Chin SF. Highly porous cellulose beads of controllable sizes derived from regenerated cellulose of printed paper wastes. *Mater Lett* 2016;164:264–266.
68. Lu Y, Sun QF, Yang DJ, She XL, Yao XD, Zhu GS, Liu Y, Zhao H, Li J. Fabrication of mesoporous lignocellulose aerogels from wood via cyclic liquid nitrogen freezing-thawing in ionic liquid solution. *J Mater Chem* 2012;22(27):13548–13557.
69. Li J, Lu Y, Yang DJ, Sun QF, Liu YX, Zhao HJ. Lignocellulose aerogel from wood-ionic liquid solution (1-allyl-3-methylimidazolium chloride) under freezing and thawing conditions. *Biomacromolecules* 2011;12(5):1860–1867.
70. Chen C, Li J. Preparation of lignocellulose aerogel from wood-ionic liquid solution. *Adv Mat Res* 2011;280:191–195.
71. Tsioptsias C, Stefopoulos A, Kokkinomalis I, Papadopoulou L, Panayiotou C. Development of micro- and nano-porous composite materials by processing cellulose with ionic liquids and supercritical CO_2. *Green Chem* 2008;10(9):965–971.
72. Lv Y, Li X, Mi Q, Wang D, Yu J, Zhang J. Cellulose aerogels prepared from cellulose/AmimCl solutions. *Scientia Sin Chim* 2011;41(8):1331–1337.
73. Mi Q-Y, Ma S-R, Yu J, He J-S, Zhang J. Flexible and transparent cellulose aerogels with uniform nanoporous structure by a controlled regeneration process. *ACS Sustain Chem Eng* 2016;4(3):656–660.
74. Yuan B, Mi QY, Yu J, Song R, Zhang JM, He JS, Zhang J. Transparent and flame retardant cellulose/aluminum hydroxide nanocomposite aerogels. *Sci China Chem* 2016;59(10):1335–1341.
75. Lv FB, Wang CX, Zhu P, Zhang CJ. Characterization of chitosan microparticles reinforced cellulose biocomposite sponges regenerated from ionic liquid. *Cellulose* 2014;21(6):4405–4418.
76. Jin CD, Han SJ, Li JP, Sun QF. Fabrication of cellulose-based aerogels from waste newspaper without any pretreatment and their use for absorbents. *Carbohydr Polym* 2015;123:150–156.

77. Sescousse R, Gavillon R, Budtova T. Aerocellulose from cellulose-ionic liquid solutions: Preparation, properties and comparison with cellulose-NaOH and cellulose-NMMO routes. *Carbohydr Polym* 2011;83(4):1766–1774.
78. Aaltonen O, Jauhiainen O. The preparation of lignocellulosic aerogels from ionic liquid solutions. *Carbohydr Polym* 2009;75(1):125–129.
79. Xiong R, Lu CH, Wang YR, Zhou ZH, Zhang XX. Nanofibrillated cellulose as the support and reductant for the facile synthesis of Fe_3O_4/Ag nanocomposites with catalytic and antibacterial activity. *J Mater Chem A* 2013;1(47):14910–14918.
80. Xu MM, Bao WQ, Xu SP, Wang XH, Sun RC. Porous cellulose aerogels with high mechanical performance and their absorption behaviors. *BioResources* 2016;11(1):8–20.
81. Xu MM, Huang QB, Wang XH, Sun RC. Highly tough cellulose/graphene composite hydrogels prepared from ionic liquids. *Ind Crop Prod* 2015;70:56–63.
82. Liu XY, Chang PR, Zheng PW, Anderson DP, Ma XF. Porous cellulose facilitated by ionic liquid BMIM Cl: Fabrication, characterization, and modification. *Cellulose* 2015;22(1):709–715.
83. Shamsuri AA, Abdullah DK, Daik R. Fabrication of agar/biopolymer blend aerogels in ionic liquid and co-solvent mixture. *Cell Chem Technol* 2012;46(1–2):45–52.
84. Deng ML, Zhou Q, Du AK, van Kasteren J, Wang YZ. Preparation of nanoporous cellulose foams from cellulose-ionic liquid solutions. *Mater Lett* 2009;63(21):1851–1854.
85. Oshima T, Sakamoto T, Ohe K, Baba Y. Cellulose aerogel regenerated from ionic liquid solution for immobilized metal affinity adsorption. *Carbohydr Polym* 2014;103:62–69.
86. Zhang HZ, Zhang ZG, Liu KX. Regenerated lignocellulose beads prepared with wheat straw. *BioResources* 2016;11(2):4281–4294.
87. Zhou ZH, Zhang XX, Lu CH, Lan LD, Yuan GP. Polyaniline-decorated cellulose aerogel nanocomposite with strong interfacial adhesion and enhanced photocatalytic activity. *RSC Adv* 2014;4(18):8966–8972.
88. Pircher N, Fischhuber D, Carbajal L, Strauss C, Nedelec JM, Kasper C, Rosenau T, Liebner F. Preparation and reinforcement of dual-porous biocompatible cellulose scaffolds for tissue engineering. *Macromol Mater Eng* 2015;300(9):911–924.
89. Demilecamps A, Beauger C, Hildenbrand C, Rigacci A, Budtova T. Cellulose-silica aerogels. *Carbohydr Polym* 2015;122:293–300.
90. Pour G, Beauger C, Rigacci A, Budtova T. Xerocellulose: lightweight, porous and hydrophobic cellulose prepared via ambient drying. *J Mater Sci* 2015;50(13):4526–4535.
91. Rein DM, Cohen Y. Novel method for manufacturing of aerocellulose. In: DeWilde WP, Brebbia CA, Mander U, (Eds.) *High Performance Structures and Materials V*, vol. 1122010. pp. 63–68.
92. Wang HQ, Shao ZQ, Bacher M, Liebner F, Rosenau T. Fluorescent cellulose aerogels containing covalently immobilized (ZnS)(x)(CuInS2)(1-x)/ZnS (core/shell) quantum dots. *Cellulose* 2013;20(6):3007–3024.
93. Duchemin BJC, Staiger MP, Tucker N, Newman RH. Aerocellulose based on all-cellulose composites. *J Appl Polym Sci* 2010;115(1):216–221.
94. Chen MJ, Zhang XQ, Zhang AP, Liu CF, Sun RC. Direct preparation of green and renewable aerogel materials from crude bagasse. *Cellulose* 2016;23(2):1325–1334.
95. Wang ZG, Liu SL, Matsumoto Y, Kuga S. Cellulose gel and aerogel from LiCl/DMSO solution. *Cellulose* 2012;19(2):393–399.
96. Liebner F, Potthast A, Rosenau T, Haimer E, Wendland M. Ultralight-weight cellulose aerogels from NBnMO-stabilized lyocell dopes. *Research Letters in Materials Science* 2007; doi:10.1155/2007/73724.
97. Liebner F, Potthast A, Rosenau T, Haimer E, Wendland M. Cellulose aerogels: Highly porous, ultra-lightweight materials. *Holzforschung* 2008;62(2):129–135.

98. Liebner F, Haimer E, Potthast A, Loidl D, Tschegg S, Neouze MA, Wendland M, Rosenau T. Cellulosic aerogels as ultra-lightweight materials. Part 2: Synthesis and properties. *Holzforschung* 2009;63(1):3–11.
99. Litschauer M, Neouze MA, Haimer E, Henniges U, Potthast A, Rosenau T, Liebner F. Silica modified cellulosic aerogels. *Cellulose* 2011;18(1):143–149.
100. Laskowski J, Milow B, Ratke L. The effect of embedding highly insulating granular aerogel in cellulosic aerogel. *J Supercrit Fluids* 2015;106:93–99.
101. Karadagli I, Schulz B, Schestakow M, Milow B, Gries T, Ratke L. Production of porous cellulose aerogel fibers by an extrusion process. *J Supercrit Fluids* 2015;106: 105–114.
102. Ganesan K, Dennstedt A, Barowski A, Ratke L. Design of aerogels, cryogels and xerogels of cellulose with hierarchical porous structures. *Mater Des* 2016;92:345–355.
103. Schestakow M, Muench F, Reimuth C, Ratke L, Ensinger W. Electroless synthesis of cellulose-metal aerogel composites. *Appl Phys Lett* 2016;108(21):4.
104. Hoepfner S, Ratke L, Milow B. Synthesis and characterisation of nanofibrillar cellulose aerogels. *Cellulose* 2008;15(1):121–129.
105. Stefelova J, Slovak V. Reproducibility of preparation of cellulose-based cryogels characterised by TG-MS. *J Therm Anal Calorim* 2015;119(1):359–367.
106. Gavillon R, Budtova T. Kinetics of cellulose regeneration from cellulose–NaOH–water gels and comparison with cellulose–N-methylmorpholine-N-oxide–water solutions. *Biomacromolecules* 2007;8(2):424–432.
107. Biganska O, Navard P. Kinetics of precipitation of cellulose from cellulose–NMMO–water solutions. *Biomacromolecules* 2005;6(4):1948–1953.
108. Dash R, Li Y, Ragauskas AJ. Cellulose nanowhisker foams by freeze casting. *Carbohydr Polym* 2012;88(2):789–792.
109. Zhang W, Zhang Y, Lu CH, Deng YL. Aerogels from crosslinked cellulose nano/micro-fibrils and their fast shape recovery property in water. *J Mater Chem* 2012;22(23):11642–11650.
110. Siro I, Plackett D. Microfibrillated cellulose and new nanocomposite materials: A review. *Cellulose* 2010;17(3):459–494.
111. Kim C, Youn H, Lee H. Preparation of cross-linked cellulose nanofibril aerogel with water absorbency and shape recovery. *Cellulose* 2015;22(6):3715–3724.
112. Zhang F, Wu WB, Sharma S, Tong GL, Deng YL. Synthesis of cyclodextrin-functionalized cellulose nanofibril aerogel as a highly effective adsorbent for phenol pollutant removal. *BioResources* 2015;10(4):7555–7568.
113. Nemoto J, Saito T, Isogai A. Simple freeze-drying procedure for producing nanocellulose aerogel-containing, high-performance air filters. *ACS Appl Mater Interfaces* 2015;7(35):19809–19815.
114. Zhang Z, Sèbe G, Rentsch D, Zimmermann T, Tingaut P. Ultralightweight and flexible silylated nanocellulose sponges for the selective removal of oil from water. *Chem Mater* 2014;26(8):2659–2668.
115. Melone L, Bonafede S, Tushi D, Punta C, Cametti M. Dip in colorimetric fluoride sensing by a chemically engineered polymeric cellulose/bPEI conjugate in the solid state. *RSC Adv* 2015;5(101):83197–83205.
116. Melone L, Rossi B, Pastori N, Panzeri W, Mele A, Punta C. TEMPO-oxidized cellulose cross-linked with branched polyethyleneimine: Nanostructured adsorbent sponges for water remediation. *ChemPlusChem* 2015;80(9):1408–1415.
117. Chen WS, Li Q, Wang YC, Yi X, Zeng J, Yu HP, Liu Y, Li J. Comparative study of aerogels obtained from differently prepared nanocellulose fibers. *ChemSusChem* 2014;7(1):154–161.
118. Li W, Zhao X, Liu SX. Preparation of entangled nanocellulose fibers from APMP and its magnetic functional property as matrix. *Carbohydr Polym* 2013;94(1):278–285.

119. Mueller S, Sapkota J, Nicharat A, Zimmermann T, Tingaut P, Weder C, Faster EJ. Influence of the nanofiber dimensions on the properties of nanocellulose/poly(vinyl alcohol) aerogels. *J Appl Polym Sci* 2015;132(13):13.
120. Nystrom G, Marais A, Karabulut E, Wagberg L, Cui Y, Hamedi MM. Self-assembled three-dimensional and compressible interdigitated thin-film supercapacitors and batteries. *Nat Commun* 2015;6:7259.
121. Carlsson DO, Nystrom G, Zhou Q, Berglund LA, Nyholm L, Stromme M. Electroactive nanofibrillated cellulose aerogel composites with tunable structural and electrochemical properties. *J Mater Chem* 2012;22(36):19014–19024.
122. Jiang F, Hsieh YL. Amphiphilic superabsorbent cellulose nanofibril aerogels. *J Mater Chem A* 2014;2(18):6337–6342.
123. Lin JY, Yu LB, Tian F, Zhao N, Li XH, Bian FG, Wang J. Cellulose nanofibrils aerogels generated from jute fibers. *Carbohydr Polym* 2014;109:35–43.
124. Koga H, Azetsu A, Tokunaga E, Saito T, Isogai A, Kitaoka T. Topological loading of Cu(I) catalysts onto crystalline cellulose nanofibrils for the Huisgen click reaction. *J Mater Chem* 2012;22(12):5538–5542.
125. Maatar W, Boufi S. Poly(methacylic acid-co-maleic acid) grafted nanofibrillated cellulose as a reusable novel heavy metal ions adsorbent. *Carbohydr Polym* 2015;126:199–207.
126. Gebald C, Wurzbacher JA, Tingaut P, Zimmermann T, Steinfeld A. Amine-based nanofibrillated cellulose as adsorbent for CO_2 capture from air. *Environ Sci Technol* 2011;45(20):9101–9108.
127. Wong JCH, Kaymak H, Tingaut P, Brunner S, Koebel MM. Mechanical and thermal properties of nanofibrillated cellulose reinforced silica aerogel composites. *Microporous Mesoporous Mat* 2015;217:150–158.
128. Fumagalli M, Sanchez F, Molina-Boisseau S, Heux L. Surface-restricted modification of nanocellulose aerogels in gas-phase esterification by di-functional fatty acid reagents. *Cellulose* 2015;22(3):1451–1457.
129. Meng YJ, Young TM, Liu PZ, Contescu CI, Huang B, Wang SQ. Ultralight carbon aerogel from nanocellulose as a highly selective oil absorption material. *Cellulose* 2015;22(1):435–447.
130. Wan CC, Lu Y, Jiao Y, Jin CD, Sun QF, Li J. Ultralight and hydrophobic nanofibrillated cellulose aerogels from coconut shell with ultrastrong adsorption properties. *J Appl Polym Sci* 2015;132(24):7.
131. Wicklein B, Kocjan A, Salazar-Alvarez G, Carosio F, Camino G, Antonietti M, Bergström L. Thermally insulating and fire-retardant lightweight anisotropic foams based on nanocellulose and graphene oxide. *Nat Nanotechnol* 2015;10(3):277–283.
132. Sajab MS, Chia CH, Chan CH, Zakaria S, Kaco H, Chook SW, Chin SX. Bifunctional graphene oxide-cellulose nanofibril aerogel loaded with Fe(III) for the removal of cationic dye via simultaneous adsorption and Fenton oxidation. *RSC Adv* 2016;6(24):19819–19825.
133. Jiang F, Hsieh YL. Self-assembling of TEMPO oxidized cellulose nanofibrils as affected by protonation of surface carboxyls and drying methods. *ACS Sustain Chem Eng* 2016;4(3):1041–1049.
134. He ZY, Zhang XW, Batchelor W. Cellulose nanofibre aerogel filter with tuneable pore structure for oil/water separation and recovery. *RSC Adv* 2016;6(26):21435–21438.
135. Zhao SY, Zhang Z, Sebe G, Wu R, Virtudazo RVR, Tingaut P et al. Multiscale assembly of superinsulating silica aerogels within silylated nanocellulosic scaffolds: Improved mechanical properties promoted by nanoscale chemical compatibilization. *Adv Funct Mater* 2015;25(15):2326–2334.
136. Wicklein B, Kocjan D, Carosio F, Camino G, Bergstrom L. Tuning the nanocellulose-borate interaction to achieve highly flame retardant hybrid materials. *Chem Mat* 2016;28(7):1985–1989.

137. Zheng QF, Cai ZY, Gong SQ. Green synthesis of polyvinyl alcohol (PVA)-cellulose nanofibril (CNF) hybrid aerogels and their use as superabsorbents. *J Mater Chem A* 2014;2(9):3110–3118.
138. Chen WS, Yu HP, Li Q, Liu YX, Li J. Ultralight and highly flexible aerogels with long cellulose I nanofibers. *Soft Matter* 2011;7(21):10360–10368.
139. Sehaqui H, Zhou Q, Berglund LA. High-porosity aerogels of high specific surface area prepared from nanofibrillated cellulose (NFC). *Compos Sci Technol* 2011;71(13):1593–1599.
140. Hamedi M, Karabulut E, Marais A, Herland A, Nystrom G, Wagberg L. Nanocellulose aerogels functionalized by rapid layer-by-layer assembly for high charge storage and beyond. *Angew Chem Int Ed* 2013;52(46):12038–12042.
141. Ishida O, Kim DY, Kuga S, Nishiyama Y, Brown RM. Microfibrillar carbon from native cellulose. *Cellulose* 2004;11(3–4):475–480.
142. He X, Cheng L, Wang YR, Zhao JQ, Zhang W, Lu CH. Aerogels from quaternary ammonium-functionalized cellulose nanofibers for rapid removal of Cr(VI) from water. *Carbohydr Polym* 2014;111:683–687.
143. Arboleda JC, Hughes M, Lucia LA, Laine J, Ekman K, Rojas OJ. Soy protein-nanocellulose composite aerogels. *Cellulose* 2013;20(5):2417–2426.
144. Javadi A, Zheng QF, Payen F, Altin Y, Cai ZY, Sabo R et al. Polyvinyl alcohol-cellulose nanofibrils-graphene oxide hybrid organic aerogels. *ACS Appl Mater Interfaces* 2013;5(13):5969–5975.
145. Cervin NT, Aulin C, Larsson PT, Wagberg L. Ultra porous nanocellulose aerogels as separation medium for mixtures of oil/water liquids. *Cellulose* 2012;19(2):401–410.
146. Gao KZ, Shao ZQ, Li J, Wang X, Peng XQ, Wang WJ, Wang F. Cellulose nanofiber-graphene all solid-state flexible supercapacitors. *J Mater Chem A* 2013;1(1):63–67.
147. Fu JJ, Wang SQ, He CX, Lu ZX, Huang JD, Chen ZL. Facilitated fabrication of high strength silica aerogels using cellulose nanofibrils as scaffold. *Carbohydr Polym* 2016;147:89–96.
148. Valo H, Arola S, Laaksonen P, Torkkeli M, Peltonen L, Linder MB, Serimaa R, Kuga S, Hirvonen J, Laaksonen T. Drug release from nanoparticles embedded in four different nanofibrillar cellulose aerogels. *Eur J Pharm Sci* 2013;50(1):69–77.
149. Gao KZ, Shao ZQ, Wang X, Zhang YH, Wang WJ, Wang FJ. Cellulose nanofibers/multi-walled carbon nanotube nanohybrid aerogel for all-solid-state flexible supercapacitors. *RSC Adv* 2013;3(35):15058–15064.
150. Fumagalli M, Ouhab D, Boisseau SM, Heux L. Versatile gas-phase reactions for surface to bulk esterification of cellulose microfibrils aerogels. *Biomacromolecules* 2013;14(9):3246–3255.
151. Jiang F, Hsieh YL. Super water absorbing and shape memory nanocellulose aerogels from TEMPO-oxidized cellulose nanofibrils via cyclic freezing-thawing. *J Mater Chem A* 2014;2(2):350–359.
152. Bendahou D, Bendahou A, Seantier B, Grohens Y, Kaddami H. Nano-fibrillated cellulose-zeolites based new hybrid composites aerogels with super thermal insulating properties. *Ind Crop Prod* 2015;65:374–382.
153. Hayase G, Kanamori K, Abe K, Yano H, Maeno A, Kajo H et al. Polymethylsilsesquioxane-cellulose nanofiber biocomposite aerogels with high thermal insulation, bendability, and superhydrophobicity. *ACS Appl Mater Interfaces* 2014;6(12):9466–9471.
154. Zhao JQ, Lu CH, He X, Zhang XF, Zhang W, Zhang XM. Polyethylenimine-grafted cellulose nanofibril aerogels as versatile vehicles for drug delivery. *ACS Appl Mater Interfaces* 2015;7(4):2607–2615.
155. Aulin C, Netrval J, Wagberg L, Lindstrom T. Aerogels from nanofibrillated cellulose with tunable oleophobicity. *Soft Matter* 2010;6(14):3298–3305.

156. Cai HL, Sharma S, Liu WY, Mu W, Liu W, Zhang XD, Deng Y. Aerogel microspheres from natural cellulose nanofibrils and their application as cell culture scaffold. *Biomacromolecules* 2014;15(7):2540–2547.
157. Silva TCF, Habibi Y, Colodette JL, Elder T, Lucia LA. A fundamental investigation of the microarchitecture and mechanical properties of tempo-oxidized nanofibrillated cellulose (NFC)-based aerogels. *Cellulose* 2012;19(6):1945–1956.
158. Liu J, Cheng F, Grenman H, Spoljaric S, Seppala J, Eriksson JE, Willför S, Xu C. Development of nanocellulose scaffolds with tunable structures to support 3D cell culture. *Carbohydr Polym* 2016;148:259–271.
159. Naseri N, Poirier JM, Girandon L, Frohlich M, Oksman K, Mathew AP. 3-Dimensional porous nanocomposite scaffolds based on cellulose nanofibers for cartilage tissue engineering: Tailoring of porosity and mechanical performance. *RSC Adv* 2016;6(8):5999–6007.
160. Jin H, Kettunen M, Laiho A, Pynnonen H, Paltakari J, Marmur A, Ikkala O, Ras RH. Superhydrophobic and superoleophobic nanocellulose aerogel membranes as bioinspired cargo carriers on water and oil. *Langmuir* 2011;27(5):1930–1934.
161. Kettunen M, Silvennoinen RJ, Houbenov N, Nykanen A, Ruokolainen J, Sainio J, Pore V et al. Photoswitchable superabsorbency based on nanocellulose aerogels. *Adv Funct Mater* 2011;21(3):510–517.
162. Lee J, Deng YL. The morphology and mechanical properties of layer structured cellulose microfibril foams from ice-templating methods. *Soft Matter* 2011;7(13):6034–6040.
163. Korhonen JT, Hiekkataipale P, Malm J, Karppinen M, Ikkala O, Ras RHA. Inorganic hollow nanotube aerogels by atomic layer deposition onto native nanocellulose templates. *ACS Nano* 2011;5(3):1967–1974.
164. Sehaqui H, Salajkova M, Zhou Q, Berglund LA. Mechanical performance tailoring of tough ultra-high porosity foams prepared from cellulose I nanofiber suspensions. *Soft Matter* 2010;6(8):1824–1832.
165. Svagan AJ, Samir MASA, Berglund LA. Biomimetic foams of high mechanical performance based on nanostructured cell walls reinforced by native cellulose nanofibrils. *Adv Mater* 2008;20(7):1263–1269.
166. Svagan AJ, Berglund LA, Jensen P. Cellulose nanocomposite biopolymer foam-hierarchical structure effects on energy absorption. *ACS Appl Mater Interfaces* 2011;3(5):1411–1417.
167. Svagan AJ, Jensen P, Dvinskikh SV, Furo I, Berglund LA. Towards tailored hierarchical structures in cellulose nanocomposite biofoams prepared by freezing/freeze-drying. *J Mater Chem* 2010;20(32):6646–6654.
168. Wang M, Anoshkin IV, Nasibulin AG, Korhonen JT, Seitsonen J, Pere J, Kauppinen EI, Ras RH, Ikkala O. Modifying native nanocellulose aerogels with carbon nanotubes for mechanoresponsive conductivity and pressure sensing. *Adv Mater* 2013;25(17):2428–2432.
169. Xiao SL, Gao RA, Lu Y, Li J, Sun QF. Fabrication and characterization of nanofibrillated cellulose and its aerogels from natural pine needles. *Carbohydr Polym* 2015;119:202–209.
170. Saito T, Uematsu T, Kimura S, Enomae T, Isogai A. Self-aligned integration of native cellulose nanofibrils towards producing diverse bulk materials. *Soft Matter* 2011;7(19):8804–8809.
171. Mulyadi A, Zhang Z, Deng YL. Fluorine-free oil absorbents made from cellulose nanofibril aerogels. *ACS Appl Mater Interfaces* 2016;8(4):2732–2740.
172. Korhonen JT, Kettunen M, Ras RHA, Ikkala O. Hydrophobic nanocellulose aerogels as floating, sustainable, reusable, and recyclable oil absorbents. *ACS Appl Mater Interfaces* 2011;3(6):1813–1816.

173. Zhai TL, Zheng QF, Cai ZY, Xia HS, Gong SQ. Synthesis of polyvinyl alcohol/cellulose nanofibril hybrid aerogel microspheres and their use as oil/solvent superabsorbents. *Carbohydr Polym* 2016;148:300–308.
174. Chen WS, Zhang Q, Uetani K, Li Q, Lu P, Cao J, Wang Q et al. Sustainable carbon aerogels derived from nanofibrillated cellulose as high-performance absorption materials. *Adv Mater Interfaces* 2016;3(10):9.
175. Jiao Y, Wan CC, Qiang TG, Li J. Synthesis of superhydrophobic ultralight aerogels from nanofibrillated cellulose isolated from natural reed for high-performance adsorbents. *Appl Phys A-Mater Sci Process* 2016;122(7):10.
176. Zhai T, Zheng Q, Cai Z, Turng L-S, Xia H, Gong S. Poly(vinyl alcohol)/cellulose nanofibril hybrid aerogels with an aligned microtubular porous structure and their composites with polydimethylsiloxane. *ACS Appl Mater Interfaces* 2015;7(13):7436–7444.
177. Zhang F, Wu WB, Zhang XD, Meng XZ, Tong GL, Deng YL. Temperature-sensitive poly-NIPAm modified cellulose nanofibril cryogel microspheres for controlled drug release. *Cellulose* 2016;23(1):415–425.
178. Cai HL, Mu W, Liu W, Zhang XD, Deng YL. Sol-gel synthesis highly porous titanium dioxide microspheres with cellulose nanofibrils-based aerogel templates. *Inorg Chem Commun* 2015;51:71–74.
179. Zheng QF, Javadi A, Sabo R, Cai ZY, Gong SQ. Polyvinyl alcohol (PVA)-cellulose nanofibril (CNF)-multiwalled carbon nanotube (MWCNT) hybrid organic aerogels with superior mechanical properties. *RSC Adv* 2013;3(43):20816–20823.
180. Kobayashi Y, Saito T, Isogai A. Aerogels with 3D ordered nanofiber skeletons of liquid-crystalline nanocellulose derivatives as tough and transparent insulators. *Angew Chem Int Ed* 2014;53(39):10394–10397.
181. Panzella L, Melone L, Pezzella A, Rossi B, Pastori N, Perfetti M, D'Errico G, Punta C, d'Ischia M. Surface-functionalization of nanostructured cellulose aerogels by solid state eumelanin coating. *Biomacromolecules* 2016;17(2):564–571.
182. Dong H, Snyder JF, Tran DT, Leadore JL. Hydrogel, aerogel and film of cellulose nanofibrils functionalized with silver nanoparticles. *Carbohydr Polym* 2013;95(2):760–767.
183. Melone L, Altomare L, Alfieri I, Lorenzi A, De Nardo L, Punta C. Ceramic aerogels from TEMPO-oxidized cellulose nanofibre templates: Synthesis, characterization, and photocatalytic properties. *J Photochem Photobiol A* 2013;261:53–60.
184. Toivonen MS, Kaskela A, Rojas OJ, Kauppinen EI, Ikkala O. Ambient-dried cellulose nanofibril aerogel membranes with high tensile strength and their use for aerosol collection and templates for transparent, flexible devices. *Adv Funct Mater* 2015;25(42):6618–6626.
185. Zhao JQ, Zhang XF, He X, Xiao MJ, Zhang W, Lu CH. A super biosorbent from dendrimer poly(amidoamine)-grafted cellulose nanofibril aerogels for effective removal of Cr(VI). *J Mater Chem A* 2015;3(28):14703–14711.
186. Yang X, Shi KY, Zhitomirsky I, Cranston ED. Cellulose nanocrystal aerogels as universal 3D lightweight substrates for supercapacitor materials. *Adv Mater* 2015;27(40):6104–6109.
187. Zhou YM, Fu SY, Pu YQ, Pan SB, Levit MV, Ragauskas AJ. Freeze-casting of cellulose nanowhisker foams prepared from a water-dimethylsulfoxide (DMSO) binary mixture at low DMSO concentrations. *RSC Adv* 2013;3(42):19272–19277.
188. He F, Chao S, Gao Y, He XD, Li MW. Fabrication of hydrophobic silica-cellulose aerogels by using dimethyl sulfoxide (DMSO) as solvent. *Mater Lett* 2014;137:167–169.
189. Huang P, Fan MZ. Development of facture free clay-based aerogel: Formulation and architectural mechanisms. *Compos Pt B-Eng* 2016;91:169–175.
190. Fumagalli M, Sanchez F, Boisseau SM, Heux L. Gas-phase esterification of cellulose nanocrystal aerogels for colloidal dispersion in apolar solvents. *Soft Matter* 2013;9(47):11309–11317.

191. Yang X, Cranston ED. Chemically cross-linked cellulose nanocrystal aerogels with shape recovery and superabsorbent properties. *Chem Mat* 2014;26(20):6016–6025.
192. Berlioz S, Molina-Boisseau S, Nishiyama Y, Heux L. Gas-phase surface esterification of cellulose microfibrils and whiskers. *Biomacromolecules* 2009;10(8):2144–2151.
193. Gawryla MD, van den Berg O, Weder C, Schiraldi DA. Clay aerogel/cellulose whisker nanocomposites: A nanoscale wattle and daub. *J Mater Chem* 2009;19(15): 2118–2124.
194. Han JQ, Zhou CJ, Wu YQ, Liu FY, Wu QL. Self-assembling behavior of cellulose nanoparticles during freeze-drying: Effect of suspension concentration, particle size, crystal structure, and surface charge. *Biomacromolecules* 2013;14(5):1529–1540.
195. Surapolchai W, Schiraldi DA. The effects of physical and chemical interactions in the formation of cellulose aerogels. *Polym Bull* 2010;65(9):951–960.
196. Majoinen J, Walther A, McKee JR, Kontturi E, Aseyev V, Malho JM, Ruokolainen J, Ikkala O. Polyelectrolyte brushes grafted from cellulose nanocrystals using Cu-mediated surface-initiated controlled radical polymerization. *Biomacromolecules* 2011;12(8):2997–3006.
197. Heath L, Thielemans W. Cellulose nanowhisker aerogels. *Green Chem* 2010;12(8): 1448–1453.
198. Pircher N, Veigel S, Aigner N, Nedelec JM, Rosenau T, Liebner F. Reinforcement of bacterial cellulose aerogels with biocompatible polymers. *Carbohydr Polym* 2014;111:505–513.
199. Yin N, Chen S, Li Z, Ouyang Y, Hu W, Tang L, Zhang W et al. Porous bacterial cellulose prepared by a facile surfactant-assisted foaming method in azodicarbonamide-NaOH aqueous solution. *Mater Lett* 2012;81:131–134.
200. Thiruvengadam V, Vitta S. Ni-bacterial cellulose nanocomposite; a magnetically active inorganic-organic hybrid gel. *RSC Adv* 2013;3(31):12765–12773.
201. Wang YG, Yadav S, Heinlein T, Konjik V, Breitzke H, Buntkowsky G, Schneider JJ, Zhang K. Ultra-light nanocomposite aerogels of bacterial cellulose and reduced graphene oxide for specific absorption and separation of organic liquids. *RSC Adv* 2014;4(41):21553–21558.
202. Wu ZY, Li C, Liang HW, Chen JF, Yu SH. Ultralight, flexible, and fire-resistant carbon nanofiber aerogels from bacterial cellulose. *Angew Chem Int Ed* 2013;52(10):2925–2929.
203. Liang HW, Guan QF, Zhu Z, Song LT, Yao HB, Lei X, Yu SH. Highly conductive and stretchable conductors fabricated from bacterial cellulose. *NPG Asia Mater* 2012;4:6.
204. Jiang YT, Yan J, Wu XL, Shan DD, Zhou QH, Jiang LL, Yang D, Fan Z. Facile synthesis of carbon nanofibers-bridged porous carbon nanosheets for high-performance supercapacitors. *J Power Sources* 2016;307:190–198.
205. Leitch ME, Li CK, Ikkala O, Mauter MS, Lowry GV. Bacterial nanocellulose aerogel membranes: Novel high-porosity materials for membrane distillation. *Environ Sci Technol Lett* 2016;3(3):85–91.
206. Sai HZ, Xing L, Xiang JH, Cui LJ, Jiao JB, Zhao CL, Li Z, Li F, Zhang T. Flexible aerogels with interpenetrating network structure of bacterial cellulose-silica composite from sodium silicate precursor via freeze drying process. *RSC Adv* 2014;4(57):30453–30461.
207. Sai H, Xing L, Xiang J, Cui L, Jiao J, Zhao C, Li Z, Li F. Flexible aerogels based on an interpenetrating network of bacterial cellulose and silica by a non-supercritical drying process. *J Mater Chem A* 2013;1(27):7963–7970.
208. Luo HL, Ji DH, Li W, Xiao J, Li CZ, Xiong GY, Zhu Y, Wan Y. Constructing a highly bioactive 3D nanofibrous bioglass scaffold via bacterial cellulose-templated sol-gel approach. *Mater Chem Phys* 2016;176:1–5.
209. Liang HW, Wu ZY, Chen LF, Li C, Yu SH. Bacterial cellulose derived nitrogen-doped carbon nanofiber aerogel: An efficient metal-free oxygen reduction electrocatalyst for zinc-air battery. *Nano Energy* 2015;11:366–376.

210. Wu ZY, Liang HW, Li C, Hu BC, Xu XX, Wang Q, Chen JF, Yu SH. Dyeing bacterial cellulose pellicles for energetic heteroatom doped carbon nanofiber aerogels. *Nano Res* 2014;7(12):1861–1872.
211. Chen LF, Huang ZH, Liang HW, Guan QF, Yu SH. Bacterial-cellulose-derived carbon nanofiber@MnO_2 and nitrogen-doped carbon nanofiber electrode materials: An asymmetric supercapacitor with high energy and power density. *Adv Mater* 2013;25(34):4746–4752.
212. Chen LF, Huang ZH, Liang HW, Gao HL, Yu SH. Three-dimensional heteroatom-doped carbon nanofiber networks derived from bacterial cellulose for supercapacitors. *Adv Funct Mater* 2014;24(32):5104–5111.
213. Chen LF, Huang ZH, Liang HW, Yao WT, Yu ZY, Yu SH. Flexible all-solid-state high-power supercapacitor fabricated with nitrogen-doped carbon nanofiber electrode material derived from bacterial cellulose. *Energ Environ Sci* 2013;6(11):3331–3338.
214. Wang Y, Zou YC, Chen J, Li GD, Xu Y. A flexible and monolithic nanocomposite aerogel of carbon nanofibers and crystalline titania: fabrication and applications. *RSC Adv* 2013;3(46):24163–24168.
215. Olsson RT, Samir M, Salazar-Alvarez G, Belova L, Strom V, Berglund LA, Ikkala O, Nogues J, Gedde UW. Making flexible magnetic aerogels and stiff magnetic nanopaper using cellulose nanofibrils as templates. *Nat Nanotechnol* 2010;5(8):584–588.
216. Menchaca-Nal S, Londono-Calderon CL, Cerrutti P, Foresti ML, Pampillo L, Bilovol V, Candal R, Martínez-García R. Facile synthesis of cobalt ferrite nanotubes using bacterial nanocellulose as template. *Carbohydr Polym* 2016;137:726–731.
217. Wang PP, Zhao J, Xuan RF, Wang Y, Zou C, Zhang ZQ, Wan Y, Xu Y. Flexible and monolithic zinc oxide bionanocomposite foams by a bacterial cellulose mediated approach for antibacterial applications. *Dalton Trans* 2014;43(18):6762–6768.
218. Wang B, Li XL, Luo B, Yang JX, Wang XJ, Song Q, Chen S, Zhi L. Pyrolyzed bacterial cellulose: A versatile support for lithium ion battery anode materials. *Small* 2013;9(14):2399–2404.
219. Wang LP, Schutz C, Salazar-Alvarez G, Titirici MM. Carbon aerogels from bacterial nanocellulose as anodes for lithium ion batteries. *RSC Adv* 2014;4(34):17549–17554.
220. Yu WD, Lin WR, Shao XF, Hu ZX, Li RC, Yuan DS. High performance supercapacitor based on Ni_3S_2/carbon nanofibers and carbon nanofibers electrodes derived from bacterial cellulose. *J Power Sources* 2014;272:137–143.
221. Hu ZX, Li SS, Cheng PP, Yu WD, Li RC, Shao XF, Lin W, Yuan D. N, P-co-doped carbon nanowires prepared from bacterial cellulose for supercapacitor. *J Mater Sci* 2016;51(5):2627–2633.
222. Long CL, Qi DP, Wei T, Yan J, Jiang LL, Fan ZJ. Nitrogen-doped carbon networks for high energy density supercapacitors derived from polyaniline coated bacterial cellulose. *Adv Funct Mater* 2014;24(25):3953–3961.
223. Haimer E, Wendland M, Schlufter K, Frankenfeld K, Miethe P, Potthast A, Rosenau T, Liebner F. Loading of bacterial cellulose aerogels with bioactive compounds by antisolvent precipitation with supercritical carbon dioxide. *Macromolecular Symposia* 2010;294:64–74.
224. Huang Y, Lin ZX, Zheng MB, Wang TH, Yang JZ, Yuan FS, Lu X, Liu L, Sun D. Amorphous Fe_2O_3 nanoshells coated on carbonized bacterial cellulose nanofibers as a flexible anode for high-performance lithium ion batteries. *J Power Sources* 2016;307:649–656.
225. Huang Y, Zheng M, Lin Z, Zhao B, Zhang S, Yang J, Zhu C, Zhang H, Sun D, Shi Y. Flexible cathodes and multifunctional interlayers based on carbonized bacterial cellulose for high-performance lithium-sulfur batteries. *J Mater Chem A* 2015;3(20):10910–10918.

226. Russler A, Wieland M, Bacher M, Henniges U, Miethe P, Liebner F, Potthast A, Rosenau T. AKD-Modification of bacterial cellulose aerogels in supercritical CO_2. *Cellulose* 2012;19(4):1337–1349.
227. Li LM, Zhou Y, Li ZQ, Ma YJ, Pei CH. One step fabrication of Mn_3O_4/carbonated bacterial cellulose with excellent catalytic performance upon ammonium perchlorate decomposition. *Mater Res Bull* 2014;60:802–807.
228. Tian LM, Luan JY, Liu KK, Jiang QS, Tadepalli S, Gupta MK, Naik RR, Singamaneni S. Plasmonic biofoam: A versatile optically active material. *Nano Lett* 2016;16(1):609–616.
229. Liebner F, Haimer E, Wendland M, Neouze MA, Schlufter K, Miethe P,, Heinze T, Potthast A, Rosenau T. Aerogels from unaltered bacterial cellulose: Application of scCO(2) drying for the preparation of shaped, ultra-lightweight cellulosic aerogels. *Macromol Biosci* 2010;10(4):349–352.
230. Thiruvengadam V, Vitta S. Interparticle interactions mediated superspin glass to superferromagnetic transition in Ni-bacterial cellulose aerogel nanocomposites. *J Appl Phys* 2016;119(24):12.
231. Sai HZ, Fu R, Xing L, Xiang JH, Li ZY, Li F, Zhang T. Surface modification of bacterial cellulose aerogels' web-like skeleton for oil/water separation. *ACS Appl Mater Interfaces* 2015;7(13):7373–7381.
232. Ye TN, Lv LB, Li XH, Xu M, Chen JS. Strongly veined carbon nanoleaves as a highly efficient metal-free electrocatalyst. *Angew Chem Int Ed* 2014;53(27):6905–6909.
233. Meng FL, Li L, Wu Z, Zhong HX, Li JC, Yan JM. Facile preparation of N-doped carbon nanofiber aerogels from bacterial cellulose as an efficient oxygen reduction reaction electrocatalyst. *Chin J Catal* 2014;35(6):877–883.
234. Xu XZ, Zhou J, Nagaraju DH, Jiang L, Marinov VR, Lubineau G et al. Flexible, highly graphitized carbon aerogels based on bacterial cellulose/lignin: Catalyst-free synthesis and its application in energy storage devices. *Adv Funct Mater* 2015;25(21):3193–3202.
235. Srinivasa P, Kulachenko A, Aulin C. Experimental characterisation of nanofibrillated cellulose foams. *Cellulose* 2015;22(6):3739–3753.
236. Ali ZM, Gibson LJ. The structure and mechanics of nanofibrillar cellulose foams. *Soft Matter* 2013;9(5):1580–1588.
237. Wang CH, Li YB, He XD, Ding YJ, Peng QY, Zhao WQ, Shi E, Wu S, Cao A. Cotton-derived bulk and fiber aerogels grafted with nitrogen-doped graphene. *Nanoscale* 2015;7(17):7550–7558.
238. Wang L, Sanchez-Soto M. Green bio-based aerogels prepared from recycled cellulose fiber suspensions. *RSC Adv* 2015;5(40):31384–31391.
239. Han SJ, Sun QF, Zheng HH, Li JP, Jin CD. Green and facile fabrication of carbon aerogels from cellulose-based waste newspaper for solving organic pollution. *Carbohydr Polym* 2016;136:95–100.
240. Bi HC, Huang X, Wu X, Cao XH, Tan CL, Yin ZY et al. Carbon microbelt aerogel prepared by waste paper: An efficient and recyclable sorbent for oils and organic solvents. *Small* 2014;10(17):3544–3550.
241. Seantier B, Bendahou D, Bendahou A, Grohens Y, Kaddami H. Multi-scale cellulose based new bio-aerogel composites with thermal super-insulating and tunable mechanical properties. *Carbohydr Polym* 2016;138:335–348.
242. Feng JD, Nguyen ST, Fan Z, Duong HM. Advanced fabrication and oil absorption properties of super-hydrophobic recycled cellulose aerogels. *Chem Eng J* 2015;270:168–175.
243. Tejado A, Chen WC, Alam MN, van de Ven TGM. Superhydrophobic foam-like cellulose made of hydrophobized cellulose fibres. *Cellulose* 2014;21(3):1735–1743.
244. Zhang JP, Li BC, Li LX, Wang AQ. Ultralight, compressible and multifunctional carbon aerogels based on natural tubular cellulose. *J Mater Chem A* 2016;4(6): 2069–2074.

245. Madani A, Zeinoddini S, Varahmi S, Turnbull H, Phillion AB, Olson JA, Martinez DM. Ultra-lightweight paper foams: Processing and properties. *Cellulose* 2014;21(3):2023–2031.
246. Liu L, Xie JP, Li YJ, Zhang Q, Yao JM. Three-dimensional macroporous cellulose-based bioadsorbents for efficient removal of nickel ions from aqueous solution. *Cellulose* 2016;23(1):723–736.
247. Liebner F, Dunareanu R, Opietnik M, Haimer E, Wendland M, Werner C, Maitz M et al. Shaped hemocompatible aerogels from cellulose phosphates: Preparation and properties. *Holzforschung* 2012;66(3lei):317–321.
248. Granstrom M, Paakko MKN, Jin H, Kolehmainen E, Kilpelainen I, Ikkala O. Highly water repellent aerogels based on cellulose stearoyl esters. *Polym Chem* 2011;2(8):1789–1796.
249. Pinnow M, Fink H-P, Fanter C, Kinize J. Characterization of highly porous materials from cellulose carbamate. *Macromolecular Symposia* 2008;262:129–139.
250. Fischer F, Rigacci A, Pirard R, Berthon-Fabry S, Achard P. Cellulose-based aerogels. *Polymer* 2006;47(22):7636–7645.
251. Grzyb B, Hildenbrand C, Berthon-Fabry S, Begin D, Job N, Rigacci A, Achard P. Functionalisation and chemical characterisation of cellulose-derived carbon aerogels. *Carbon* 2010;48(8):2297–2307.
252. Luong ND, Lee Y, Nam JD. Facile transformation of nanofibrillar polymer aerogel to carbon nanorods catalyzed by platinum nanoparticles. *J Mater Chem* 2008;18(36):4254–4259.
253. Luong ND, Lee Y, Nam JD. Highly-loaded silver nanoparticles in ultrafine cellulose acetate nanofibrillar aerogel. *Eur Polym J* 2008;44(10):3116–3121.
254. Guilminot E, Fischer F, Chatenet M, Rigacci A, Berthon-Fabry S, Achard P, Chainet E. Use of cellulose-based carbon aerogels as catalyst support for PEM fuel cell electrodes: Electrochemical characterization. *J Power Sources* 2007;166(1):104–111.
255. Jin MM, Luo N, Li GP, Luo YJ. The thermal decomposition mechanism of nitrocellulose aerogel. *J Therm Anal Calorim* 2015;121(2):901–908.
256. Patel AR, Schatteman D, Lesaffer A, Dewettinck K. A foam-templated approach for fabricating organogels using a water-soluble polymer. *RSC Adv* 2013;3(45):22900–22903.
257. Lin RJ, Li A, Lu LB, Cao Y. Preparation of bulk sodium carboxymethyl cellulose aerogels with tunable morphology. *Carbohydr Polym* 2015;118:126–132.
258. He F, Sui C, He XD, Li MW. Facile synthesis of strong alumina-cellulose aerogels by a freeze-drying method. *Mater Lett* 2015;152:9–12.
259. Lu TH, Li Q, Chen WS, Yu HP. Composite aerogels based on dialdehyde nanocellulose and collagen for potential applications as wound dressing and tissue engineering scaffold. *Compos Sci Technol* 2014;94:132–138.
260. Zheng QF, Cai ZY, Ma ZQ, Gong SQ. Cellulose nanofibril/reduced graphene oxide/carbon nanotube hybrid aerogels for highly flexible and all-solid-state supercapacitors. *ACS Appl Mater Interfaces* 2015;7(5):3263–3271.
261. Yao XL, Yu WJ, Xu X, Chen F, Fu Q. Amphiphilic, ultralight, and multifunctional graphene/nanofibrillated cellulose aerogel achieved by cation-induced gelation and chemical reduction. *Nanoscale* 2015;7(9):3959–3964.
262. Smith SW, Buesch C, Matthews DJ, Simonsen J, Conley JF. Improved oxidation resistance of organic/inorganic composite atomic layer deposition coated cellulose nanocrystal aerogels. *J Vac Sci Technol A* 2014;32(4):041508.
263. Baillis D, Coquard R, Moura LM. Heat transfer in cellulose-based aerogels: Analytical modelling and measurements. *Energy* 2015;84:732–744.
264. Wang ZP, Ma YJ, He HL, Pei CH, He P. A novel reusable nanocomposite: FeOOH/CBC and its adsorptive property for methyl orange. *Appl Surf Sci* 2015;332:456–462.

265. Wang CH, Ding YJ, Yuan Y, He XD, Wu ST, Hu S, Zou M et al. Graphene aerogel composites derived from recycled cigarette filters for electromagnetic wave absorption. *J Mater Chem C* 2015;3(45):11893–11901.
266. Alatalo SM, Pileidis F, Makila E, Sevilla M, Repo E, Salonen J, Sillanpää M, Titirici MM. Versatile cellulose-based carbon aerogel for the removal of both cationic and anionic metal contaminants from water. *ACS Appl Mater Interfaces* 2015;7(46):25875–25883.
267. Alatalo SM, Qiu KP, Preuss K, Marinovic A, Sevilla M, Sillanpaa M, Guo X, Titirici MM. Soy protein directed hydrothermal synthesis of porous carbon aerogels for electrocatalytic oxygen reduction. *Carbon* 2016;96:622–630.
268. Li YQ, Samad YA, Polychronopoulou K, Liao K. Lightweight and highly conductive aerogel-like carbon from sugarcane with superior mechanical and EMI shielding properties. *ACS Sustain Chem Eng* 2015;3(7):1419–1427.
269. Li Y-Q, Samad YA, Polychronopoulou K, Alhassan SM, Liao K. Carbon aerogel from winter melon for highly efficient and recyclable oils and organic solvents absorption. *ACS Sustain Chem Eng* 2014;2(6):1492–1497.
270. Men Y, Siebenbuerger M, Qiu X, Antonietti M, Yuan J. Low fractions of ionic liquid or poly(ionic liquid) can activate polysaccharide biomass into shaped, flexible and fire-retardant porous carbons. *J Mater Chem A* 2013;1(38):11887–11893.
271. Bi HC, Yin ZY, Cao XH, Xie X, Tan CL, Huang X, Chen B et al. Carbon fiber aerogel made from raw cotton: A novel, efficient and recyclable sorbent for oils and organic solvents. *Adv Mater* 2013;25(41):5916–5921.
272. Falco C, Baccile N, Titirici MM. Morphological and structural differences between glucose, cellulose and lignocellulosic biomass derived hydrothermal carbons. *Green Chem* 2011;13(11):3273–3281.

12 Production of Cellulose Nanocrystals at InnoTech Alberta

Tri-Dung Ngo, Christophe Danumah and Behzad Ahvazi

CONTENTS

12.1 Introduction 269
12.2 From Cellulose to Cellulose Nanocrystals 270
12.3 Growing Market for Cellulose Nanocrystals 271
12.4 A Versatile Range of Applications 272
12.5 InnoTech Alberta's Cellulose Nanocrystals Pilot Plant 273
12.5.1 Feedstock for Cellulose Nanocrystals Production 275
12.5.2 Acid Hydrolysis Reactions 277
12.5.3 Centrifuge 278
12.5.4 Purification—Ultrafiltration 279
12.5.5 Spray Dryer 280
12.5.6 Storage and Shelf-Life 282
12.6 Characterisation of Cellulose Nanocrystals Materials 282
12.7 Remarks 286
Acknowledgements 286
References 286

12.1 INTRODUCTION

In recent decades, the conversion of renewable lignocellulosic biomass and natural biopolymers into chemicals, liquid fuels and feed supplements has gained considerable attention. This has been driven by the instability of petroleum prices and the high-energy intensity involved in the production of chemicals and synthetic polymers [1–3]. With appropriate conversion and extraction technologies, including modification and characterisation, biopolymers such as cellulose nanocrystals (CNCs) can be successfully integrated into biobased products. Their use in novel materials and various applications suggests a future of cellulose and cellulose-based biomass components offering substantial environmental and economic benefits [4–7].

InnoTech Alberta plays a strategic role in bridging Alberta's industries by transforming biomass, including wood and crop waste, into chemicals and materials, with real-world commercial applications. The goal is to grow Alberta's economy by pursuing higher-value opportunities for the province's natural resources, while decreasing our environmental footprint. Dedicated research scientists at InnoTech Alberta are experimenting with CNCs produced from residual crop and wood fibre to find new environmentally sustainable products. This could help diversify and strengthen Canada's agriculture and forestry industries, paving the way for novel, biobased, 'greener' materials for use in the energy sector and many other industries in Alberta.

In the fall of 2013, Alberta's one-of-a-kind CNC pilot plant was commissioned at InnoTech Alberta's Edmonton facility. With the capacity to produce up to 100 kg of CNC per week, the plant is helping to advance new CNC-based products that have the potential to further diversify and increase the global competitiveness of Alberta's economy.

The production of pilot-scale samples of CNC represents a critical step in the introduction of the material to industrial markets and in providing a platform for the development of novel, high-value applications. The $6.7 million CAD pilot plant is the product of a collaboration of the Governments of Canada and Alberta, with funding from Western Economic Diversification Canada and Alberta Enterprise and Advanced Education and financial contributions and 'in kind support' from Alberta-Pacific Forest Industries Inc. (Al-Pac). The plant is part of InnoTech Alberta's mandate to provide technical services and funding support to facilitate the commercialisation of technologies, to develop new knowledge-based industry clusters and to foster the entrepreneurial culture in Alberta. Two objectives motivated this significant investment: first, to produce CNC from various forestry and agricultural feedstock, with an eye to improve consistency, yield, cost efficiency and chemical recovery from the current approaches, and, second, to develop an Alberta-based and industry-focussed, CNC applications' development program for CNC.

12.2 FROM CELLULOSE TO CELLULOSE NANOCRYSTALS

Cellulose is the most important and naturally abundant organic biopolymer in the biosphere. It is the basic structural component of plant cell walls. As such, it is distributed in all plants, from highly developed trees to primitive organisms such as seaweeds, flagellates and bacteria. Cellulose is the main constituent of wood (40%–45%), cotton fibre (90%) and dried hemp (45%) [8,9]. It is composed of carbon, hydrogen and oxygen bioengineered as a linear homopolysaccharide chain, with cellobiose as the repeating building block. Each block is composed of two adjacent anhydro-glucose rings, connected via a β-1,4 glycosidic linkage, as illustrated in Figure 12.1.

The number of glucose units in cellulose varies from 7,000 to 10,000 [10]. Cellulose molecules are linked laterally into linear bundles by hydrogen bonds. The extremely

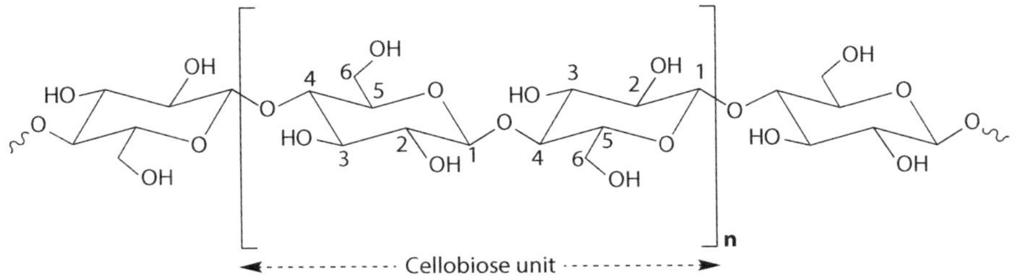

FIGURE 12.1 Schematic diagram of partial molecular structure of a cellulose chain with numbering for carbon atoms and n = number of cellobiose repeating unit.

large number of hydrogen bonds results in a strong lateral association of linear cellulose molecules. This strong association and almost-perfect alignment of cellulose molecules give rise to their crystallinity. The degree of crystallinity has a great influence on a material's hardness, density, transparency and diffusion. X-ray measurements have shown that the crystalline regions of cellulose are interrupted every 600 Å with non-crystalline or 'amorphous' regions. In these amorphous regions, molecules are arranged in an irregular and non-periodic manner. Therefore, the cellulose molecules can be considered highly oriented (crystalline) for a distance of about 600 Å. Then, they pass through an area of poor orientation (amorphous) and re-enter a crystalline region. This pattern repeats throughout the length of the cellulose molecule.

Cellulose is odourless, flavourless, hydrophilic and insoluble in most solvents, including strong alkali. It is biodegradable and can be broken down chemically, amongst other techniques, into its monomeric building blocks to produce biofuels. By treating cellulose with concentrated acids (a process known as acid hydrolysis), the amorphous regions can be broken up, thereby isolating the nanosized cellulose-based crystals or CNCs. The CNCs (also known as nanocrystalline cellulose or NCC, or cellulose whiskers) are elongated, rigid, rod-like or whisker-shaped particles with a rectangular cross-section. The CNCs can be prepared from any cellulose source material, including wood pulp, recycled paper and paperboard, cotton fibres, hemp, flax and other agro-biomass [11].

12.3 GROWING MARKET FOR CELLULOSE NANOCRYSTALS

Nanotechnology is a rapidly evolving area of development, as science, engineering and technology have merged to bring nanoscale materials much closer to reality. As an emerging technology, CNC has gained considerable interest from those seeking increased use of biobased materials, leading many countries such as Israel, Germany, France, Sweden, Finland, Switzerland, Norway, Iran and Japan to build various-sized pilot or demonstration plants. However, the most well-known CNC research and development centres, along with their various industrial partners and development clusters, are located in the United States and Canada. CelluForce Inc. is the only demonstration-scale pilot plant currently operating in Canada. This facility, built as a joint venture between FPInnovations and Domtar, can produce up to one tonne per day of CNC. Canadian CNC units are pilot plants located at FPInnovations (Pointe-Claire, Québec) and InnoTech Alberta (Edmonton), with production capacities of 10 kg/week and up to 100 kg/week, respectively [12]. As CNC becomes more widely known, there has been increasing interest from different corporations, governments, research organisations, universities and the public to use CNC in various applications.

12.4 A VERSATILE RANGE OF APPLICATIONS

Cellulose nanocrystals are considered to have many desirable industrial and commercial properties. They have a high aspect ratio (length-to-width ratio), with typical lateral dimensions of 100–200 nm and longitudinal dimensions of 5–20 nm. Their dimensions vary depending on the native cellulose source, extraction methods and recovery processes (e.g. hydrolysis time and temperature). Cellulose nanocrystals offer high tensile strength and can be compared to Kevlar fibre in stiffness, making it an excellent material for reinforcement of natural or synthetic matrix polymers [13,14]. Recently, CNC was used as the filler phase in polymer matrices to produce a biobased nanocomposite with superior thermal and mechanical properties [6,7].

Cellulose nanocrystals have unique rheological properties. Aqueous dispersions of CNC, even at low concentrations (8–10 wt.%), generate gels, making CNC a great ingredient as a non-caloric stabiliser for food industry. In addition, CNCs are highly shear-thinning, a property particularly important in different coating applications; this makes CNC an ideal value-added biomaterial for use in stimulation, fracturing and completion fluids [15–18]. Cellulose nanocrystals have low density and exhibit properties that have the potential to be altered or controlled [19]. Cellulose nanocrystals, when used in composite films, is considered to be gas-impermeable and has been suggested as a barrier material for perishable foods and materials sensitive to air and oxidation [20]. Recently, InnoTech Alberta was awarded U.S. patent number 8,105,430 [21], which describes the use of CNC as a water-based viscosity control, specifically as an additive for anti-icing fluids. In oil and gas sector applications, CNC could improve drilling fluid, fracturing fluid and enhanced oil recovery [15–18]. Demonstrating its ongoing potential, CNC is currently being exploited beyond the original patent claims, and a continuation patent may be filed.

The expanding interest in CNC applications from a number of major industrial sectors is the driving force needed to facilitate CNC production at a large scale. These sectors include packaging, aerospace, automotive, coating and consumer goods such as electronics and appliances. Other sectors such as energy, paper and paperboard, hygiene and sorbents, medical industry and cosmetic industry stand to benefit from CNC [22]. Cellulose nanocrystals, with their unique physical and chemical properties, will directly impact and improve the quality and performance of end use, biobased products. Research scientists at InnoTech Alberta have developed methods and processes for the production of CNCs with controlled morphology, structure and properties. The chemical and physical properties of CNCs produced from several feedstock and production lines at InnoTech Alberta's pilot plant are fully characterised to reveal their complete profiles to inform specific applications' development. Ultrastructure, morphology, chemical composition and purity are constantly tested by using the state-of-the-art instrumentations located at InnoTech Alberta and at partnering facilities. Meanwhile, national efforts to develop experimental data and standardisation terms, including lifecycle analysis to better understand the impact of CNCs on human health and the environment, are underway. Alberta is uniquely positioned to build the CNC value chain because of its distinguished nanotechnology research institutes, well-known forest products' companies and proximity to end-use markets such as the energy and chemical sectors. Alberta's major advantage in CNC and CNC-based product development is the innovation capabilities of research centred at the National Institute of Nanotechnology and the University of Alberta, Edmonton, Canada, coupled with InnoTech Alberta's research network partners. Collectively, they have joined forces to help transform Alberta's forest products' industry by providing a new value proposition to enhance manufacturing, processing and design technologies. Consequently, this will accelerate the production and implementation of 'smart' biobased materials, engineered intrinsically and functionalised with specific properties for advanced applications.

12.5 INNOTECH ALBERTA'S CELLULOSE NANOCRYSTALS PILOT PLANT

The pilot capacity at InnoTech Alberta is helping foster and advance new CNC-based products that have the potential to further diversify and increase the global competitiveness of Alberta's economy. The overview of InnoTech Alberta's pilot plant is shown in Figure 12.2. The production of pilot-scale CNC samples represents a critical step in providing a development platform for high-value CNC applications

FIGURE 12.2 Overview of the InnoTech Alberta CNC pilot plant.

and in introducing this novel material to industrial markets. Key process optimisation factors remain at the pilot plant. We are constantly driving to increase yields and improve the quality, while maintaining consistent production, all of which decrease production costs and open up opportunities for CNCs in new markets. The goal is to produce cost-effective, sustainable, high-quality CNCs from Alberta feedstock, complete with an extensive CNC sample inventory for application developers, supported by data informed by material characterisation and standardisation.

The flowchart of major CNC production activities at InnoTech Alberta's pilot plant is presented in Figure 12.3. Before the construction of any large-scale CNC production, essential production factors (e.g. moisture content of the feedstock, feedstock type and quality, acid concentration, reaction time and temperature, acid-to-pulp ratio and mixing conditions) need to be tightly controlled. Optimisation of CNC production processes and the resulting reduction in CNC production costs are essential to CNC's incorporation into commercial products. Increasing CNC yield and monitoring the amount of unreacted cellulosic material (not converted to CNC)

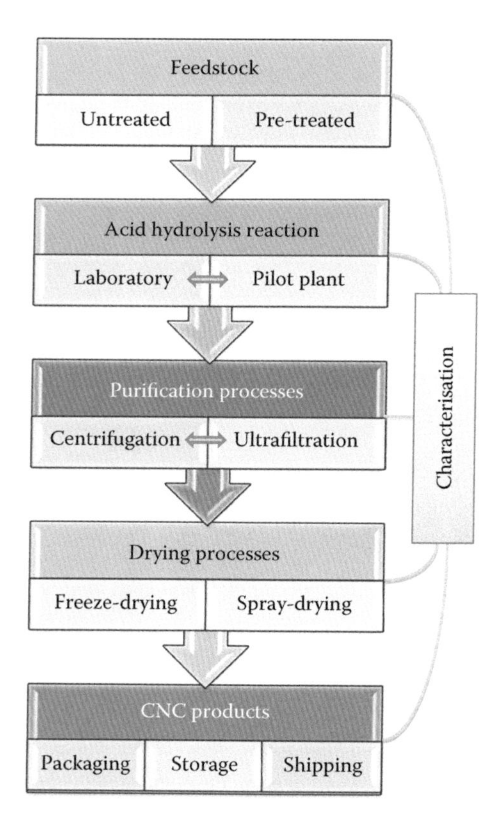

FIGURE 12.3 Flowchart for the major CNC pilot plant activities.

during the acid hydrolysis are also critical to process economics at demonstration and commercial scales. Unreacted cellulosic material, also known as 'CNC rejects', can be post-processed to generate other value-added nanocelluloses. The recovery of excess spent acid remaining after the quenching of the acid hydrolysis reaction, the extraction of sugars from the recycling of acid and even different morphologies created by alternative approaches to drying, all are being investigated.

The overall objective of InnoTech Alberta's CNC pilot plant production is the development of strategies that allow the optimisation, at large scale, of CNC yield, consistency, purity and quality, all while respecting the needs of economics. In a dedicated effort to achieve these strategies, any important process parameter change is first validated at the laboratory scale and then implemented in the pilot plant with economic simulation modelling completed, in order to track changes in resultant production costs. Maintaining quality control and characterisation throughout the production at each step of the pilot plant is paramount.

Overcoming technical challenges and pursuing optimisation are fundamental characteristics of operating any pilot plant. Not surprisingly, InnoTech Alberta's pilot plant efforts include (1) constant tuning and troubleshooting of different pilot plant units; (2) the maintenance of tight control on feedstock, reaction conditions and the final CNC product to achieve homogeneous, clean, high-yield and cost-effective CNC; (3) the development of standard test methods for characterisation and standardisation and (4) active applications' development.

12.5.1 Feedstock for Cellulose Nanocrystals Production

In general, CNC can be produced from any cellulosic-based material. This includes wood pulp, recycled paper and paperboard, cotton fibres, hemp, flax, bamboo, sugarcane bagasse and other agro-biomass. The conversion of cellulosic-based material to CNC can be a chemical (acid hydrolysis or oxidation) or biological (enzymatic) treatment. The objective of the conversion is to cleave the amorphous regions of the cellulose fibres to isolate the nanocrystals.

As mentioned earlier, different feedstocks can be utilised to produce CNC. At InnoTech Alberta's pilot plant, several different types of wood pulp have been used to produce CNC, including Neucel Specialty's dissolving pulp received from Cellulose Ltd. British Colombia (BC) and fully bleached unpressed northern bleached softwood kraft (NBSK) and northern bleached hardwood kraft (NBHK) pulp grades received from Al-Pac. The elucidation of any cellulosic feedstock profiles is crucial for effective and efficient CNC production. The CNC yield is very dependent on the α-cellulose content and purity of the cellulosic feedstock. Because of their high α-cellulose content, the dissolving pulp and a new generation of chemical treated pulp developed at InnoTech Alberta have consistently shown higher CNC production yields during laboratory and large-scale pilot plant trials. The resulting CNC materials appear to be cleaner, more homogeneous and brighter than other feedstock derivatives. Examples of the feedstock forms and their scanning electron microscopy (SEM) images are shown in Figures 12.4 and 12.5, respectively.

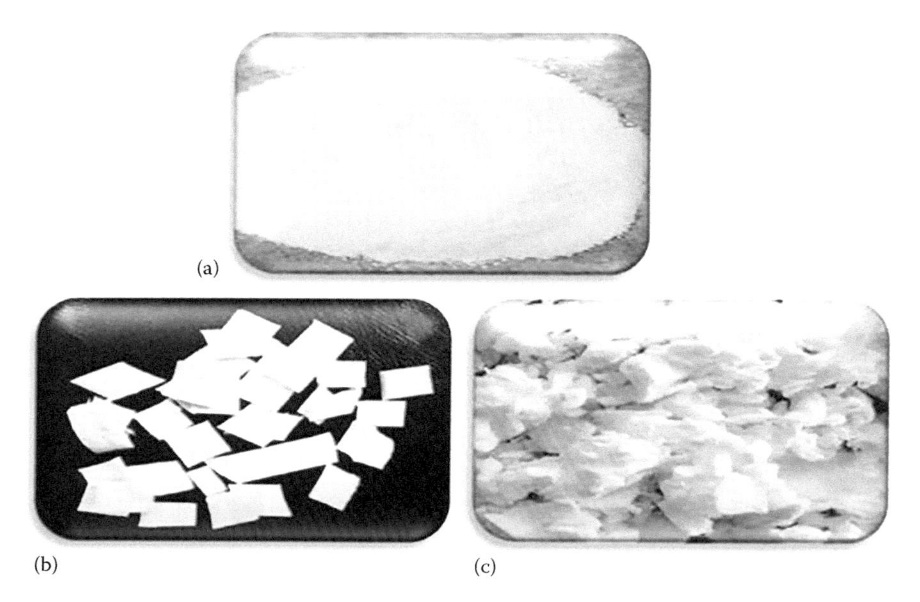

FIGURE 12.4 Types of cellulosic feedstocks used in CNC production: (a) microcrystalline cellulose powder (MCC); (b) pierret chopped beached pulp and (c) fluffed bleached pulp.

FIGURE 12.5 Scanning electron microscopic (SEM) images of cellulosic feedstocks used in CNC production: (a) microcrystalline cellulose powder (MCC); (b) pierret chopped bleached pulp and (c) fluffed bleached pulp.

12.5.2 Acid Hydrolysis Reactions

Acid hydrolysis is a process related to the breakage of dominantly ether bonds in the cellulose. Acid hydrolysis, using concentrated acid, is the predominant and most efficient chemical method to produce CNC, with minimal energy consumption. Commonly, hydrolysis is performed in the presence of mineral acids (e.g. 50–70 wt.% H_2SO_4) to break β-1,4 glycosidic bonds and depolymerise cellulose. The low-density amorphous regions present in native cellulose are more accessible to acid and more susceptible to hydrolytic action than the crystalline domains, and therefore, these amorphous regions of cellulose will break down and release individual nanocellulose crystallites when subjected to acid treatment. Cellulose nanocrystals produced by sulphuric acid hydrolysis are electrostatically stabilised in aqueous suspension with the introduction of charged sulphate ester groups onto the crystallites' surface during reaction. A particular CNC's sulphate group content and the overall surface charges may hinder the desirable formulation of CNC-based biomaterials. Therefore, a thorough understanding of sulphur content is crucial to understanding material properties.

In InnoTech Alberta's pilot plant, hydrolysis reactions occur in a Pfaudler 50-gallon glass-lined reactor, which is shown in Figure 12.6. Typically, 110–155 kg of 63.5 or 64 wt.% H_2SO_4 is pumped into the reactor from the acid storage tank. The acid is then stirred at 200 rpm and heated up to ~45°C through the reactor jacket with a low-pressure steam. Thereafter, 10–13.5 kg of cellulosic feedstock is added into reactor and mixed at 200 rpm. The reaction persists for 2 h at a mixing rate of 200 rpm. After 2 h of reaction, 50 kg of water is pumped into the reactor to begin

FIGURE 12.6 Pfaudler 50-gallon glass-lined reactors.

quenching the reaction. The hydrolysate mixture is then transferred from the reactor into a 7,500 L storage tank containing approximately 1,200 kg reverse osmosis (RO) water to finalise the reaction quenching, followed by neutralisation through the slow addition of sodium hydroxide.

12.5.3 Centrifuge

In the CNC production process, centrifugation is used as a first-stage purification process to separate CNC from the waste stream. The centrifuge used in InnoTech Alberta's CNC pilot plant is a GEA Westfalia SC-35 disk stack centrifuge with a bowl speed of 6,500 rpm; it is shown in Figure 12.7. The unit separates the liquid from the solid, ideally at up to 90 L of material per minute. Through centrifugation, CNC is collected as a paste-like cake and the centrate is collected as the waste stream. The centrate is sent to sewage, whereas the solids' discharge is pumped to a storage tank and diluted with 1,500 L of water. At this stage, CNC particles begin to suspend or disperse in water. Owing to the high viscosity of CNC cakes, it is very easy to plug the transferring pump and lines. Therefore, the cake discharge frequency and the frequency and duration of hood water spray were optimised to avoid plugging. Separation efficiency is determined by the material's residence time in the centrifuge bowl, with residence time controlled by the flow rate of the feed into the bowl. Decreasing the flow rate will increase the residence time, resulting in better separation efficiency and higher yield. Conversely, a low flow rate will increase the

FIGURE 12.7 GEA Westfalia SC-35 disk stack centrifuge.

time of operation. An optimal flow rate of 25 L/min for CNC product purification is in place at InnoTech Alberta's CNC pilot plant.

The centrifugation is also used to remove high-molecular-weight cellulosic material, large particles, dirt and unreacted materials. After the purification step, the CNC suspension is pumped back into the centrifuge and then processed at a bowl speed of 6,500 rpm to remove unacceptable materials that may affect the quality of the CNC.

12.5.4 Purification—Ultrafiltration

After centrifugation, the CNC materials will suspend in the liquid stream. However, there are still large amounts of sodium sulphate salts, glucose and oligomers present in the liquid. To further purify the CNC, an ultrafiltration system (tangential flow filtration) is used. The ultrafiltration system used in InnoTech Alberta's CNC pilot plant was built by GEA-Niro, as shown Figure 12.8, whereas the membrane modules are produced by pall corporation (PALL). The membrane separation technique has 11–13 modules arranged in parallel and designed to process about 1,500 L of material in 8 h. The system is used to remove salt, simple sugar, oligomers and others impurities in the CNC suspension and also used to concentrate the product up to 3 wt.% consistency. The membrane is made of hollow fibres with a diameter of 800 μm and has a molecular weight cut-off of 50,000 Da. The CNC suspension is circulated through a parallel series of Pall hollow fibre tube modules, where the dilute, low-molecular-weight salt or sugar contaminates are passed through the membrane, whereas the CNC particles are retained within the tubes. Reverse osmosis water is

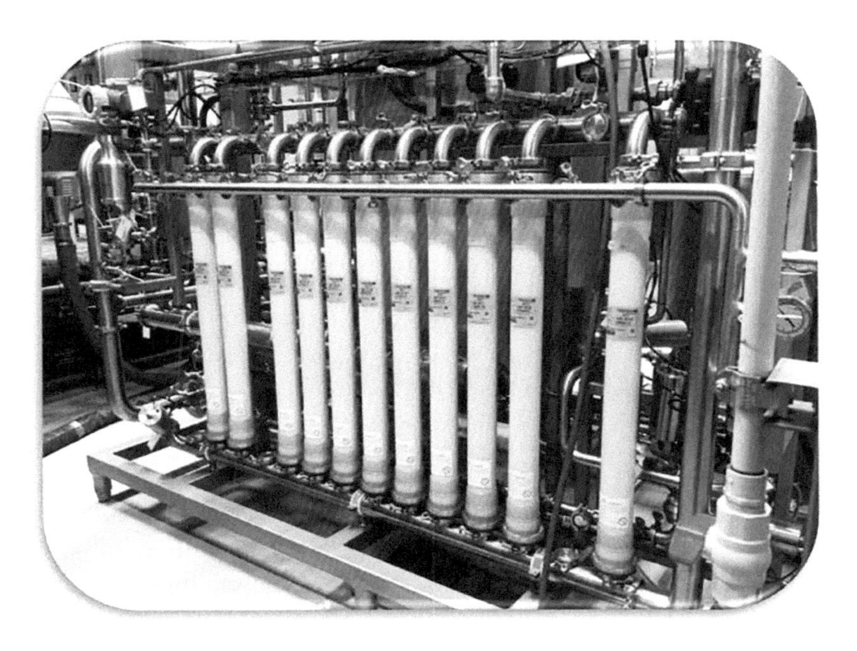

FIGURE 12.8 **(See colour insert.)** GEA-Niro ultrafiltration unit.

added, as required, to maintain the CNC concentration at approximately 0.5 wt.%. This diafiltration continues until the conductivity of the suspension is reduced to less than 300 μS/cm. The small diameter of the hollow fibres provides a large surface area for an effective separation process. However, when using the same membrane to concentrate CNC suspensions, the membrane fouls very easily due to the high viscosity of CNC suspensions. The ultrafiltration system is an integral part of the pilot plant operation, and identifying efficient cleaning agents and developing cleaning protocols have a significant impact on process operations. Without proper cleaning agents, it can take months to clean membranes fouled by CNC products. Using a side product of CNC production as a cleaning agent, the InnoTech Alberta team has developed a convenient cleaning protocol in which about 80% of membrane separation efficiency was recovered in 2 days. An additional washing with RO water allows the recovery of membrane efficiency.

Once filtered, the purified CNC suspension is centrifuged to remove high-molecular-weight cellulosic material, large particles, dirt and unreacted materials. The colloidal CNC suspension is then filtered using a 10-μm cartridge-style filter to remove dirt and small unreacted cellulosic materials (so-called 'off-spec' materials). The resulting 'clean' CNC suspension is transferred to the ultrafiltration system for second-stage purification, until the conductivity of the suspension is less than 100 μS/cm. This purification technique employs the same filtration system as is used for diafiltration. The purified CNC suspension is then concentrated up to 3 wt.% by using the ultrafiltration system and pumped to a 300 L transfer vessel.

12.5.5 Spray Dryer

The purified and concentrated CNC suspension is dried to powder by using a spray dryer. Spray drying is presently one of the most common technologies used by the food and pharmaceutical industries and has been identified as the ideal method when the end product must comply with precise quality standards (e.g. particles size distribution, residual moisture content and bulk density). The physicochemical properties of the final product depend mainly on inlet and outlet temperatures, air flow rate, feed flow rate, atomiser speed, types of carrier agent and their concentration. Flow rate is a critical parameter owing to its influence on the outlet temperature. The drying equipment used in the CNC pilot plant is an SPX-Anhydro MS-400 spray dryer (Figure 12.9). Water removal can occur at up to 37 kg/h for some materials. Inlet temperatures can be varied from 170°C to 230°C, whereas outlet temperatures can range from 80°C to 100°C. Inlet and outlet temperatures affect the spray-dried CNC output. At higher inlet air temperatures, there is a greater temperature gradient between the atomised feed and drying air, which results in the greatest driving force for water evaporation. The use of higher inlet air temperature leads to the production of larger particles and causes increased swelling. If the inlet air temperature is low, the particles remain shrunken and smaller in size. Higher inlet and lower outlet temperatures yield greater product output. Inlet and outlet temperatures at InnoTech Alberta's pilot scale spray dryer are 220°C and 90°C, respectively.

FIGURE 12.9 SPX13-Anhydro MS-400 spray dryer.

FIGURE 12.10 CNCs before and after spray drying.

Typically, InnoTech Alberta uses a 3 wt.% CNC suspension in the pilot plant spray dryer. Figure 12.10 shows the images for the CNC suspension before spray drying and the CNC powder after spray drying.

There are several concerns about spray drying CNC, including the long cleaning time (approximately 5–7 h), which involves several steps (i.e. removing the atomiser, washing it with water for 60–90 min, scrubbing the front door with soap for about

30–40 min, rinsing it with water again for about 60 min to rinse any soap residue and removing the pipes for cleaning and drying). The other spray-dryer issue is related to sample collection and the risk of losing a large amount of processed CNC sample from the collection container. Improving spray-dryer efficiency, maximising the operating capacity of the spray dryer and improving the sample collection system in order to collect samples directly to avoid losing sample, all are recommended. In addition, atomising and cleaning the spray dryer immediately after operating the dryer will reduce the cleaning time.

12.5.6 Storage and Shelf-Life

In InnoTech Alberta's pilot plant, CNC is extracted using sulphuric acid. This process generates unmodified CNCs, with anionic sulphate half-ester groups on their surface. In addition, as organic nanomaterials, CNCs are composed primarily of highly crystal native cellulose. As such, CNCs are subject to chemical degradation, such as the acid catalysed hydrolysis of sulphate half-ester groups or the cellulose chains, and the potential of bacterial and fungal growth, resulting in the biodegradation of CNC materials. The shelf-life of CNCs is an essential information for manufacturers and consumers. The monitoring of chemical and physical stability of CNCs during long-term storage under a variety of conditions is critical to understand the changes in sulphate half-ester content, cellulose chain length, thermal stability, brightness and their re-dispersability in water over time [23]. It is therefore of great interest to determine the mechanisms and kinetics of such degradation and its ultimate effect on CNCs being used in different applications [24]. Stable CNCs products, with reproducible and predictable properties, are critically dependent on storage conditions. Receiving a stable product is of great importance to many members of the cellulosic nanomaterial community, from researchers to producers and companies.

12.6 CHARACTERISATION OF CELLULOSE NANOCRYSTALS MATERIALS

Characterisation is an essential part of the larger CNC program underway at InnoTech Alberta [22,24–26]. The importance of characterisation is highlighted by its contribution to achieving the following:

- Study or reveal CNC features relevant for (specific) applications' development
- Develop or establish testing protocols for CNC quality assurance and quality control
- Optimise CNC pilot plant processes and equipment to achieve high yield and quality and consistent CNC
- Establish scientific and technical collaborations with partners
- Sustain applications' developments
- Focus efforts to decrease production costs
- Develop health and safety procedures for bulk handling of CNC materials

TABLE 12.1
Some Features of CNC

Purity	Morphology	Chemical and Physical Properties	
Moisture content	Length and diameter	Total sulphur content	Suspension conductivity
Crystallinity	Aspect ratio	Thermal stability	Specific surface area
Colour (brightness)	Particle size distribution	Sulphate ester content	Viscosity
Residual-free sulphate content	Degree of agglomeration	Zeta potential	Turbidity
Metal ion profile	Degree of aggregation	Dispensability in water	pH

Source: Bouchard, J., The cellulosic nanomaterials value-chain and standardization challenges, *1st Meeting – CSA Technical Committee on Cellulose Nanomaterials*, March 25–26, Mississauga, ON, 2013.

Characterisation is also crucial for optimisation of individual operating units, feedstock and CNC quality; assessment of performance and grade; scientific development of process optimisation and development of viable CNC production technologies. The full exploitation of CNC as an end product is directly dependent on the availability of the best knowledge of CNC features, such as those outlined in Table 12.1.

The characterisation of the features of unmodified CNC is necessary to provide CNC samples in sufficient quantity and quality to recipients (e.g. universities, government laboratories and industry) for applications' development. Most importantly, more accurate, faster and robust standardised testing methods for characterisation are required to avoid the creation of barriers to commercialising and promoting CNCs. Therefore, InnoTech Alberta's CNC scientists have created a set of potential tests methods, including standard approaches for the characterisation of unmodified CNC and CNC produced from pre-treated cellulosic feedstocks.

Using different state-of-art characterisation techniques shown in Table 12.2, the InnoTech Alberta CNC team were able to identify key parameters influencing the crystalline structure, degree of crystallinity, thermal stability, morphology, dimensions and yields of CNC and validate, develop and improve physical and chemical characterisation methods for CNCs. Some of these works were executed in collaboration with others (i.e. researchers at FPInnovations; the University of Alberta, Edmonton, Canada; the Ingenuity Lab and the Measurement Science and Standard Institute of the National Research Council of Canada) (Figure 12.11).

Table 12.2 identifies analytical techniques required for accessing specific properties of the CNC materials. Of these listed techniques, only dynamic light scattering, zeta potential, SEM, X-ray diffraction spectroscopy, elemental analysis (EA) [inductively coupled plasma optical emission spectrometry (ICP-OES)] by acid digestion, EA by combustion for sulphur, thermogravimetric analysis (TGA/N_2) and rheology were utilised for samples' characterisation. These tests assess the ultrastructure, morphology, chemical composition, viscosity and purity of CNC materials

TABLE 12.2
A List of Some Techniques for CNC Characterisation

Technique	Rank	Information	Physical State of Sample
Scanning electron microscopy	1	Detailed nanostructural information (shape, size and aspect ratio) Dispersion state	Dry or suspended material
Transmission electron microscopy	1	Detailed nanostructural information (shape, size and aspect ratio) Dispersion state Aspect ratio	Dry or suspended material
Atomic force microscopy	1	Structural evaluation (both lateral dimensions and particle height)	Dry material
Dynamic light scattering	1	Particle size and distribution Aggregation state	Colloidal suspension
X-ray diffraction spectroscopy	1	Crystalline structure Crystal size Crystallinity index or degree of crystallinity	Dry material
Zeta potential	1	Nature of surface charge and density Stability Type of interactions	Colloidal suspension
Elemental analysis by combustion for sulphur	1	Total sulphur content	Dry material
Elemental analysis (ICP-OES) by acid digestion	1	Sulphate half-ester content Traces of impurities (Na, etc.)	Dry material
^{13}C Solid-state nuclear magnetic resonance	2	Structural evaluation and crystallinity index or degree of crystallinity	Dry material
Fourier transform infrared spectroscopy	3	Spectroscopic evaluation (functional groups)	Dry material
Thermogravimetric analysis (TGA/N_2)	2	Thermal resistance Moisture content	Dry material
Rheology	2	Viscosity Degree of polymerisation	Wet material (suspension, hydrogel or gel)
Physisorption of gas (Brunauer–Emmett–Teller model)	4	Specific surface area, pore size and pore volume	Dry material

Ranking 1 to 4 to rate the levels of importance: (1) very important, (2) fairly important, (3) important and (4) slightly important.

and of their starting feedstocks. In Table 12.2, the measurement techniques are ranked from the very important (1) to the less important (4) to know. Because CNC materials' properties are highly dependent on the cellulosic feedstock quality, pulp properties (i.e. α-cellulose, acetone extractives and ash) are tested in accordance with technical association of the pulp and paper industry (TAPPI) standard test methods

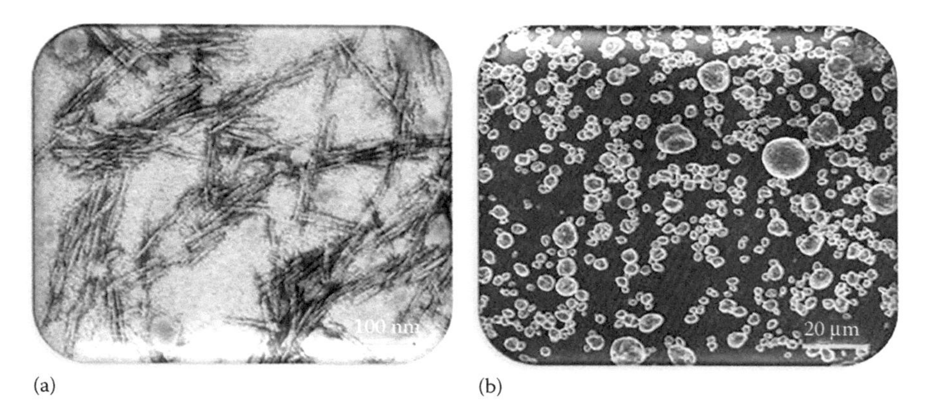

FIGURE 12.11 Morphologies of CNC in (a) suspension and (b) after spray drying.

T203 cm-99, T204 om-97 and T211 om-93, respectively, before the feedstock's use in CNC production. Pulp viscosity, which is also critical, is tested according to American Society for Testing and Materials (ASTM) D1795–1796. Through these analytical techniques, the InnoTech Alberta team has demonstrated the importance of the pre-treatment of cellulosic feedstock on CNC production yields and quality by increasing the α-cellulose content of the cellulosic feedstock through pre-treatment.

Overall, characterisation helps in the development of novel applications through improved engagement and collaboration with researchers and industry as well as in the improvement of CNC pilot plant production process, yield and quality. It also helps build capability, knowledge and expertise in the area of CNC characterisation as well in the development of improved characterisation methods and the establishment of protocols for CNC quality control. In addition, the work is relevant for other team members' CNC-related research activities and enables InnoTech Alberta to scientifically and technically support and troubleshoot client needs.

Scientists at InnoTech Alberta's CNC pilot plant, along with other members of the Canadian Standards Association Technical Committee (CSA TC), have worked to establish standard test methods for use in the characterisation of unmodified CNC. Collectively, they developed the first standard in 2015 ('CSA Z5100 Cellulosic Nanomaterials Test Methods for Characterization of Cellulosic Nanomaterials') and helped compile the International Cellulosic Nanomaterials (CNM) Activities under the CSA TC and the International Organization for Standardization Technical Committee 6 Task Group 1 (ISO TC6 TG1), respectively [12,24–26].

Standard CNC characterisation test methods will foster partnerships and collaborations and also support the commercialisation of CNC by communicating consistent quality characteristics based on reliable and repeatable measurement methods. For some applications, such as packaging, where CNC has direct contact with foods, health and safety testing requires international standardisation to ensure that possible negative health impacts are avoided [28]. This, in turn, will help speed up CNC's use in high-value applications.

12.7 REMARKS

Successful pilot-scale CNC production is critical for the commercialisation of CNC. At pilot scale, the focus is to optimise CNC production, while reducing associated production costs. Cellulose nanocrystals samples produced at pilot scale are provided to clients for applications' development and ultimately commercialisation. The CNC team at InnoTech Alberta has worked on various aspects of CNC production from feedstock preparation, improving processing conditions, recycling waste and unreacted materials to optimise production and reduce costs and finally to characterise CNC materials. Computer simulation and economic modelling have been used to identify cost efficiencies at pilot plant scale. It is crucial to provide quality CNC samples, in sufficient quantity, to universities, government laboratories and industry for applications' development. Scientists at InnoTech Alberta's CNC pilot plant continue to be actively involved in national and international nanocellulose work (including ISO TC6 TG1 and CSA TC) to develop standardised testing methods for characterisation of CNC to help avoid the creation of barriers to commercialise and promote Canadian CNCs.

ACKNOWLEDGEMENTS

We thank NanoAlberta and AI-Bio for their financial supports, as well as the Alberta CNC Steering and Working Committees for providing direction and coordinating efforts. We also thank Al-Pac for their partnership, financial support and collaboration. We would also like to extend our appreciation to all scientists and technologists from the Biomass Conversion and Processing team at InnoTech Alberta for their continuing and valuable contributions.

REFERENCES

1. C.J. Chirayil, L. Mathew and S. Thomas, Review of recent research in nanocellulose preparation from different lignocellulosic fibers, *Rev. Adv. Mater. Sci.*, **2014**, 37, 20–28.
2. I.M.N. Iqbal, G. Kyazze and T. Keshavarz, Advances in the valorization of lignocellulosic materials by biotechnology: On overview, *BioResources*, **2013**, 8(2), 3157–3176.
3. L.L. Da Silva, Adding value to agro-industrial waste, *Ind. Chem.*, **2016**, 2(2), 1000e103.
4. G. Siqueira, J. Bras and A. Dufresne, Cellulosic bionanocomposites: A review of preparation, properties and applications, *Polymers*, **2010**, 2, 728–765.
5. S. Rebouillat and F. Pla, State of the art manufacturing and engineering of nanocellulose: A review of available data and industrial applications, *J. Biomater. Nanobiotechnol.*, **2013**, 4, 165–188.
6. L. Brinchi, F. Cotana, E. Fortunati and J.M. Kenny, Production of nanocrystalline cellulose from lignocellulosic biomass: Technology, and applications, *Carbohydr. Polym.*, **2013**, 94, 154–169.
7. A. Sinha, E.M. Martin, K.-T. Lim, D.J. Carrier, H. Han, V.P. Zharov and J.-W. Kim, Cellulose nanocrystals as advanced 'Green' materials for biological and biomedical engineering, *Biosyst. Eng.*, **2015**, 40(4), 373–393.
8. S. Kobayashi, K. Kashiwa, J. Shimada, T. Kawasaki and S. Shoda, Enzymatic polymerization: The first in vitro synthesis of cellulose via non-biosynthetic path catalyzed by cellulase, *Makromol. Chem. Macromol. Symp.*, **1992**, 54/55, 509–518.
9. S. Piotrowski and M. Carus, *Multi-Criteria Evaluation of Lignocellulosic Niche Crops for Use in Biorefinery Processes*, Nova-Institut GmbH, Hürth, Germany, **2011**.

10. NIIR Board of Consultants & Engineers, *The Complete Technology Book on Wood and Its Derivatives*, National Institute of Industrial Research, Delhi, India, **2005**.
11. Y. Habibi, L.A. Lucia and O.J. Rojas, Cellulose nanocrystals: Chemistry, self-assembly, and applications, *Chem. Rev.*, **2010**, 110 (6), 3479–3500.
12. Summary of international activities on cellulosic nanomaterials. http://www.tappinano.org/media/1096/tc6-world-cnm-activities-summary-july-29-2015.pdf (accessed on July 29, 2015).
13. R.J. Moon, A. Martini, J. Nairn, J. Simonsen and J. Youngblood, Cellulose nanomaterials review: Structure, properties and nanocomposites, *Chem. Soc. Rev.*, **2011**, 40, 3941–3994.
14. C.-J. Zhou and Q.-L. Wu, Recent development in applications of cellulose nanocrystals for advanced polymer-based nanocomposites by novel fabrication strategies, In *Nanocrystals – Synthesis, Characterization and Applications*, S. Neralla (Ed.), Intech, **2012**, (Chapter 6), pp. 103–120, Rijeka, Croatia.
15. Canadian Forest Nanoproducts Network, *Building a Canadian Advantage with Nanocellulose*, Arboranano, Canada, Spring, **2014**.
16. M.-C. Li, Q. Wu, K. Song, S. Lee, Y. Qing and Y. Wu, Cellulose nanoparticles: Structure-morphology-rheology relationships, *ACS Sustain. Chem. Eng.*, **2015**, 3, 821–832.
17. M.-C. Li, Q.-L. Wu, K.-L. Song, C.F. De Hoop, S.-Y. Lee, Y. Qing and Y.-Q. Wu, Cellulose nanocrystals and polyanionic cellulose as additives in bentonite water-based drilling fluids: Rheological modeling and filtration mechanisms, *Ind. Eng. Chem. Res.*, **2016**, 55 (1), 133–143.
18. M.-C. Li, Q.-L. Wu, K.-L. Song, Y. Qing and Y.-Q. Wu, Cellulose nanoparticles as modifiers and fluid loss in bentonite water-based fluids, *ACS Appl. Mater. Interf.*, **2015**, 7, 5006–5016.
19. R. Bardet, C. Sillard, N. Belgacem and J. Bras, Self-assembly of cellulose nanocrystals with fluorescent agent in iridescent films, *Cellulose Chem. Technol.*, **2015**, 49 (7–8), 587–595.
20. S.S. Nair, J.Y. Zhu, Y. Deng and A.J. Ragauskas, High performance green barriers based on nanocellulose, *Sustain. Chem. Process.*, **2014**, 2:23, 1–7.
21. Y. Boluk and L. Zhao, Aircraft anti-icing fluids formulated with nanocrystalline cellulose, US Patent 8,105,430 B2, **2012**.
22. O.J. Rojas, Cellulose nanomaterials-a path towards commercialization, Workshop report, **2014**, http://www.nano.gov/ncworkshop.
23. S. Beck and J. Bouchard, Effect of storage conditions on cellulose nanocrystal stability, *TAPPI J.*, **2014**, 13 (5), 53–61.
24. C.S. Davis, R.J. Moon, S. Ireland, E.J. Foster, L. Johnston, J.A. Shatkin, K. Nelson et al., NIST-TAPPI workshop on measurement needs for cellulose nanomaterials, *TAPPI International Conference on Nanotechnology for Renewable Materials*, **2014**, Vancouver, Canada.
25. CSA Group, Cellulose nanomaterials-Test methods for characterisation, **2014**. http://shop.csa.ca/en/canada/sustainable-forest-management/z5100-14/invt/27036672014.
26. CSA Group, CSA announces first cellulosic nanomaterials standard: z5100-14, cellulosic nanomaterials-test methods for characterization, Canada, July, **2014**.
27. J. Bouchard, The cellulosic nanomaterials value-chain and standardization challenges, *1st Meeting – CSA Technical Committee on Cellulose Nanomaterials*, March 25–26, **2013**, Mississauga, ON.
28. M. Roman, Toxicity of cellulose nanocrystals: A review, *Indus. Biotechnol.*, **2015**, 11 (1), 25–33.

Index

Note: Page numbers followed by f and t refer to figures and tables respectively.

2,2,6,6-tetramethylpiperidine-1-oxyl (TEMPO), 150, 188, 222
- mediated oxidation, 37, 48–54
- oxidised materials, 94
- oxidised pulps, 46
- radical-catalyzed process, 21
- TOCNF, 151–152, 151f
- TOCNs, 50

2,3-dialdehyde cellulose (DAC), 57
2-Azaadamantane *N*-oxyl (AZADO) oxidation, 54–55, 54f
4-Acetamido-TEMPO (4-AcNH-TEMPO), 52–53

A

Abrasion resistance, 199
Absorption, cellulose-based aerogels, 242, 242f, 243t–246t
Acetobacter, 1–2
- species, 68

Acetobacter xylinum BPR2001, 4, 210
Acetylation, 38, 164
Acid hydrolysis reactions, 271, 277–278
Acrylic resins, 153
Adsorption
- mechanism, nanocellulose and emulsions, 185–186, 186f
- membranes, nanocellulose as, 136

Aerocellulose, 218
Aerogels, 116, 218
- ambient dried, 218
- carbon, cellulose-based, 251–252
- with catalytic activity, cellulose composite, 251t
- Ce-D, 224, 225f
- Ce-I, 221–224
- Ce-II, 219–221, 219t
- cellulose, 236, 237f
  - after formation, 231–232, 233t–234t
  - AmimCl solution, 235, 235f
  - antibacterial activity, 237–238
  - preparation, 218
- cellulose-based. *See* Cellulose-based aerogels
- CL, 222
- inorganic NP, 232
- lignocellulose, 221
- magnetic cellulose composite, 238–239, 238t
- monolithic, 235
- PMSQ, 236
- regenerated cellulose, 219

AFM (atomic force microscopy), 148–149
Airlift reactor, 3
Alberta-Pacific Forest Industries Inc. (Al-Pac), 270
Alkaline TEMPO/NaClO/NaBr oxidation, 49, 49f
Ambient dried aerogels, 218
American Society for Testing and Materials (ASTM), 285
AmimCl (1-allyl-3-methylimidazolium chloride), 221
Amorphous, 29
Antibacterial activity, cellulose-based aerogels, 237–238, 238f
ASTM (American Society for Testing and Materials), 285
Atomic force microscopy (AFM), 148–149
AZADO (2-Azaadamantane *N*-oxyl) oxidation, 54–55, 54f

B

Bacterial cellulose (BC), 1–3, 34, 68, 69f, 176, 201, 218
- applications, 1–2
- cellulose nanopapers, 158
- composites, 163–168, 163f, 165f, 166f, 167f
- degree of crystallinity, 74
- fermentation process, 10
  - cost estimations, 3–4
  - cost structure and profitability analysis, 11t–12t
  - overview, 1–3
  - process simulation, 3
- paper, 158
- pellicles, 210, 211f
- PLA nanocomposites, 165–166
- production, process simulation, 4–12
  - with different strains and culture conditions, 5t–7t
  - equipment specifications and costs, 8, 10t
  - raw materials, 8, 8t
- properties, 1
- PVA nanocomposites, 166
- thermal degradation behaviour, 77, 77f
- *vs*. CNF, crystallinity, 74–75
- XRD pattern, 74, 74f

BET (Brunauer–Emmet–Teller), 92, 93f
Bi-functional esterifying reagents, 232
Binder-less fibreboards manufacturing, 200, 200f
Bio-mediated oxidation, 53

Bottom–up approach, nanocellulose, 68
Branched polyethyleneimine (b-PEI), 242, 242f
Broido–Shafizadeh mechanism, 77–78, 77f
Brunauer–Emmet–Teller (BET), 92, 93f
Bulk methods, 108

C

CAB (cellulose acetate butyrate), 163
Carbon aerogels, cellulose-based, 251–252
Carbon nanotube (CNT), 249
Cartridge housing, 142, 142f
Cationic functionalisation, 58
CBM (cellulose-binding module), 185
Ce-D aerogels, 224, 225f
Ce-I aerogels, 221–224
Ce-II aerogels, 219–221, 219f
Cellouronic acid, 45, 51
CelluForce Inc., 271
Cellulon®, 2
Cellulose, 17, 67, 175, 217, 270–271
activation, 78
aerogels, 231–232, 233t–234t
amphiphilic character, 242
biocompatible biomass, 250
chain, partial molecular structure, 270, 271f
chemical structure, 67, 68f
CI, 73
composite aerogels with catalytic activity, 251t
crystal structures, 69–76, 70t
lattice parameters, 69, 69t
projections, 71, 71f
degree of crystallinity, 73
derivatisation, 224
esters, 224
foams/sponges, 218
gelation, 218, 220
gels formation, 225–230
hydrophilic, 185
microcrystalline, 37
microfibrils, 67, 68f
overview, 67–69
pulp, 148
pyrolysis process, 78
thermal degradation pathway, 77–78, 79t
thermal stability, 77–81
unit cells structures, 70, 71f
whiskers, 271
Cellulose acetate butyrate (CAB), 163
Cellulose–AH composite aerogels, 236, 237f
Cellulose-based aerogels
carbon, 251–252
carbonisation, 251
composite, 225–234
functional
absorption, 242, 243t–246t
antibacterial activity, 237–238, 238f
biomedical features, 250–251
catalytic activity, 250
electrical properties, 249
flame retardancy, 236, 237f
hydrophobicity, 239–241
magnetic properties, 238–239
optical properties, 235–236, 235f
oxidation resistance, 247
sound absorption, 247
thermal conductivity, 247, 248t–249t
overview, 217–218
preparation
Ce-D, 224, 225f
Ce-I, 221–224
Ce-II, 219–221, 219t
Cellulose-binding module (CBM), 185
Cellulose-degrading enzymes, 36
Cellulose Iα, 69–70
Cellulose Iβ, 69–70
Cellulose nanocrystals (CNCs), 28f, 89, 111, 176, 221–222, 269
applications, 272–273
aqueous dispersions, 272
from cellulose to, 270–271
characterisation, 282–285
techniques for, 283, 284t
esterification routes, 38–39
features, 283, 283t
as filler phase, 272
growing market for, 271–272
at InnoTech Alberta, 273–275
acid hydrolysis reactions, 277–278
centrifuge, 278–279
feedstock, 275–276
purification—ultrafiltration, 279–280
spray dryer, 280–282
storage and shelf-life, 282
isolated and aggregated, 184, 184f
lengths, 32, 32t
morphologies, 285, 285f
overview, 27–29
oxidation routes, 36–38, 37f
pilot plant activities flowchart, 274, 274f
preparation, 29–30, 29f
enzymatic hydrolysis, 36
with other mineral acids, 35–36
rejects, 275
shape, impact of, 187–188, 188f
shear-thinning, 272
spray drying, before and after, 281, 281f
sulphuric acid hydrolysis, 30–31, 277
reaction conditions effect, 31–32, 33f
source material effect, 34–35, 34t
suspension, 279
tensile strength, 272
θ content, 275
vs. CNF, crystallinity, 75–76

as water-based viscosity control, 272
XRD pattern, 76, 76f
yield, 39, 39f
Cellulose nanofibres (CNFs), 18, 69f, 148, 175, 176f, 201, 221–222
compression moulding method, 159
crystallinity
vs. BC, 74–75
vs. CNC, 75–76
extraction processes
developments, 21–23
mainstream, 17–21
impregnation method, 159
melt compounding method, 159
production of, 149–150
in situ biosynthesis method, 159
solvent casting method, 159
surface selective dissolution method, 159
thermal degradation behaviour, 77, 77f
XRD pattern, 74, 74f
Cellulose nanopapers, 149
enzymatic treatment, 155, 155f
fabrication methods, 149–151
generalities, 148–149
NFC, 151–157
light transmittance, 152f
optical properties, 154f, 156, 157f
papermaking process, 151, 152f
ultraviolet–visible transmittance, 151, 151f
Cellulose nanoparticles (CNs), 177
oxidation on
2,2,6,6-tetramethylpiperidine-1-oxyl-mediated, 48–54
AZADO, 54–55, 54f
industrial possibility, 58
NHPI, 55
by nitrogen oxides, 46–48, 48f
overview, 45–46
periodate, 56–58, 56f
Cellulose nanowhiskers, 221–222
Cellulosic feedstocks types, 275, 276f
Centrifugation, 181–182, 278–279
Chitosan, 156
CI (crystallinity index), 73
Cladophora nanocrystals, 187
CL (cross-linked) aerogels, 222
CL-CNF, 222–223
CNCs. *See* Cellulose nanocrystals (CNCs)
CNF-rGO aerogel, strain-sensitive conductivity of, 250f
CNFs. *See* Cellulose nanofibres (CNFs)
CNs. *See* Cellulose nanoparticles (CNs)
CNT (carbon nanotube), 249
Collenchyma cells, 23
Composite aerogels, cellulose-based, 225–234, 227t–229t
Composites, optically transparent nanocellulose
BC, 163–168, 163f, 165f, 166f, 167f
fabrication methods, 158–159
generalities, 158
nanofibrillated cellulose, 160–163, 160f, 162f
Compression moulding method, CNFs, 159
Contact angle, 178
Counterion exchange, 31
Cross-linked (CL) aerogels, 222
Cryocrushing method, 21
Cryogel, 218
Cryoscopic transmission electron micrographs, 32, 34f
Crystallinity index (CI), 73
Cu(II) adsorption capacities, 132, 132f

D

DAC (2,3-dialdehyde cellulose), 57
DCNC (dialdehydemodified nanocrystalline cellulose), 57
Degree of crystallinity, 271
BC, 74
Ce-II aerogels, 220
cellulose, 73
Densely coated neat sisal (DCNS) fibres, 211, 212f
Density function theory (DFT), 138
Depreciation, 12
Dialdehydemodified nanocrystalline cellulose (DCNC), 57
Dicarboxylic acids, 38–39
Dimethyl sulphoxide (DMSO), 221
Direct chemical modification methods, 106
Directed water channels, 135
Donnan effect, 112
Drying and/or hot-pressing, 150
Dyes removal, 136, 137f

E

Electrokinetic phenomena, 100
Electrophoresis, 101–102
Electrophoretic deposition (EPD), 111
Electrospun fibrous membranes, 138
Electrostatic interactions, 177
Electrostatic stabilisation, 46
EmimCl (1-ethyl-3-methylimidazolium chloride), 221
Emulsions
characteristics control, 180
destabilisation, 180
HIPE, 188–189, 188f
nanocellulose and, 180–185, 181f, 183f, 184f, 185f
adsorption mechanism, 185–186, 186f
CNCs shape, impact of, 187–188, 188f
pickering, 177–180, 178f, 179f
stability parameters, 182

Enzymatic hydrolysis, 21
Experimental paper-making machine (XPM), 139–140, 140f
Extrusion papermaking process, 150

F

Fabricated nanofibrillated cellulose film, 140, 141f
Fabrication methods, nanocellulose
  paper, 149–151
  polymer composites, 158–159
FD. *See* Freeze-drying (FD)
Fermentation methods, 2–3
Ferromagnetic material, 190
Filtration efficiency, 199
Fixed capital investment components, 4
Flame retardancy, cellulose-based aerogels, 236, 237f
Fluffed bleached pulp, 275, 276f
Fluorinated silane, 239
Freeze-drying (FD), 218
  Ce-I aerogels, 224
  hydrogels, 223
Freundlich model, 114
Functional cellulose-based aerogels
  absorption, 242, 242f, 243t–246t
  antibacterial activity, 237–238, 238f
  biomedical features, 250–251
  catalytic activity, 250, 251t
  flame retardancy, 236, 237f
  hydrophobicity, 239–241, 240t–241t
  oxidation resistance, 247
  properties
    electrical, 249, 250f
    magnetic, 238–239
    optical, 235–236, 235f
  sound absorption, 247
  thermal conductivity, 247, 248t–249t
'Functional material,' 135

G

Gas-phase esterification, 232, 232f
Gas-phase methods, 231
GEA-Niro ultrafiltration unit, 279, 279f
GEA Westfalia SC-35 disk stack centrifuge, 278, 278f
Gengiflex®, 2
Gluconacetobacter xylinus, 1, 167
Grafting functional groups, 242
Grafting reactions, 106
Grinder process, 19–20, 19f

H

Hairy fibres of neat sisal fibres (HNSF), 212f, 213
Hairy nanocrystalloids, 57
HCl, 35
HCNF (holocellulose nanofibres), 155
Hemicelluloses, 53, 99
  in nanofibrillation, 20
Hestrin and Schramn (HS) culture medium, 4
High internal phase emulsion (HIPE), 188–189, 188f
High-performance disperser, 20
High-pressure homogeniser, 18–19, 19f, 22–23
High-pressure water-jet system, 153
High-shear fibrillation process, 68
HIPE (high internal phase emulsion), 188–189, 188f
HNSF (hairy fibres of neat sisal fibres), 212f, 213
Holocellulose nanofibres (HCNF), 155
Homogenisation process, 18
Hornification, 205
HPMC (hydroxyl propyl methylcellulose), 224, 225f
HS (Hestrin and Schramn) culture medium, 4
Hydroentanglement-treated flax fibre fabric, 199
Hydrophobic cellulose-based aerogels, 240t–241t
Hydrophobicity, 239–241
Hydrophobisation method, 239
Hydroxyl propyl methylcellulose (HPMC), 224, 225f

I

IEP (isoelectric point), 101, 103
IFFS (interfacial shear strength), 210
Impregnation method, CNFs, 159
Inhibition zones, CNF–Fe_3O_4 aerogel, 237, 238f
InnoTech Alberta, 270
  CNC pilot plant at, 273–275, 273f
    acid hydrolysis reactions, 277–278
    centrifuge, 278–279
    feedstock, 275–276
    objective, 275
    purification—ultrafiltration, 279–280
    spray dryer, 280–282
    storage and shelf-life, 282
  pilot capacity, 273
Inoculum propagation, 8
Inorganic NP aerogels, 232
In situ biosynthesis method, CNFs, 159
Interfacial shear strength (IFFS), 210
Ionic liquids, 220–221
Isoelectric point (IEP), 101, 103

K

Kevlar fibre, 272
Kitchen blender, 22, 150
Komagataeibacter xylinus, 1, 68

L

Langmuir – Flory–Huggins – clustering model, 111
Langmuir isotherm, 91–92, 114, 132
Langmuir–Schäfer technique, 112
Layer-by-layer (LBL)
  filtration process, 201, 203f
  method, 232
Levelling-off degree of polymerisation (LODP), 30
Lignocellulose aerogels, 221
(ligno)cellulosic fibres, surface fibrillation of, 198–201, 199t
Limited coalescence
  phenomenon, 179
  process, 182, 183f
Liquid-phase method, 232
LODP (levelling-off degree of polymerisation), 30
'Loose' reverse osmosis (RO) membranes, 134

M

Magnetic cellulose composite aerogels, 238–239, 238t
Materials of nanocellulose, 87–88
  amphiphilic character, hydrophilic nanomaterial, 104–106, 105f
  chemical composition building blocks, 96–98, 97f
  contaminant adsorption mechanism, 113–116
  different grades of, 88f
  as emulsion stabilisers, 117–118
  interfacial properties application, 116–117
  reactivity and surface modification, 106–108
  SSA, 88–94, 90f, 94f, 95t–96t
  surface charge building blocks, 98–104, 101f
  water interactions, 108–113, 109f
MCC (microcrystalline cellulose powder), 275, 276f
Mechanical properties
  cellulosic materials, 199, 199t
  natural fibres, 202t, 203
Melt compounding method, CNFs, 159
Membrane fouling, 117
Membrane module design concepts, 141–142, 143f
Membrane scale-up, 139–141
Membrane separation technique, 279
Mercerisation, 70–71
Methyltrimethoxysilane (MTMS), 236
Microcrystalline cellulose, 37
Microcrystalline cellulose powder (MCC), 275, 276f
Microfibrillated cellulose (MFC), 18–19, 46, 148, 187–188, 221
Microfibrils, 18, 29
Microfluidizer, 20, 20f
Molecular structure, cellulose chain, 270, 271f
Monolayered membranes/adsorbents, 136, 137f
Montmorillonite (MMT), 236
MTMS (methyltrimethoxysilane), 236
Multilayered membranes, 137–138, 138f

N

Nano banana cellulose, 132–133
Nanocellulose, 67, 147
  approaches, 67–68
  as binder
    for plant-based natural fibres, 201–205
    for wood pulp fibres, 205–209, 206t–207t
  bonding, 209
  coated natural fibres, 209–213, 210f
  crystallinity types, 74–76, 75f
  fabrication methods
    paper, 149–151
    polymer composites, 158–159
  filter cake, 150
  materials. *See* Materials of nanocellulose
  papers. *See* Nanocellulose papers, optically transparent
  physical properties, 148
  surface fibrillation of (ligno)cellulosic fibres, 198–201, 199t
  thermal expansion coefficient, 148
  thermal stability, 78–81, 80t
  for tissue engineering applications, 149
  water purification and. *See* Water purification and nanocellulose
  XRD pattern, 72–74, 72f, 73t
Nanocellulose and emulsions, 180–185, 181f
  adsorption mechanism, 185–186, 186f
  CNCs shape, impact of, 187–188, 188f
  HIPE, 188–189, 188f
Nanocellulose papers, optically transparent, 149
  BC, 158
  benefits, 153
  conductive properties, 153
  fabrication methods, 149
    CNFs production, 149–150
    drying and/or hot-pressing, 150
    extrusion papermaking process, 150
    solvent exchange, 150
    surface treatments, 150–151
    vacuum filtration, 150
  generalities, 148–149
  NFC, 151–157
  transmittance, 156
  using paracetic acid, 155
Nanocrystalline cellulose (NCC), 271
Nano-enhanced membranes (NEMs), 135

Nanofibrillated cellulose (NFC), 148, 188, 221
cellulose nanopapers, 151–157
composites, 160–163, 160f, 162f
vacuum impregnation, 160
Nanostructured membrane (NSM), 134
Nanotechnology, 129, 271
Nata de coco, 2
Natural fibres, 197
into fabrics/mats, 198
nanocellulose
binder for plant-based, 201–205
coated, 209–213, 210f
wood pulp fibres, 205–209, 206t–207t
properties, 198, 205
NBHK (northern bleached hardwood kraft), 275
NBSK (northern bleached softwood kraft), 275
NCC (nanocrystalline cellulose), 271
NEMs (nano-enhanced membranes), 135
Net profitability, 3
NFC. *See* Nanofibrillated cellulose (NFC)
N-hydroxypthalimide (NHPI) oxidation, 55, 55f
Nippon Paper, 58
N-methylmorpholine-*N*-oxide (NMMO) monohydrate, 219
Non-crystalline/amorphous regions, 271
Non-selective oxidation methods, 46
Non-wovens fibers, 203, 204f
Northern bleached hardwood kraft (NBHK), 275
Northern bleached softwood kraft (NBSK), 275

O

OLED (organic light-emitting diode) display, 161, 165
Optically transparent nanocellulose, 147–148
composites, 148
BC, 163–168, 163f, 165f, 166f, 167f
fabrication methods, 158–159
generalities, 158
nanofibrillated cellulose, 160–163, 160f, 162f
papers, 148
BC, 158
fabrication methods, 149–151
generalities, 148–149
nanofibrillated cellulose, 151–157
Organic light-emitting diode (OLED) display, 161, 165
Oxidation resistance, 247

P

Pall corporation (PALL), 279
Paper-making approach, 139
Parenchyma cells, 23
Periodate oxidation, 37, 56–58, 56f
PET (polyethylene terephthalate), 153
Pfaudler 50-gallon glass-lined reactors, 277, 277f
PHB (poly(3-hydroxybutyrate)), 166–167
Physical gel, 218
Physisorption process, 91
Pickering emulsions, 117, 177–180, 178f, 179f
Pierret chopped beached pulp, 275, 276f
Pilot-scale method, 140, 141f
PLA (polylactic acid) nanocomposites, 165–166
PMSQ (polymethylsilsesquioxane) aerogels, 236
Poly(3-hydroxybutyrate) (PHB), 166–167
Polyelectrolyte titration, 99
Polyethylene terephthalate (PET), 153
Polyglucuronic acids, 45
Polylactic acid (PLA) nanocomposites, 165–166
Polymethylsilsesquioxane (PMSQ) aerogels, 236
Polypyrrole (PPy), 237
Polyvinyl alcohol (PVA), 104, 166
PrimaCels, 2
Pulp viscosity, 285
Purification mechanism, 135
Purification—ultrafiltration, 279–280
PVA (polyvinyl alcohol), 104, 166

Q

Q^{-4} decay, 185
Quartz crystal microbalance with dissipation (QCM-D), 109, 111

R

Regenerated cellulose aerogels, 219
Rieter Perfojet pilot spun-lace machine, 198
Rietveld refinement method, 74

S

SANS (small-angle neutron scattering), 183–184
Scanning electron microscopy (SEM), 89
images, 275, 276f
polymerised styrene/water emulsions, 185, 185f
Sclerenchyma cells, 23
SE (spectroscopic ellipsometry), 110
Segal method, 73
Selective oxidation methods, 46
Semi-emperical Segal's method, 73, 76
Separation efficiency, 278
Silanols, 239
Single-step filtration process, 201, 203f
Sips isotherm model, 133
Sisal fibres, 201, 209, 210f
Small-angle neutron scattering (SANS), 183–184
Sodium nitrate, 47
Solid acid catalysts, 36
Solvent casting method, CNFs, 159
Sonication process, 182
Sound absorption, cellulose-based aerogels, 247

Specific surface area (SSA), 88–94, 90f, 94f, 95t–96t
Spectroscopic ellipsometry (SE), 110
Spiral wound structure, 142, 142f
SPR (surface plasmon resonance), 110
Spray dryer, CNCs, 280–282
Spray-FD method, 224
SPX-Anhydro MS-400 spray dryer, 280, 281f
SSA (specific surface area), 88–94, 90f, 94f, 95t–96t
Steric stabilisation, 46
Stirred tank reactor, 3
Storage and shelf-life, CNCs, 282
Sulphuric acid hydrolysis, 30–31, 131, 277
  reaction conditions effect, 31–32, 33f
  source material effect, 34–35, 34t
Super-Pro process flowsheet, 8, 9f
Surface plasmon resonance (SPR), 110
Surface selective dissolution method, CNFs, 159
Surface-sensitive methods
  swelling behaviour analysed, 111–113
  water vapour sorption analysed, 111
Surface-sensitive techniques, 108
Surfactants, 117, 177, 189

T

TBA (tetrabutyl ammonium) fluoride, 242, 242f
Technical association of the pulp and paper industry (TAPPI), 284
TEMPO. *See* 2,2,6,6-tetramethylpiperidine-1-oxyl (TEMPO)
TEMPO-mediated oxidation, 37, 37f, 48–54, 151
TEMPO/NaClO/$NaClO_2$ oxidation, reaction scheme, 52, 52f
TEMPO-oxidised cellulose nanofibrils (TOCNFs), 50, 50f, 131, 222
TEMPO-oxidised materials, 94
Tetrabutyl ammonium (TBA) fluoride, 242, 242f
Thermal conductivity, cellulose-based aerogels, 247, 248t–249t
Top–down approach, nanocellulose, 67–68, 89
Transmittance, 156
Trichoderma Reesei, 185

U

Ultrasonication, 22

V

Vacuum filtration, NFC, 150
Valonia, 67
Van der Waals interactions, 176
Voids/porosities, 158
VTT method, 140

W

Water-in-oil emulsions, 189, 224
Water purification and nanocellulose, 129–131, 130f
  impregnated electrospun mats, 138–139, 139t
  membrane module design concepts, 141–142, 143f
  membranes, 134–136, 136t
    characterisation, 134, 134f
  membrane scale-up, 139–141
  monolayered membranes/adsorbents, 136, 137f
  multilayered membranes, 137–138, 138f
  overview, 131–134
Wood powder, 161
Wood pulp, 67–68
Wood pulp fibres
  cell wall layered structure, 18, 18f
  nanocellulose as binder, 205–209, 206t–207t
Woody biomass, 67
Working capital components, 4

X

Xerogels, 218
X-ray diffraction (XRD) pattern, nanocellulose, 72–74, 72f, 73t
  BC, 74, 74f
  CNC, 76, 76f
  CNF, 74, 74f, 76, 76f
X-ray photoelectron spectroscopy (XPS), 105, 107, 211
  C 1s and Si 2p high-resolution regions, 108, 108f

Z

Zelfo Technology GmbH®, 205
Zeta potential (ξ), 100–104, 101f
  electrophoresis, 102, 102f
  streaming potential measurements, 103, 103f

#0001 - 030118 - C4 - 229/152/17 [19] - CB - 9781498761031